# Studies in Computational Intelligence

Volume 915

**Series Editor**

Janusz Kacprzyk, Polish Academy of Sciences, Warsaw, Poland

The series "Studies in Computational Intelligence" (SCI) publishes new developments and advances in the various areas of computational intelligence—quickly and with a high quality. The intent is to cover the theory, applications, and design methods of computational intelligence, as embedded in the fields of engineering, computer science, physics and life sciences, as well as the methodologies behind them. The series contains monographs, lecture notes and edited volumes in computational intelligence spanning the areas of neural networks, connectionist systems, genetic algorithms, evolutionary computation, artificial intelligence, cellular automata, self-organizing systems, soft computing, fuzzy systems, and hybrid intelligent systems. Of particular value to both the contributors and the readership are the short publication timeframe and the world-wide distribution, which enable both wide and rapid dissemination of research output.

Indexed by SCOPUS, DBLP, WTI Frankfurt eG, zbMATH, SCImago.

More information about this series at http://www.springer.com/series/7092

Patricia Melin · Oscar Castillo · Janusz Kacprzyk
Editors

# Recent Advances of Hybrid Intelligent Systems Based on Soft Computing

Springer

*Editors*
Patricia Melin
Division of Graduate Studies
and Research
Tijuana Institute of Technology
Tijuana, Mexico

Oscar Castillo
Division of Graduate Studies
and Research
Tijuana Institute of Technology
Tijuana, Mexico

Janusz Kacprzyk
Systems Research Institute
Polish Academy of Sciences
Warsaw, Poland

ISSN 1860-949X ISSN 1860-9503 (electronic)
Studies in Computational Intelligence
ISBN 978-3-030-58727-7 ISBN 978-3-030-58728-4 (eBook)
https://doi.org/10.1007/978-3-030-58728-4



This Springer imprint is published by the registered company Springer Nature Switzerland AG
The registered company address is: Gewerbestrasse 11, 6330 Cham, Switzerland

# Preface

We describe in this book the recent advances on fuzzy logic, neural networks and optimization algorithms, as well as their hybrid combinations, and their application in areas, such as intelligent control and robotics, pattern recognition, medical diagnosis, time series prediction, and optimization of complex problems. In particular, the enhancements of neural and optimization algorithms are given particular attention. There are papers with the main theme of type-1 and type-2 fuzzy logic, which basically consists of papers that propose new concepts and algorithms based on type-1 and type-2 fuzzy logic and their applications. There are also papers that present theory and practice of meta-heuristics in different areas of application. There are interesting papers on diverse applications of fuzzy logic, neural networks, and hybrid intelligent systems in medical applications. In addition, we can find papers describing applications of fuzzy logic, neural networks, and meta-heuristics in robotics problems. Another set of papers is presenting theory and practice of neural networks in different areas of application. There are also a group of papers that present theory and practice of optimization and evolutionary algorithms in different areas of application. Finally, we can find a set of papers describing applications of fuzzy logic, neural networks, and meta-heuristics in pattern recognition problems.

In conclusion, the edited book comprises papers on diverse aspects of fuzzy logic, neural networks, and nature-inspired optimization meta-heuristics for forming hybrid intelligent systems and their application in areas, such as intelligent control and robotics, pattern recognition, time series prediction, and optimization of complex problems. There are theoretical aspects as well as application papers.

Tijuana, Mexico — Patricia Melin
Tijuana, Mexico — Oscar Castillo
Warsaw, Poland — Janusz Kacprzyk
May 2020

# Contents

# Optimization of Fuzzy Logic Controllers with Distributed Bio-Inspired Algorithms

**Alejandra Mancilla, Oscar Castillo, and Mario García Valdez**

**Abstract** Optimization problems are present in many fields of science, engineering, and industry. In most cases, optimization problems involve a large number of complex decision variables. Often, the application of fuzzy systems to traditional control strategies requires the use of optimization techniques to define the parameters of the membership functions capturing the designer's experience. Currently, bioinspired algorithms are one of the most widely used methods for fuzzy system optimization. In this paper, we review the literature on the application of different bioinspired techniques such as Genetic Algorithms, Ant Colony Optimization, Particle Swarm Optimization, Galaxy Swarm Optimization, Harmony Search, Firefly, Differential Evolution, to optimize the parameters of the fuzzy controllers. We found particular trends in the current literature, with many methods using dynamic parameter adaptation, hybrids, distributed algorithms, and type-2 fuzzy logic.

**Keywords** Bioinspired algorithms · Fuzzy systems · Fuzzy control

## 1 Introduction

Systems incorporating fuzzy logic are currently widely used, and the type of applications is also very diverse. However, perhaps the field in which it has been applied more is in control, where we can find extensive publications about theoretical and practical aspects of what is called fuzzy control. We can attribute the success and industrial acceptance of this type of system to the advantage it presents to model design desitions by using words. A fuzzy inference system makes it possible to trans-

A. Mancilla (✉) · O. Castillo · M. García Valdez
Instituto Tecnológico de Tijuana, Tijuana, BC, Mexico
e-mail: alejandra.mancilla@tectijuana.edu.mx

O. Castillo
e-mail: ocastillo@tectijuana.mx

M. García Valdez
e-mail: mario@tectijuana.edu.mx



P. Melin et al. (eds.), *Recent Advances of Hybrid Intelligent Systems Based on Soft Computing*, Studies in Computational Intelligence 915,
https://doi.org/10.1007/978-3-030-58728-4_1

fer the knowledge of an expert to a set of fuzzy rules, where he can easily express the relationships between numeric inputs and outputs, using words to model the ambiguity. Fuzzy logic, together with other nature-inspired techniques such as genetic algorithms or neural networks, cooperate in many applications to solve complex problems. The incorporation of fuzzy logic in a control system also has the objective of giving a simple solution to complex problems in a natural way [1]. Fuzzy control systems have been applied to many types of applications, such as articulated arms and autonomous vehicles, in which the mathematical models are very complex [2]. In robotics, they are used to control the speed and the position of mobile robots [3–5]. Other applications are the navigation of mobile robots [6], flight control [7–9].

In this work, we are interested in the optimization of fuzzy controllers by using bioinspired algorithms. In particular, we are interested in which are the optimization techniques employed in the current literature, and if there are works that employ hybrid approaches, distributed systems, dynamic adaptation or other techniques that have the objective of reducing the computing time of the optimization process.The main contribution of the paper is to present a brief review and discussion of state of the art. The organization of the paper is as follows. First, we present the main concepts and design issues of fuzzy control in Sect. 2, then an overview of Bioinspired algorithms is presented in Sect. 3, followed by a review of related work in Sect. 4, and distributed bioinspired algorithms in Sect. 5, finally Conclusions are presented in Sect. 6.

## 2 Fuzzy Inference Systems for Control

Most applications implement fuzzy control as a rule-based system termed Fuzzy Inference System (FIS) as a way of representing inaccurate knowledge and data-values similarly as a human does. A FIS defines a nonlinear correspondence between one or more input variables and one output variable; From this input-output view, Babuvska [10] describes that these systems as "flexible mathematical functions which can approximate other functions or just data (measurements) with a desired accuracy". This provides a basis from which decisions can be made or patterns defined. Next we define the fuzzy production rules, that are the main components of a FIS:

Fuzzy rules or fuzzy IF-THEN rules are expressions of the form IF antecedent THEN consequent, where the antecedent is a proposition of the form "$x$ is $A$" where $x$ is a linguistic variable, and $A$ is a linguistic term. We use fuzzy logic sets to characterize the variables and terms used in these propositions. The truth value of this proposition is given by the the matching degree between $x$ and $A$. Linguistic Variables (LVs) are variables that can be assigned linguistic terms as values, i.e., if we define a linguistic variable *POSITION* we can assign it the linguistic terms *LOW*, *MEDIUM*, or *HIGH*. We can express the meaning of these linguistic terms by defining their membership functions (MFs). LVs can be defined as a 5-tuple $LV=<v,T,X,g,m>$ where $v$ is the name of the variable, $T$ is the set of linguistic terms of $v$, $X$ is the domain (universe) of $v$, $g$ is a syntactic rule to generate linguistic terms, m is a semantic rule that assigns to each term $t$ its meaning m($t$), which is a fuzzy

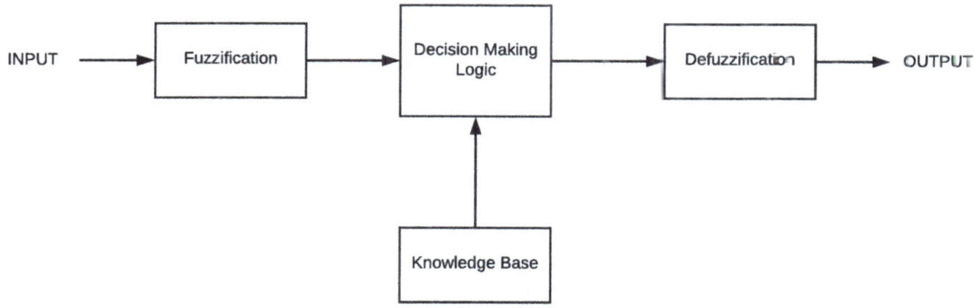


**Fig. 1** Fuzzy inference system block diagram

set defined in $X$. Depending on the form of the consequent, two main types of fuzzy production systems are distinguished:

- Linguistic fuzzy model: where both the antecedent and consequent are fuzzy propositions.
- Takagi-Sugeno fuzzy model: the antecedent is a fuzzy proposition; the consequent is a crisp function.

In fuzzy control, the input receives the current information in the feedback loop from the device as it operates, and the output controls the operation of the device. The stages that must be met to implement a Fuzzy Inference System are shown in Fig. 1 [11].

- Knowledge Base. A set of fuzzy rules, defining the relationship between input and output variables. Membership Functions are also defined here.
- Desition Making Logic or Fuzzy Inference Engine. This module executes the fuzzy inference operations.
- Fuzzifier. This interface transforms the inputs of the systems (numerical data) into linguistic values.
- Defuzzifier. This interface transforms the fuzzy results into numerical data.

From receiving the input to generating an output, a fuzzy inference follows these steps [12]:

1. Compare the input variables with the membership functions in the antecedent, to obtain the membership values of each linguistic term. This step is frequently called fuzzification.
2. Compose through a specific T-Norm operator (mainly max-min or max-product) the membership values to obtain the degree of support of each rule.
3. Generate the qualified consequence (fuzzy or numeric) of each rule depending on the degrees of support. These outputs are then aggregated to form a unified output.
4. Then the output fuzzy set is resolved or defuzzified to a single numeric value

Although the design of a basic fuzzy system could seem like an easy task, because it is only a matter of expressing our knowledge as fuzzy-rules, in reality, fuzzy control design is an engineering process. The main steps for designing a fuzzy system are:

1. Identify the type of problem and the type of fuzzy system that best fits the data.
2. We must define the system's input and output variables. Also, their fuzzy values and their membership functions, this is called the parameterization of variables.
3. Definition of the knowledge base or fuzzy rules.
4. Obtaining system outputs by input variable information using the fuzzy inference system.
5. Defuzzification of the output, to a crisp value.
6. Adjust the system, validating the results.

As we can see, the design of a FIS requires knowledge about the behavior of the system. In the case of the controller design, the adjustment step involves running simulations or testing on real devices. The adjustment, in this case, can be done on modifying the parameters of the membership functions, adding more membership functions to a variable, changing the rules, or even considering more input variables. Changing the parameters of membership functions is an easy way of improving the performance of a fuzzy controller, and this technique has been used extensively [13]. Another decision can be the selection of FIS, based on the type of fuzzy logic employed. When there is more noise in the system, a type-2 fuzzy logic system could be a better choice [14]. A drawback is that a type-2 fuzzy system increases the number of parameters and search space for optimization. Adjusting the system or optimizing the FIS parameters is a time consuming iterative process. From the earlier works, researchers have employed some optimization algorithms for searching the parameter space, trying to find a sub-optimal but good enough fuzzy controller. Next, we review the subject of bioinspired algorithms that are common in fuzzy controller optimization.

## 3 Bioinspired Algorithms

Bioinspired algorithms are a branch of Artificial Intelligence in which the behavior of natural systems is emulated to design nondeterministic heuristic search methods for optimization, learning, pattern recognition, and classification [15, 16]. A bioinspired system can be developed by using hardware and software whose objective is to simulate the way biological systems process information for/or solving problems [17]. Bioinspired algorithms focus on finding the optimal solution to a problem, in which all of its possible solutions can be represented as points in space. Each position or point in this space is a candidate solution. Algorithms have a function that evaluates the quality or fitness of a solution. Solutions with higher fitness are at a local optimum point in the solution space; bioinspired algorithm uses this fitness to guide the search, following a stochastic process. Due to the random nature of these algorithms, they may lose some local optimums, or even the global optimum, which would make the algorithm get stuck in a local optimum, avoiding it to find a better solution. Because of that, algorithms include procedures to ensure that further exploration of the search space can be possible, endowing the algorithm with what is known as a global search ability [17].

Among the first bioinspired algorithms, we can mention Genetic Algorithms (GAs) and Particle Swarm Optimization (PSO) that have proven to be very efficient in optimizing parameters.

### *3.1 Genetic Algorithms*

We can define a genetic algorithm as"a problem-solving technique that imitates biological evolution as a strategy to solve problems, encompassing what we have previously called population-based techniques" [18]. GAs are inspired by natural selection in the way individuals can reproduce them selfs naturally, combining their gens and or mutating some of their characteristics. In GAs we have as an input a collection of candidate solutions to a given problem. Candidate solutions are coded typically as a string of bits, and there is a fitness function, to quantitatively evaluate each candidate. Among these solutions, there may be some local optimums, and the genetic algorithm aims to improve on them. For this, there is a process of selection of the fittest, and they have a better chance of reproduction. The process repeats iteratively (generations) that are needed until we get a solution ones good enough for our problem. GA is a population-based search method that reproduces the life cycle: the selection of individuals, reproduction of these using genetic crossover and mutation operators to generate offspring, and a replacement mechanism to select the best individuals [19].

### *3.2 Particle Swarm Optimization*

Particle Swarm Optimization (PSO) is an optimization technique that is inspired by the behavior of insect swarms in nature. Specifically, we can think of a swarm of bees, since when searching for pollen they look for the region of space with the highest density of flowers, because the probability of finding pollen it is higher. The PSO algorithm borrowed this idea, and currently, it is one of the most used in the optimization of different types of systems [20]. The algorithm works by randomly placing particles in the search space, and giving them the possibility to move through it according to rules that take into account the personal knowledge of each particle and the global knowledge of the swarm. Here the knowledge is related to the quality of the solution similar to the fitness in GAs. One difference with GAs, is that in a PSO particles can move between iterations, and the algorithm does not replace them When they move around this landscape and to communicate with each other, they can discover local or global minimums. A general term used for describing swarm-based algorithms is swarm-intelligence. They refer to algorithms inspired by the behavior of swarms of bees, schools of fish or flocks of birds. In them, each member of a swarm it is represented as a particle with at least two parameters: the position, representing the current candidate solution, and the speed at which it moves on the solution space. Furthermore, each particle is capable of "remembering" the position with the best fitness in which it has been found (personal optimum) and the position with the best fitness of all neighboring particles (global optimum) [17]. In a new iteration, the new position of a particle considers its previous position and velocity and considers local and global knowledge. Similar to a flock of birds, where the position of a single bird is influenced by other members and leaders of the flock.

### 3.3 *Other Bioinspired Algorithms*

More bio-inspired optimization algorithms have been proposed in the last decades. These algorithms are based on the previous ideas of evolutionary algorithms (EAs) [21] and swarm intelligence (SI) [22]. Borrowing from Evolutionary algorithms we can mention Differential Evolution (DE) [23], and Genetic Programming (GP) [21], while examples of SI algorithms [22] are artificial bee colony (ABC) [24], Grey Wolf Optimization (GWO) [25], Firefly [26] ,and ant colony algorithms (ACO) [27], and Harmony Search [28].

## 4 Related Work

In this section, we consider some of the work that has been done recently, in terms of optimizing parameters of fuzzy controllers using bioinspired techniques:

Oscar Castillo [29] compared the performance of two bioinspired methods, Ant Colony Optimization (ACO) and the Gravitational Search Algorithm (GSA). They implemented a dynamic adaptation of parameters to enhanced both algorithms. Dynamic adaptation means that specific parameters of the algorithm are changed dynamically on each iteration, to increase or decrease the exploitation or exploration in the search space. To make these adaptations, they considered some metrics about the algorithm execution, like the percentage of iterations elapsed or the diversity of the population. Then by using a type-2 fuzzy logic decision system, they adapted the parameters on each iteration. They compered these algorithms to adjust a fuzzy controller optimized to minimize the error in the simulation with complex nonlinear plants. In this work, the parameters of membership functions of the input variables of the controller were optimiezed.

Balcucho et al. [30], addresses the problem of improving the performance of fuzzy controllers through the improvement of their membership rules and functions, for which they use a GA and simulated annealing.

In a work of Arostegui et al. [31], they propose the implementation of an algorithm based on the biological behavior of bees to a PID controller and a non-optimized Fuzzy PD + I controller also improving its membership functions. Subsequently, they applied this algorithm to control the outlet temperature of a heat exchanger in a PCT32-MKII pasteurization plant and compared the results with a classic model of a PID controller and a non-optimized Fuzzy PD + I controller.

In a recent work, Lara Lagunes et al. [4] proposed a methodology for optimizing the parameters of the membership functions of a fuzzy tracker controller for a mobile autonomous robot using the firefly algorithm. In this work, the optimization of the parameters of the membership functions in the fuzzy control systems allows a better performance of the actuators that control an autonomous robot.

Bernal et al. [32], present a fuzzy method for dynamic parameter adjustment using the galactic swarm optimization algorithm (GSO). GSO is based on the movement of stars and galaxies in the universe, as well as on their attractive influence that allows the use of multiple cycles of exploration and exploitation to solve complex optimization problems. In this work, they dynamically adjust two of the GSO parameters by using a fuzzy system. They tested the fuzzy approach with a set of mathematical benchmark functions and with the fuzzy controller of the water tank problem. Finally, they presented a comparison of the results between the proposed method and other metaheuristics.

Peraza e al. [33] present the application of the fuzzy harmony search algorithm (FHS). In this case, they also adapt the parameters of the original harmony search algorithm (HS) by using a fuzzy system for the dynamic adaptation of parameters The parameters they adapt are Harmony Memory Acceptance (HMR) and Pitch Adjustment (PArate), with the Fuzzy System rules controlling the intensification and diversification of the search space. They applied this method to the mathematical functions provided by CEC 2017, which are unimodal, multimodal, hybrid, and composite functions to verify the efficiency of the proposed method. They presented a comparison to verify the results obtained with the original harmony search algorithm and the fuzzy harmony search algorithm.

Castillo et al. [34] perform a comparison between three optimization algorithms Bee Colony Optimization (BCO), Differential Evolution (DE), and Harmony Search (HS). Furthermore, for each algorithm, they propose a type 1 fuzzy logic system (T1FLS) for the dynamic modification of the main parameters. They analyze and compare each algorithm (modified and original versions) based on the optimal design of fuzzy systems for benchmark control problems, especially in the design of fuzzy controllers. The simulation results provide evidence that the algorithm FDE outperforms the FBCO and FHS algorithms in optimizing fuzzy drivers. Statistically, they show that the best errors are found with the implementation of fuzzy systems to improve each proposed algorithm.

Martinez-Soto et al. [5] propose the use of a hybrid PSO-GA optimization method for the design of fuzzy logic controllers (FLC). They control the trajectory tracking of autonomous mobile robots They have used the PSO-GA algorithms to find the parameters of the membership functions of the FLC to obtain the optimal controller.

Hernández et al. [35]. propose to use the Grey Wolf Optimizer (GWO), inspired by the hunting behavior and leadership hierarchy of grey wolves, for the optimization of fuzzy controllers for a two-wheeled autonomous mobile robot to validate the efficiency of the proposed approach.

| References | Bioinspired algorithms | Parameter adaptation |
|---|---|---|
| Castillo [29] | ACO, GSA | Yes-FL |
| Balcucho et al. [30] | GA, SA | No |
| Arostegui et al. [31] | ABC | No |
| Lara Lagunes et al. [4] | Firefly | Yes |
| Bernal et al. [32] | GSO | No |
| Peraza e al. [33] | HS | Yes |
| Martinez-Soto et al. [5, 34] | PSO, GA | Yes |
| Hernández et al. [35] | GWO | No |
| Castillo et al. [34] | BCO, DE, HS | Yes |

In this brief review, we can see the following trends. Many works are related to applying novel bio-inspired metaheuristics, to optimize parameters of fuzzy controllers, mainly adjusting the parameters of membership functions. There is also work related to improving the search algorithm, and the most common is the dynamic adaptation of parameters that control the degree of exploration and exploitation. Also, some papers present combinations of metaheuristics. Currently, there is a lack of distributed algorithms or parallel implementations, using a multi-population approach.

## 5 Distributed Bioinspired Algorithms

This work Muelas et. to the. [muelas2009algoritmos] analyze the behavior of two heterogeneous distributed algorithms that combine evolutionary techniques: Genetic Algorithms (GA), Distribution Estimation Algorithms (EDA), and Differential Evolution (DE) on the set of functions of the CEC'05. The results obtained showed that the interaction of these algorithms, by means of the distributed model, obtains very competitive results, improving or equalizing in various functions, the best result obtained by the three best algorithms that were proposed in the CEC'05. They showed that the interaction of GA, EDA, and DE algorithms could be very beneficial in solving functions with different degrees of difficulty in a continuous search space. The two-island model with the GA-DE algorithms has obtained better overall results than the three-island model, mainly due to the higher number of generations available and the ballast derived from the low performance of the EDA algorithm in some of the functions analyzed.

Currently, several evolutionary pool-based algorithms (PEAs) have been proposed [36, 37], which asynchronously distributes an evolutionary search among heterogeneous devices, using nodes and nodes controlled outside the local network, through web browsers or cloud services. In a PEA, the population is stored in a shared group, while distributed processes called workers execute the evolutionary search.

This approach allows researchers to use low-cost computing power that might not otherwise be available.

Based on these references, our work proposes evaluating various bioinspired algorithms to optimize a fuzzy controller [2, 38, 39], since these allow us to apply different techniques where we try to optimize their parameters or find the ones that work best. The optimization of controllers using bioinspired algorithms has a reasonably high computational cost, so we propose to implement the algorithms in a distributed and asynchronous way.

## 6 Conclusions

As we have seen so far, many of the works are applied to the optimization of fuzzy controllers, and they do so through centralized, monolithic bioinspired techniques, also using some other techniques for classic controllers or a combination of various bioinspired algorithms. Most of the works employ some dynamic adaptation techniques of parameters of the bioinspired algorithm, functions, or membership rules, trying to obtain better results. However, we have seen that few works have tried to do it in the cloud or only in a distributed way but not doing a combination of these strategies. So we see the opportunity to work on the optimization of the parameters of fuzzy controllers by applying different bioinspired algorithms but in a distributed, asynchronous, and cloud-based manner.

## References

1. C. De Los Ríos, W.I. Alama, L.A.S. Dedios, Evaluación de estructuras y métodos de ajuste de reguladores pid difusos. (Universidad de Piura, Perú, 2004)
2. M.L. Lagunes, O. Castillo, J. Soria, Methodology for the optimization of a fuzzy controller using a bio-inspired algorithm, in *North American Fuzzy Information Processing Society Annual Conference* (Springer, 2017), pp. 131–137
3. O. Castillo, H. Neyoy, J. Soria, P. Melin, F. Valdez, A new approach for dynamic fuzzy logic parameter tuning in ant colony optimization and its application in fuzzy control of a mobile robot. Appl. Soft Comput. **28**, 150–159 (2015)
4. M.L. Lagunes, O. Castillo, J. Soria, Optimization of membership function parameters for fuzzy controllers of an autonomous mobile robot using the firefly algorithm, in *Fuzzy Logic Augmentation of Neural and Optimization Algorithms: Theoretical Aspects and Real Applications* (Springer, 2018), pp. 199–206
5. R. Martinez-Soto, O. Castillo, L.T. Aguilar, I.S. Baruch, Bio-inspired optimization of fuzzy logic controllers for autonomous mobile robots, in *2012 annual Meeting of the North American Fuzzy Information Processing Society (NAFIPS)* (IEEE, 2012), pp. 1–6
6. M.P. Garcia, O. Montiel, O. Castillo, R. Sepúlveda, P. Melin, Path planning for autonomous mobile robot navigation with ant colony optimization and fuzzy cost function evaluation. Appl. Soft Comput. **9**(3), 1102–1110 (2009)
7. L. Cervantes, O. Castillo, Design of a fuzzy system for the longitudinal control of an f-14 airplane, in *Soft computing for intelligent control and mobile robotics* (Springer, 2010), pp. 213–224

8. L. Cervantes, O. Castillo, Type-2 fuzzy logic aggregation of multiple fuzzy controllers for airplane flight control. Inf. Sci. **324**, 247–256 (2015)
9. P. Melin, O. Castillo, Adaptive intelligent control of aircraft systems with a hybrid approach combining neural networks, fuzzy logic and fractal theory. Appl. Soft Comput. **3**(4), 353–362 (2003)
10. R. Babuška, in *Fuzzy Systems, Modeling and Identification*, vol. 4. (Delft University of Technology, Department of Electrical Engineering Control Laboratory, Mekelweg, 1996)
11. M.G. Simoes, Introduction to fuzzy control. Colorado School of Mines Eng. Div. Gold. Colorado **8**, 18–22 (2010)
12. D.J. Dubois, *Fuzzy Sets and Systems: Theory and Applications*, vol. 144. (Academic Press, 1980)
13. O. Castillo, P. Melin, Soft computing for control of non-linear dynamical systems. Appl. Soft Comput. **3**, 301 (2003)
14. O. Castillo, L. Amador-Angulo, J.R. Castro, M. Garcia-Valdez, A comparative study of type-1 fuzzy logic systems, interval type-2 fuzzy logic systems and generalized type-2 fuzzy logic systems in control problems. Inf. Sci. **354**, 257–274 (2016)
15. C. Caraveo, F. Valdez, O. Castillo, Optimization of fuzzy controller design using a new bee colony algorithm with fuzzy dynamic parameter adaptation. Appl. Soft Comput. **43**, 131–142 (2016)
16. A. Cárdenas Cardona, Inteligencia artificial, métodos bio-inspirados: un enfoque funcional para las ciencias de la computación. B.S. thesis, Universidad Tecnológica de Pereira, 8 2012
17. I. Riquelme Medina, *Revisión de los Algoritmos Bioinspirados*. Ph.D. thesis, 07 2014
18. F.S. Caparrini, Algoritmos genéticos. Technical report, Universidad de Sevilla, 2019
19. S. Muelas, J. Pena, A. LaTorre, V. Robles, *Algoritmos distribuidos heterogéneos para problemas de optimización continua* (In VI Congreso Español sobre Metaheurísticas, Algoritmos Evolutivos y Bioinspirados, MAEB, 2009), pp. 425–432
20. F.S. Caparrini, Pso: Optimización por enjambre de partículas. Technical report, Universidad de Sevilla, 2019
21. T. Back, *Evolutionary algorithms in theory and practice: evolution strategies, evolutionary programming, genetic algorithms* (Oxford University Press, 1996)
22. J. Kennedy, Swarm intelligence, in *Handbook of Nature-Inspired and Innovative Computing* (Springer, 2006), pp. 187–219
23. D. Karaboğa, S. Ökdem, A simple and global optimization algorithm for engineering problems: differential evolution algorithm. Turkish J. Electr. Eng. Comput. Sci. **12**(1), 53–60 (2004)
24. D. Karaboga, An idea based on honey bee swarm for numerical optimization. Technical report, Technical report-tr06, Erciyes University, Engineering Faculty, 2005
25. S. Mirjalili, S.M. Mirjalili, A. Lewis, Grey wolf optimizer. Adv. Eng. Softw. **69**, 46–61 (2014)
26. X.-S. Yang et al., Firefly algorithm, in *Nature-Inspired Metaheuristic Algorithms*, vol. 20, pp. 79–90 (2008)
27. M. Dorigo, G. Di Caro, Ant colony optimization: a new meta-heuristic, in *Proceedings of the 1999 congress on evolutionary computation-CEC99 (Cat. No. 99TH8406)*, vol. 2, (IEEE, 1999), pp. 1470–1477
28. X.-S. Yang, Harmony search as a metaheuristic algorithm, in *Music-Inspired Harmony Search Algorithm* (Springer, 2009), pp, 1–14
29. O. Castillo, Dynamic parameter adaptation based on using interval type-2 fuzzy logic in bio-inspired optimization methods, in *International Conference on Innovations in Bio-Inspired Computing and Applications* (Springer, 2018), pp. 1–12
30. J.A. Balcucho Mogollón, O.E. Gualdrón Guerrero. Optimización de la base de reglas de un controlador difuso, mediante técnicas estocásticas como algoritmos genéticos y el simulated annealing. (Universidad de Pamplona, Revista Colombiana de Tecnologías de Avanzada, Pamplona, Norte de Santander, Colombia, 2010)
31. A.V. Arosteguí Arias et al., Optimización de un esquema de control difuso mediante algoritmos evolutivos (artificial bee colony). B.S. thesis, Quito, 2019

32. E. Bernal, O. Castillo, J. Soria, F. Valdez, Fuzzy galactic swarm optimization with dynamic adjustment of parameters based on fuzzy logic. SN Comput. Sci. **1**(1), 59 (2020)
33. C. Peraza, F. Valdez, O. Castillo, Harmony search with dynamic adaptation of parameters for the optimization of a benchmark set of functions, in *Hybrid Intelligent Systems in Control, Pattern Recognition and Medicine* (Springer, 2020), pp. 97–108
34. O. Castillo, F. Valdez, J. Soria, L. Amador-Angulo, P. Ochoa, C. Peraza, Comparative study in fuzzy controller optimization using bee colony, differential evolution, and harmony search algorithms. Algorithms **12**(1), 9 (2019)
35. E. Hernández, O. Castillo, J. Soria. Optimization of fuzzy controllers for autonomous mobile robots using the grey wolf optimizer, in *2019 IEEE International Conference on Fuzzy Systems (FUZZ-IEEE)* (IEEE, 2019), pp. 1–6
36. M. García-Valdez, A. Mancilla, L. Trujillo, J.-J. Merelo, F. Fernández-de Vega, Is there a free lunch for cloud-based evolutionary algorithms? in *2013 IEEE Congress on Evolutionary Computation*, (IEEE, 2013), pp. 1255–1262
37. M. García-Valdez, L. Trujillo, J. J. Merelo-Guérvos, and F. Fernández-de Vega. Randomized parameter settings for heterogeneous workers in a pool-based evolutionary algorithm, in *International Conference on Parallel Problem Solving from Nature* (Springer, 2014), pp. 702–710
38. M. Guerrero, O. Castillo, M. García, Cuckoo search algorithm via lévy flight with dynamic adaptation of parameter using fuzzy logic for benchmark mathematical functions, in *Design of Intelligent Systems Based on Fuzzy Logic, Neural Networks and Nature-Inspired Optimization* (Springer, 2015), pp. 555–571
39. M. Guerrero, O. Castillo, M. García, Fuzzy dynamic parameters adaptation in the cuckoo search algorithm using fuzzy logic, in *2015 IEEE Congress on Evolutionary Computation (CEC)* (IEEE, 2015), pp. 441–448

# Parallel-Machine Scheduling Problem: An Experimental Study of Instances Difficulty and Algorithms Performance

**Octavio Ramos-Figueroa, Marcela Quiroz-Castellanos, Guadalupe Carmona-Arroyo, Betsabé Vázquez, and Rupak Kharel**

**Abstract** Parallel-machine scheduling problems are well-known NP-hard combinatorial optimization problems with many real-world applications. Given the increasing appearance of these problems, over the last few years, several algorithms have been developed to solve them, and different instances of these problems have been designed to be used as the benchmark for testing the algorithms. However, there is limited knowledge about why the algorithms show some performance and the difficulty degree of the instances. This chapter provides a starting point for the characterization of Parallel-machine scheduling problems with an emphasis on the case of unrelated machines, jobs with non-preemptions, and the reduction of the maximum completion time ($R||C_{max}$). For this purpose, we calculate seventeen measures for characterizing the structure of the instances, and we apply data analysis techniques to study how the features of the instances affect the performances of eleven constructive heuristics. The algorithmic behavior explanations obtained by the study suggest that the difficulty of the $R||C_{max}$ instances is related to: (1) the number of jobs and machines; (2) the dispersion of the processing times of the jobs; (3) characteristics generated by considering only the minimum times in which each job can be processed; and (4) the difference between the processing times of the two fastest machines.

O. Ramos-Figueroa (✉) · M. Quiroz-Castellanos · G. Carmona-Arroyo · B. Vázquez
Artificial Intelligence Research Center, Universidad Veracruzana, Xalapa, Mexico
e-mail: oivatco.rafo@gmail.com

M. Quiroz-Castellanos
e-mail: qc.marcela@gmail.com

G. Carmona-Arroyo
e-mail: lupita_0111@outlook.es

B. Vázquez
e-mail: jennybey13@gmail.com

R. Kharel
School of Engineering, Manchester Metropolitan University, Manchester, UK
e-mail: R.Kharel@mmu.ac.uk


P. Melin et al. (eds.), *Recent Advances of Hybrid Intelligent Systems Based on Soft Computing*, Studies in Computational Intelligence 915,
https://doi.org/10.1007/978-3-030-58728-4_2

**Keywords** Algorithm behavior analysis · Characterization · Unrelated parallel machine scheduling

## 1 Introduction

Applications of Combinatorial Optimization Problems (COPs) arise in engineering, science, economics, industry, and everyday life. Given the importance of COPs, several studies have been performed, providing solutions that allow minimizing their costs [1]. It is well-known that the solution process required by these problems represents high complexity because, in most cases, real-world COPs belong to a special class of problems called NP-hard [2], which implies there is no efficient algorithm to find an optimal solution for every finite-size instance of a COP. For the solution of NP-hard COPs, a wide variety of heuristic algorithms have been proposed with satisfactory performance. However, despite the efforts of the scientific community in the development of new strategies, until now, no algorithm is the best option for all possible situations. Given the above, the design of algorithms suited to specific conditions is usually the best option to address this type of problem.

One of the main challenges in the design of heuristic algorithms is to identify which strategies make an algorithm show better performance and under what conditions they obtain it. Much of the recent progress, in the development of algorithms, has been aided by a better understanding of the properties of problem instances and the performance of the algorithms that solve them. An example of that, are the numerous studies about the complexity of instances and the performance of algorithms for the Propositional Satisfiability Problem and the Traveling Salesman Problem [3, 4]. However, the percentage of studies carried out to understand how and why the algorithms follow particular behaviors is much lower than the works that create such algorithms. Furthermore, the characteristics that explain the degree of difficulty of the instances of many COPs have not yet been studied [5, 6].

This chapter addresses the characterization of the Parallel-machine scheduling problem with unrelated machines. We propose some measures to describe properties of the problem instances, and we analyze its impact with the relative performance of a set of algorithms. The main objective of this work is to promote the characterization and analysis of Parallel-machine scheduling problems and heuristics used to solve them as well as understanding the degree of difficulty of the instances. The knowledge gained from this study can be useful to design better algorithms and generate new hard-to-solve instances that fairly evaluate the performance of the algorithms discovering their strengths and weaknesses. To the authors' knowledge, no similar works, aimed at studying the hardness of Parallel-machine scheduling instances, have been presented and analyzed.

The rest of the chapter continues as follows. Section 2 summarizes previous works dealing with the characterization of instances difficulty for COPs. In Sect. 3, we outlined the theoretical foundations of parallel machine scheduling problems, with a focus on the $R||C_{max}$ variant, and we also describe the benchmark instances and algorithms considered in this research. In Sect. 4, we apply the experimental approach proposed by Quiroz et al. [7] to study the hardness of the instances of the $R||C_{max}$ Parallel-machine scheduling problem. First, the properties of the instances and the performance of the algorithm are quantified using indexes. Next, we discard incorrect and redundant indexes. Then, we analyze the indexes to obtain performance relations that eventually will become instance difficulty explanations. Finally, Sect. 5 contains the conclusions of this work.

## 2 Approaches for Characterization of Instance Difficulty for COPs

Real implementations of heuristic algorithms exploit a lot of ideas and strategies related to the characteristics of the problems to solve. However, in most cases, there is limited knowledge about why the heuristics show some performance and there is no theoretical basis on how to characterize and analyze the factors of the problem and the algorithm that lead to successful performance.

The analysis of state-of-the-art heuristics algorithms for COPs also has revealed that most of the research has focused on demonstrating that the proposed heuristics are effective or superior to others, to solve some optimization problem, but the reason, for their competitive or inadequate performance, has not been demonstrated. Several authors have agreed there is still a lack of knowledge about the exact way the performance of different heuristics is related to the characteristics of the problem instances to be solved, highlighting the need for a guide for the description and performance analysis of real implementations of these strategies [5, 6, 8]. Likewise, many NP-hard COPs have not yet been studied. Consequently, there is little knowledge about the difficulty of their test instances and the behavior of the algorithms that solve them.

The related works about the characterization and analysis of the algorithmic optimization process include different approaches, Quiroz et al. present a summary of the main works in this area [8].

In 1976, Rice introduced the algorithm selection problem [9], the key idea was to select an effective or best algorithm given some indepeendent characteristics of the problems which are important for the algorithm selection and performance.

Since the seminal work of Rice, there have been important efforts to predict the performance of algorithms for COPs via the application of statistical methods, multivariate analysis, and machine learning techniques [10].

The field of meta-learning, especially concerning algorithm selection and configuration, has been an active area of research [11, 12]. Another successful approach has been the analysis of the problem search space structure; fitness landscape analysis has allowed the study metrics to characterize the search space and its properties. many of these metrics have been used to analyze a variety of COPs [13–15].

On the other hand, studies of the relative hardness of problem instances for various NP-hard COPs have shown a phase transition property around which the most difficult problems occur. The identification of these properties has allowed characterizing the instances as easy or hard, and a large number of experimental results have demonstrated the effect of these properties for both exact and heuristics algorithms [16, 17].

The difficulty of the NP-hard COPs instances can also be characterized by studying the structure of optimal solutions. For some COPs, it has been possible to identify sets of backbone variables that have fixed values amongst all optimal solutions, showing that there is a relation between the magnitudes of these sets and the degree of the difficulty of the instances [13, 18].

Although various studies have focused on measuring the instance difficulty for COPs, the complexities of NP-hard COPs and heuristic algorithms have shown that a single characterization approach is not sufficient to understand the performance of the algorithms and the difficulty of the instances. Highlighting the need for more complete analysis, combining all the proposed characterization techniques to obtain better explanations regarding the heuristic algorithms performance and the difficulty of the instances solved are also other challenges. This type of study is important since it can provide a solid basis for the analysis and design of algorithms.

A good experimental analysis of heuristic algorithms should enable us to understand the relationships between features of instances and algorithmic performance, to explain the strengths and weaknesses of each algorithm. The definition of the optimal subset of features that adequately measure the relative difficulty of the instances is a critical task, the way instances structure affects the performance of the algorithms is complex and the explanations of the algorithm behavior depend in an unstable way on the features actually used. The construction of features to characterize the structure of the instances of an optimization problem can be made via problem-independent metrics, considering the search space and its properties, and via problem-specific features. Smith-Miles and Lopes review some of the problem-specific features that have been constructed to characterize the problem difficulty of various COPs [19]. However, as far as we know, there is not any work related to the characterization of Parallel-machine scheduling problems. The motivation behind the work presented in this chapter is to contribute to the construction of suitable problem-specific features that allow us to characterize and understand the hardness of Parallel-machine scheduling problems.

## 3 The Parallel-Machine Scheduling Problem

Parallel-machine scheduling problems are classical NP-hard combinatorial optimization problems. Currently, there exists a great interest in the scientific community to generate practical and theoretical knowledge related to these problems, mainly because of their extensive number of industrial, planning, and logistics applications.

The statement of the classical Parallel-machine scheduling problem can be generalized as scheduling a collection of $n$ jobs $N = \{j_1, \ldots, j_n\}$ in a set of $m$ machines $M = \{i_1, \ldots, i_m\}$, in such a way that each machine $i$ can process one job at the time, and every job is assigned to exclusively one machine. The aim is to find a schedule that optimizes a certain performance measure [20].

Parallel-machine scheduling is, in fact, a family of problems. Indeed, several parameters intervene in the problem definition: resource environment, job characteristics, optimization criteria, and scheduling environment, among others. To differentiate the problem variants identified in the literature, the $\alpha|\beta|\gamma$ notation was proposed [21, 22]. The first variable, $\alpha$ is used to indicate the machine environment (i.e., a problem with a single machine ($\circ$), identical parallel machines ($P$), uniform parallel machines ($Q$), unrelated parallel machines ($R$) or an open shop ($O$)). The second one, $\beta$ specifies the job characteristics (i.e., no preemption ($\circ$), limited resources ($res$), a precedence relation ($r_j$), and the processing time ($p_j$ or $p_{ij}$)). Finally, $\gamma$ defines the goal of interest (i.e., the completion time ($C_i$), the lateness ($L_j$), the tardiness ($T_j$), and the unit penalty ($U_j$)).

Efficiently solving large-scale instances of this kind of COPs with exact algorithms is almost impossible or takes a considerably long time. Therefore, the construction of efficient and accurate heuristics is often necessary to solve these problems. At the same time, the potential of these algorithms must be studied, evaluating their performance by solving different classes of test instances.

The most general variant of Parallel-machine scheduling problems is the $R||C_{max}$ problem, consisting of unrelated machines, jobs without preemption, and minimization of the maximum completion time, commonly referred to as makespan. $R||C_{max}$ can be considered an assignment problem since the ordering process of the jobs in the machines does not affect their performance. Given the above, $R||C_{max}$ is stated as a formulation of Mixed Integer Linear Programming (MILP) as follows [23]:

$$\min C_{max} \tag{1}$$

$$\sum_{i=1}^{m} x_{ij} = 1 \ \forall j \in N \tag{2}$$

$$\sum_{j=1}^{n} p_{ij} \cdot x_{ij} \leq C_{max} \ \forall i \in M \tag{3}$$

$$x_{ij} \in \{0, 1\} \ \forall j \in N, \forall i \in M, \tag{4}$$

where $C_{max} = max(C_i)$, $C_i$ indicates the total completion time for each machine $i$ to process its assigned jobs; $p_{ij}$ is the processing time of job $j$ on machine $i$; and $x_{ij} = 1$ if the job $j$ is assigned to the machine $i$, otherwise $x_{ij} = 0$.

According to the state-of-the-art, $R||C_{max}$ holds a NP-hard complexity since it was shown that $P||C_{max}$, a more simple scheduling problem, belongs to that class [24].

In addition to this, in 1990 Lenstra *et al.* showed that it is not possible to develop an algorithm with a better worst-case ratio approximation than 3/2 to solve $R||C_{max}$ unless $P = NP$ [25].

This chapter presents a study of instances difficulty and algorithms performance for the $R||C_{max}$ problem. Using a knowledge discovery approach, we seek insights into the relationships between the properties of test instances and the effectiveness of some scheduling heuristics. We argue for these characterization measures to be applied to other Parallel-machine scheduling problems, to contribute to the characterization and analysis of instances and algorithms.

## 3.1 $R||C_{max}$ benchmark of instances

The general structure of the $R||C_{max}$ test instances can be defined using a matrix of size $(m \times n)$. Each cell in the matrix contains the processing time $p_{ij}$ required by machine $i$ to process job $j$. Figure 1 shows an $R||C_{max}$ instance example.

According to the specialized literature on $R||C_{max}$ instances, in the first related works, each author generated their test instances (i.e., Martello et al. [26], Mokotoff and Jimeno [27, 28] or Ghirardi and Potts [29]). None of those instances are publicly available, nor their best-known solutions. Nevertheless, it is well-known that have been studied instances with values of $p_{ij}$ generated with a uniform distribution in intervals like $U(1, 100)$ and $U(10, 100)$, as well as with job and machine correlated. That is short (long) processing times of jobs $p_{ij}$ for all machines and slow (fast) machines for all jobs, respectively.

It was until 2010 that Fanjul-Peyro and Ruiz introduced a collection of 1400 test instances available publicly at http://soa.iti.es, and also used in this work [23]. This benchmark groups the 1400 instances in seven groups concerning the criteria used to generate the values of $p_{ij}$ in each instance. Its characteristics are the following. The first five group includes instances with values of $p_{ij}$ uniformly distributed in the intervals $U(1, 100)$ $U(10, 100)$, $U(100, 120)$, $U(100, 200)$, and $U(1000, 1100)$, respectively; while the sixth and seventh group include instances with correlated jobs (Jobcorre) and correlated machines (Machcorre), respectively. Each group (sub-set)

**Fig. 1** Example of a test instance of $R||C_{max}$

Machines

| Jobs | $N \backslash M$ | $i_1$ | $i_2$ | $\cdots$ | $i_m$ |
|---|---|---|---|---|---|
| | $j_1$ | $p_{11}$ | $p_{21}$ | $\cdots$ | $p_{m1}$ |
| | $j_2$ | $p_{12}$ | $p_{22}$ | $\cdots$ | $p_{m2}$ |
| | $\vdots$ | $\vdots$ | $\vdots$ | $\ddots$ | $\vdots$ |
| | $j_n$ | $p_{1n}$ | $p_{2n}$ | $\cdots$ | $p_{mn}$ |

holds ten instances for each of the twenty possible combinations generated with $m = 10, 20, 30, 40, 50$ and $n = 100, 200, 500, 1000$. Thus, the size of the smaller instances is $100 \times 10$, and of the larger ones is $1000 \times 50$.

## 3.2 Algorithms for the $R||C_{max}$

$R||C_{max}$ holds hard complexity; consequently, the state-of-the-art includes methods designed to solve it under different approaches, such as deterministic heuristics [30–33], exact methods [26–29, 34], two-phase algorithms or rounding methods [25, 35–41], local searches [23, 34, 42, 43, 43, 43, 44, 44, 45, 45–47, 47, 48], evolutionary algorithms [34, 43, 49], and hybrid algorithms [34, 43]. All these approaches require initial solutions to perform their search process, which is the main motivation of this experimental study, involving the characterization of $R||C_{max}$ and the algorithmic behavior of eleven simple heuristics to generate solutions, five deterministic, and six non-deterministic.

In deterministic algorithms, given a particular input, the same output is always produced, passing through the same sequence of states. Algorithms with these characteristics are far functional so that they can be run on real devices (e.g., machines, computers, robots, etc.) efficiently. This chapter comprises four classical deterministic algorithms: Lowest [23], Lowest min [33], Highest min [33], Mean min [31], and one deterministic heuristic proposed in this work, called Diff_fastest min.

In contrast, non-deterministic algorithms can generate different outputs on different runs, even for the same input. In most cases, such behavior is related to the use of random number generators. Algorithms with these characteristics are far recurrent to address complex problems, with large and hard search spaces. In this work, we analyze the algorithmic behavior of six non-deterministic constructive heuristics, one taken from the specialized literature known as Random min [33], and five more introduced in this work, called: Random, Random lowest bound min, Lowest 4 g min, Highest 4 g min, Diff_fastest 4 g min. Most of them use the scheduling strategy Min heuristic, presented in Algorithm 1. Min heuristic allocates each job $j$ to the machine $i$ that produces the lowest value of $S_i = S_i + p_{ij}$, where $S_i$ is the time demanded by machine $i$ to process all its assigned jobs $j$, and $p_{ij}$ is the time demanded by machine $i$ to process job $j$. However, some of these constructive heuristics differ in the strategy they use to sort the jobs. Figure 2 contains an example of an instance with $n = 8$ jobs and $m = 4$ machines, as well as the detail of four properties used to classify the jobs before schedule them: (1) *lowest*, the processing time $p_{ij}$ required by the fastest machine $i$ to process job $j$; (2) *highest*, the processing time $p_{ij}$ required by the slowest machine $i$ to process job $j$; (3) *mean*, the average processing time required by all the machines in $M = \{i_1, \ldots, i_m\}$ to process job $j$; and (4) *diff_fastest*, the difference between the two fastest machines to process the job $j$.

| $N \backslash M$ | $i_1$ | $i_2$ | $i_3$ | $i_m$ | *lowest* | *highest* | *mean* | *diff_fastest* |
|---|---|---|---|---|---|---|---|---|
| $j_1$ | 20 | 15 | 10 | 12 | 10 | 20 | 14.25 | 2 |
| $j_2$ | 10 | 12 | 16 | 18 | 10 | 18 | 14 | 2 |
| $j_3$ | 18 | 15 | 11 | 10 | 10 | 18 | 13.5 | 1 |
| $j_4$ | 15 | 17 | 16 | 11 | 11 | 17 | 14.75 | 4 |
| $j_5$ | 10 | 20 | 11 | 19 | 10 | 20 | 15 | 1 |
| $j_6$ | 20 | 12 | 15 | 16 | 12 | 20 | 15.75 | 3 |
| $j_7$ | 11 | 19 | 20 | 13 | 11 | 20 | 15.75 | 2 |
| $j_n$ | 12 | 10 | 18 | 20 | 10 | 20 | 15 | 2 |

**Fig. 2** Instance characteristics used by constructive heuristicsInstance characteristics used by constructive heuristics

**Algorithm 1** Min heuristic

**Input:** A job $j$, available machines $M = \{i_1, \ldots, i_m\}$, and their current workload $S_i$.
**Output:** The machine $i$ where the job $j$ must be assigned.

```
for machine i in M do
    calculate S_i = S_i + p_ij.
end for
select the machine i that produces the lowest S_i.
```

According to the scope of the literature review, the earliest efforts to solve $R||C_{max}$ were devoted to the design of simple heuristics. The first related work was presented in 1977 by Ibarra et al. [33], who proposed five well-known heuristics differing in the criterion used to establish the scheduling order of the jobs, the heuristic for allocating them, and the current status of the schedule before allocating a new job. From that study, we consider three heuristics, referred to in this chapter as Lowest min, Highest min, and Random min. Three years later, De and Morton studied the strengths and weaknesses of the heuristics proposed by Ibarra et al.; as a result, they introduced novel heuristic algorithms [31]. In this chapter, we also consider one of them, referred to as heuristic Mean min. The last deterministic constructive heuristic taken from the specialized literature is Lowest, used to generate the initial solution of different neighborhood searches, designed by Fanjul-Peyro and Ruiz in 2010 [23]. Finally, in this work, the remaining heuristics were proposed, called Diff_fastest min, Random, Random lowest bound min, Lowest 4 g min, Highest 4 g min, and Diff_fastest 4 g min. The following paragraphs contain the description of the eleven constructive heuristics considered in this chapter.

### 3.2.1 Lowest

This method is the simplest since it not requires ordering the jobs. Instead, jobs are scheduled directly considering the property *lowest* (see Fig. 2) without any other criteria. That is, each job is assigned to the machine $i$ that requires the lowest time $p_{ij}$ to process the job $j$.

### 3.2.2 Lowest Min

In the heuristic Lowest min, jobs are considered, in non-increasing order, concerning the instance property *lowest* to allocate them using the Min heuristic presented in Algorithm 1. Figure 2 details on how to get this information for an instance example. As can be seen, this characteristic refers to the processing time $p_{ij}$ required by the fastest machine $i$ to process the job $j$.

### 3.2.3 Highest Min

Similar to Lowest min, this heuristic allocates the jobs in non-increasing order based on the property *highest*, using the Min heuristic presented in Algorithm 1. As can be seen from Fig. 2, this property *highest* refers to the processing time $p_{ij}$ required by the slowest machine $i$ to process the job $j$.

### 3.2.4 Mean Min

In this algorithm, jobs are scheduled employing the Min heuristic detailed in Algorithm 1, in non-increasing order, according to the instance property *mean*. In Fig. 2 can be seen that this characteristic refers to the average processing time of each job $j$.

### 3.2.5 Diff_fastest Min

Finally, we propose a deterministic heuristic called: Diff_fastest min. This algorithm considers the jobs in non-increasing order according to the instance property *diff_fastest* to schedule them using the Min heuristic detailed in Algorithm 1. Figure 2 shows the result of extracting the characteristic *diff_fastest* from an instance example. As can be seen, it consists of identifying the two machines ($i_a$ and $i_b$) that require the lowest time $p_{ij}$ to process the job $j$, to then calculate the difference between the processing times of these machines ($p_{i_aj} - p_{i_bj}$).

### 3.2.6 Random

In this non-deterministic heuristic, referred to as Random, jobs are allocated randomly on machines without considering any other criteria. The main benefits of the heuristic Random relies on its applicability since it can be used to generate solutions for problems with different characteristics. In contrast, effectiveness is the central drawback of this heuristic since there is no control to generate the solutions. Therefore, generated solutions can be very good, but also very bad.

### 3.2.7 Random Min

Unlike Random, the heuristic Random min combines randomness and knowledge of the problem domain to generate the solutions. In this way, jobs are selected randomly as part of the randomness, to later allocate them using the Min heuristic presented in Algorithm 1. As its name indicates, this algorithm assigns each job, evaluating all the possibilities to choose the best option at the moment, assigning each job to the machine that affects as little as possible the maximum completion time.

### 3.2.8 Random Lowest Bound Min

This heuristic is an extension to the Random min that incorporates the use of the lower bound $lb$, consisting of summing the processing times $p_{ij}$ required by the fastest machines to process every job $j$ and dividing the result by the number of available machines $m$. Given the above, *Random lowest min bound* works as follows. First, it attempts to assign every job, randomly selected, in the machine $i$ with the lowest processing time $p_{ij}$ considering the following criterion. If $S_i + p_{ij} <= lb$, job $j$ is assigned to the fastest machine $i$. Otherwise, job $j$ remains unassigned. Next, unassigned jobs are allocated with the Min heuristic.

### 3.2.9 Lowest 4g Min

This heuristic is an extension to the deterministic constructive heuristic Lowest min. Therefore, the constructive process of Lowest 4 g min also starts ordering the jobs in non-increasing order based on the instance characteristic *lowest*; nevertheless, it does not assign them immediately. Instead, divide them into four groups, from $G_1$ to $G_4$ with the same number of jobs equal to $n/4$, preserving the order of the jobs during group construction. Subsequently, the jobs in each group are permuted, modifying their order. Finally, the jobs are allocated with the Min heuristic, presented in Algorithm 1, following the order of the groups: $G_1$, $G_2$, $G_3$, and $G_4$.

#### 3.2.10 Highest 4g Min

Like Lowest 4 g min, Highest 4 g min is an extension of the deterministic constructive heuristic Highest min. Hence, it starts arranging the jobs in non-increasing order based on the instance characteristic *highest*. Next, it divides them into four groups, from $G_1$ to $G_4$ with the same number of jobs equal to $n/4$, preserving the order of the jobs during group construction. Subsequently, the jobs of each group are permuted, changing their order. Finally, the jobs are allocated with the Min heuristic, presented in Algorithm 1, following the order of the groups: $G_1$, $G_2$, $G_3$, and $G_4$.

#### 3.2.11 Diff_fastest 4g Min

Besides the heuristics Lowest 4 g min and Highest 4 g min, we also used this approach with the instance property *Diff_fastest*. As can be inferred, in this heuristic, jobs are first ordered in non-increasing order based on the difference of the processing time $p_{ij}$ required by the two fastest machines ($i_a$ and $i_b$) for processing each job $j$. Next, its process continues the same as that of the heuristics Lowest 4 g min and Highest 4 g min. That is, dividing the jobs into four groups, to later permute them, and finally assign them with the Min heuristic.

## 4 Experimental Study of the Optimization Process of $R||C_{max}$

This section details the results obtained by applying the approach proposed by Quiroz et al. [7] to perform the experimental analysis of heuristics algorithms. This approach holds four phases: characterization, characteristics refining, relations study, and algorithm behavior explanations. The characterization phase consists of identifying the most relevant properties of the instances of the problem studied, and measure them by means of characterization functions (indexes), that will provide valuable information that could enable us to explain the performance of the heuristics that solve them. On the other hand, the characteristics refining stage involves the use of exploratory data analysis techniques to discard incorrect, redundant, or irrelevant indexes; if necessary, new indexes are incorporated, in order to characterize all the relevant factors that allow discriminating between instances with different features. Next, in the phase of relations study, an exploratory analysis of the final set of indexes is carried out to look for characteristics of the test instances explaining the algorithmic behavior of the heuristics under research. Finally, as the last phase of the approach, the knowledge resulting from this study conducts an understanding of the behavior of the heuristic algorithms, explaining how their final performance is affected by several factors that cause the features of the instances.

## 4.1 Phase 1: Characterization

The main objective of the characterization phase is to identify and measure relevant factors to describe the instances structure and the final performance of the algorithms. In this section, we describe the set of indexes used to describe the most relevant properties of the 1400 test instances of the $R||C_{max}$ problem and the final performance of the eleven constructive heuristics described in Sect. 3.2.

### 4.1.1 Measuring $R||C_{max}$ instances hardness

Knowing and understanding the structure of the instances of $R||C_{max}$ plays an important role in predicting the quality of the solutions generated by the constructive heuristic algorithms that address them. Characterizing the structure of an $R||Cmax$ instance is a key issue to predict the performance of a heuristic algorithm when solving it. It is well-known that factors such as the number of jobs and machines, the central tendency of their processing times, and their distribution, influence on the difficulty level that an instance may have for a solution algorithm. However, the challenge is the formulation of indexes able to characterize these factors.

According to the scope of the literature review, until now, no work related to the characterization of a Parallel-machine scheduling problem has been introduced. The above motivates this work, where we apply several general-purpose indexes, based on descriptive statistics, to analyze different characteristics of the test instances of the $R||C_{max}$ Parallel-machine scheduling problem to get relevant information about the parameters of every instance.

In Sect. 3.1 the proposed standard $R||C_{max}$ instances utilized to evaluate the solution algorithms performance were described. The characteristics used to generate this benchmark are the first that allow us to discriminate between instances with different structures. In this sense, given an instance of $R||C_{max}$, we use $m$ and $n$ (the number of machines and the number of jobs, respectively) to measure the size of the problem. Another important factor to characterize is the distribution of the processing times $p_{ij}$, which can be studied measuring its form, centralization, dispersion, as well as the correlation concerning the jobs and the machines.

Recalling from Sect. 3.2, constructive heuristics for the Parallel-machine scheduling problem consider different properties to classify the jobs before scheduling them (see Fig. 2), like the processing time $p_{ij}$ required by the fastest machine to process each job (*lowest*) and the difference between the two fastest machines to process each job (*diff_fastest*). In order to explore the impact of these properties in the performance of the algorithms, we generate plots for each instance to observe the way these characteristics can vary among different instances. For example, the *multiplicity* of the instance property *lowest*, represented as a vector that contains the processing times required by the fastest machine to process each job, is an aspect that could incise on the difficulty of the instances. Figure 3 holds information about it, which allows us to analyze how the *multiplicity*(*lowest*) varies from one instance to another.

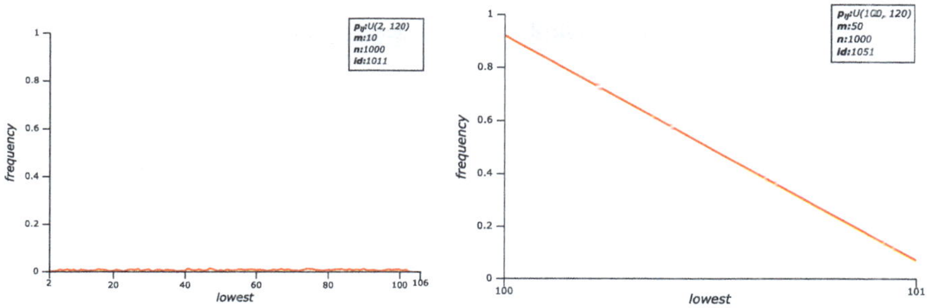


(a) Instance 1011 of the set *Jobcorre*

(b) Instance 1051 of the set $U(100, 120)$

**Fig. 3** *Multiplicity* of the lowest processing times of jobs (*lowest*)

To generate this figure, first, for each instance $I$, the set *lowest* was created with the minimum processing time $p_{ij}$ required to process every job $i$, to later identify the different values in *lowest*, and count its frequency. In this way, Fig. 3 contains the instances of the extremes, that is the instance where the most frequent processing time is the highest and the one with the lowest of the 1400 cases. In each graph, the $x-$axis depicts the processing time of the jobs, and the $y-$axis indicates its frequency as a percentage of the total number of jobs. Besides, in the upper right corner, each figure holds the general characteristics of the plotted instance, including its *id*, the range used to generate its values of $p_{ij}$, number of jobs $n$, and number of machines $m$. As can be seen in these graphs, the benchmark considered in this work contains instances where the lowest processing times are not very repeated, such as the one shown in Fig. 3a with 0.16% incidence. In contrast, Fig. 3b indicates that this benchmark also includes instances with the lowest processing times repeated until 92.5 percent.

Another aspect that could be related to the difficulty of the instances is the difference between the processing times of the two fastest machines to process each job (*diff_fastest*), mainly for constructive heuristics that sort the jobs utilizing this information. Figure 4 shows how the values of this feature vary from one instance to another. To generate this figure, for each instance $I$, the set *diff_fastest* was created, to later sort the jobs in increasing order according to this property. In this way, Fig. 4 holds the plots for the instances with the greatest and the smallest value in *diff_fastest* of the 1400 cases, respectively. In each graph, $x-$axis depicts the jobs in increasing order according to its values in *diff_fastest*, and the $y-$axis indicates the difference between the processing times of the two fastest machines to process each job. Besides, the upper right corner of each graph presents the general characteristics of the plotted instance: the criterion used to generate its values of $p_{ij}$, the number of machines $m$, the number of jobs $n$, and identifier *id*. From these figures can be observed that, in some instances, the difference between the two fastest machines to process each job is small, like the one plotted in Fig. 4a with differences of 0 and 1 only. In contrast, Fig. 4b shows that the benchmark also includes instances with values of *diff_fastest* that can vary between 0 and 70. This type of information can be

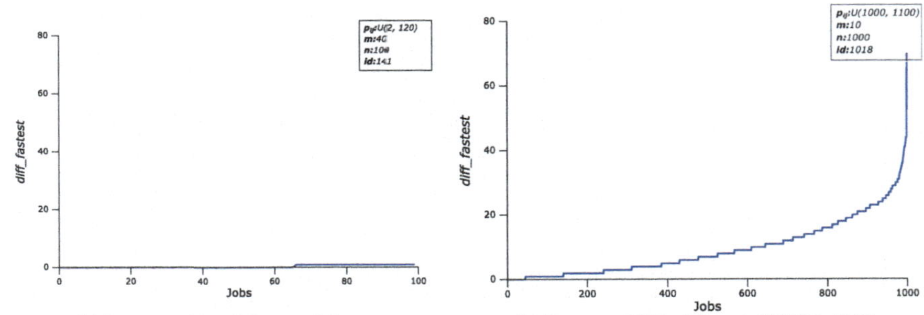


(a) Instance 141 of the set *Jobcorre* (b) Instance 1018 of the set $U(1000, 1100)$

**Fig. 4** Difference between the processing times of the two fastest machines to process each job (*diff_fastest*)

helpful to build a solution since it indicates that, in some cases, the fastest machines to process all jobs have similar speeds. Consequently, placing the jobs in the fastest or the second-fastest machine affects less than when there exists a wide variation among the speed of the two fastest machines.

Table 1 contains the general indexes proposed to analyze the 1400 instances, where the first column includes the name of each index, and the second one its description. In this sense, for each instance $I$, we collect the number of jobs $n$ and machines $m$; the quotient of the number of jobs by the number of machines $\frac{n}{m}$; the minimum $min(p_{ij})$ and the maximum $min(p_{ij})$ processing time; the average $mean(p_{ij})$ and the coefficient of variation of the processing times $cv(p_{ij})$; the difference $range(p_{ij})$ and the quotient $(q)$ between the maximum and minimum processing times; and the level of correlation among the processing times of jobs (*job_corre*) and machines (*mach_corre*). These indexes allow measuring the general characteristics of the structure of the instances. Equation 5 shows how to calculate the job correlation of an instance $I$, where $range(p_j)$ indicates the subtraction of the largest and minor processing time of the job $j$, $n$ represent the number of jobs, and $range(p_{ij})$ depicts the difference between the highest and the lowest processing time of $I$, respectively. In this way, as the value of *job_corre* of $I$ approaches 1, its jobs are more correlated, indicating that its jobs have similar processing times for all machines (either short, medium, or long). Similarly, Eq. 6 is used to measure the machine correlation (*mach_corre*) of an instance $I$, where $range(p_i)$ collects the subtraction of the highest and the lowest processing time of the machine $i$, $m$ indicates the number of machines, and $range(p_{ij})$ depicts the difference between the largest and minor processing time of $I$, respectively. Like *job_corre*, as the value of *mach_corre* of $I$ approaches 1, its machines are more correlated, indicating that its machines have similar speeds to process all the jobs (either fast or slow).

$$job\_corre = 1 - \frac{\frac{\sum_{j=1}^{n} range(p_j)}{n}}{range(p_{ij})} \tag{5}$$

**Table 1** General characteristics for $I$

| Index | Description |
|---|---|
| $n$ | Number of jobs to be scheduled |
| $m$ | Number of machines available |
| $\frac{n}{m}$ | Quotient of $n$ between $m$ |
| $min\ (p_{ij})$ | Minimum processing time in $I$ |
| $max\ (p_{ij})$ | Maximum processing time in $I$ |
| $mean\ (p_{ij})$ | Average processing time in $I$ |
| $cv\ (p_{ij})$ | Coefficient of variation of the processing times in $I$ |
| $range\ (p_{ij})$ | Difference between $G_{min}\ (I)$ and $G_{max}(I)$ in $I$ |
| $q$ | Quotient of $max\ (p_{ij})$ between $min(p_{ij})$ |
| $job\_corre$ | Level of correlation among jobs in $I$ |
| $mach\_corre$ | Level of correlation among machines in $I$ |

$$mach\_corre = 1 - \frac{\frac{\sum_{i=1}^{m} range(p_i)}{m}}{range(p_{ij})}, \tag{6}$$

Additionally, different features of jobs and machines are independently analyzed using the seven descriptive measures defined in Table 2. In this table, the first column includes the name of each index, and the second one holds its description. As can be seen from this table, each index requires an argument $(X)$ that can represent either a set of processing times of a job, a machine, or any specific feature extracted from a given instance $I$. For instance, to analyze the general structure of jobs, the $m$ processing times $p_{ij}$ of each job are used to create the set $X$. In this way, the seven measures listed in Table 2 (i.e., *mean*, *cv*, *uniformity*, *min*, *max*, *range*, and *multiplicity*) are applied to each job independently, to later calculate the average of each measure regarding the $n$ jobs under study.

The information collected with these measures and the process mentioned above allow examining the general distribution, uniformity, form, and size of the jobs in a given instance. Furthermore, these measures also can be used to study more specific characteristics of jobs, like the behavior followed by the shortest, the longest, and the average processing time of jobs. To collect this information, first, the characteristic of interest is extracted from each job and then used as the argument $(X)$ of the measures to be calculated. Thereby, data related to the characteristic of interest, like its central tendency, dispersion, and uniformity, can be explored using the indexes listed in Table 2.

Similar to jobs, several machine characteristics are studied using the seven descriptive measures detailed above. In this sense, to analyze its general structure, the measures are applied to each machine independently. That is forming the set $X$ with the $n$ possible processing times $p_{ij}$ for each machine $i$, from $p_{i1}$ to $p_{in}$. Furthermore, more specific features of machines are explored, like the time required by each

**Table 2** descriptive measures

| Index | Description |
|---|---|
| *mean*($X$) | Mean of the data set $X$. |
| *cv*($X$) | Coefficient of variation of the data set $X$ |
| *uniformity*($X$) | Uniformity of the data set $X$ [7] |
| *min*($X$) | Minimum value in the data set $X$ |
| *max*($X$) | Maximum value in the data set $X$ |
| *range*($X$) | Difference between the maximum and the minimum value in $X$ |
| *multiplicity*($X$) | Multiplicity of the data set $X$ [7] |

machine to process the job with the shortest processing time (*shortest*), the number of jobs that each machine process faster than the other ones (*n_fastest*), the average speed of machines considering the $n$ jobs (*average_speed*), and the average speed of machines considering only the jobs that each machine process faster than the others (*average_speed_fastest*).

In addition to this, we used indexes with specific-purpose. For example, to analyze relations jobs-machines we use four indexes: *above* $\frac{n}{m}$, *equal* $\frac{n}{m}$, and *below* $\frac{n}{m}$. To calculate these indexes, first, the vector *n_fastest* with length $m$ is created where each position of the vector includes the number of jobs that each machine process faster than the other ones. In this way, the characteristics of this property can be studied, applying the indexes of the Table 2. Additionally, we measure *above* $\frac{n}{m}$ that counts the number of machines processing faster a lower number of jobs that $\frac{n}{m}$, *equal* $\frac{n}{m}$ that indicates the number of machines processing faster an equal number of jobs that $\frac{n}{m}$, and *below* $\frac{n}{m}$ that measures the number of machines processing faster a lower number of jobs that $\frac{n}{m}$. However, it is important to note that these indexes are not standardized.

On the other hand, we also use the descriptive measures in Table 2 to collect information related to the difference between the processing times of the two fastest machines to process each job (see Figs. 2 and 4). The information collected from this property is important since it allows detecting the contrast among processing the jobs by the machine that processes them fastest or the second one. Recalling from Fig. 4b, this difference can be considerably large. To generate this data, first, the vector *diff_fastest* is created with the difference between the two fastest machines to process each job, to later, using the seven indexes listed in Table 2 to analyze its central tendency, uniformity, form, and multiplicity.

Finally, we use the indexes in Table 2 to analyze data of the instances related to the instance property *lowest* (see Figs. 2 and 3). The study of this property (lowest) is very useful since it provides information related to the shortest time in which the jobs of an instance $I$ can be processed. In this way, indexes of Table 2 bring information respect to its central tendency, distribution, form, and multiplicity. Additionally, we apply the indexes *max_repe*, introduced by Quiroz *et al.* in [7] as well as *no_multiplicity*, introduced in this work. *Max_repe* indicates the frequency of the most recurrent processing time, and *no_multiplicity* counts the number processing times that only appear one time. Knowing the characteristics of *lowest* is important since one of the first ideas that can arise at the moment of solving an $R||C_{max}$ problem is to assign the jobs to the machine that processes them faster. In this sense, the information collected from this property can help to validate the use of said heuristic or discard it and guide the selection of a new heuristic.

### 4.1.2 Measuring Performance

The performance of the eleven heuristics was measured using the Relative Percentage Deviation (*RPD*), presented in Eq. 7 that, for each instance ($I$), determines the error rate among the solutions generated by each constructive heuristic and the solutions generated by 2 hours run of the commercial solver CPLEX from IBM ILOG, a high-performance solver based on branch and cut methods designed to address Linear Programming, Mixed Integer Programming, and Quadratic Programming problems.

$$\text{RPD} = \frac{C_{max}(I) - C^*_{max}(I)}{C_{max}(I)} \tag{7}$$

where $C_{max}(I)$ indicates the solution found for each heuristic and $C^*_{max}$(I) is the solution found by CPLEX for the instance $I$, presented in [23], and available at http://soa.iti.es.

## 4.2 Phase 2: Characteristics Refining

Once $R||C_{max}$ interest indexes have been defined, it is necessary to perform an exploratory analysis of the variables to look for different indexes measuring the same feature of the structure of the problem, indicators measuring no relevant information, and indexes without any relation to the others. During this stage, first, we analyzed the relationships among each pair of variables to identify redundant, irrelevant, and incorrect indexes. The correlation matrix revealed that several pares of indexes measuring features related to the central tendency, dispersion, and form of different characteristics of the $R||C_{max}$ processing times are strongly correlated. Moreover, it allowed identifying that the lineal association among some variables do not make sense or collects irrelevant information. This study enabled observing the contribution and consistency of each index, detecting those that do not hold meaningful information for the characterization of the structure of the $R||C_{max}$ instances. Thus, we discarded inconsistent, irrelevant, and redundant indexes to avoid the noise provided by them to the analysis of the instance structure. Table 3 includes the detail of the final set of indexes, used to analyze the structure of the $R||C_{max}$ instances. The first column indicates the type of information that collects, while the second one holds the name of every index.

**Table 3** Final set of indexes for $R||C_{max}$ characterization

| Type | Index |
|---|---|
| Size | $n$ |
| | $m$ |
| | $q$ |
| Centralization | $mean\ (p_{ij})$ |
| | $mean\ (diff\_fastest)$ |
| Dispersion | $range\ (p_{ij})$ |
| | $cv\ (p_{ij})$ |
| | $cv\ (min(p_j))$ |
| | $cv\ (n\_fastest)$ |
| Form | $uniformity\ (p_i)$ |
| | $uniformity(p_j)$ |
| | $uniformity\ (n\_fastest)$ |
| Location | $max\ (p_{ij})$ |
| | $min\ (p_{ij})$ |
| Relations job-machine | $\frac{n}{m}$ |
| | $below\ \frac{n}{m}$ |
| Machine | $max\ (diff\_fastest)$ |
| | $mach\_corre$ |
| Jobs | $multiplicity\ (lowest)$ |
| | $job\_corre$ |

**Table 4** Characteristics of the eight groups of instance. $p_{ij}$: processing time distribution. $n$: number of jobs. $m$: number of machines. $id$ identifier of each collection of instances

| $p_{ij}$ | $n$ | $m$ | $id$ |
|---|---|---|---|
| $U(1, 100)$ | 100 | 10 | 1 |
| | | 50 | 2 |
| | 1000 | 10 | 3 |
| | | 50 | 4 |
| $U(1000, 1100)$ | 100 | 10 | 5 |
| | | 50 | 6 |
| | 1000 | 10 | 7 |
| | | 50 | 8 |

Finally, a Principal Components Analysis (PCA) allowed plotting eight subsets of instances, arranged according to Table 4. In this table, the first column indicates the processing time distribution of each collection of instances, followed by its number of jobs, its number of machines, and its identifier from 1 to 8.

Figure 5 contains the instances of the eight groups plotted according to the first three principal components. As can be seen in this figure, the instances of each group

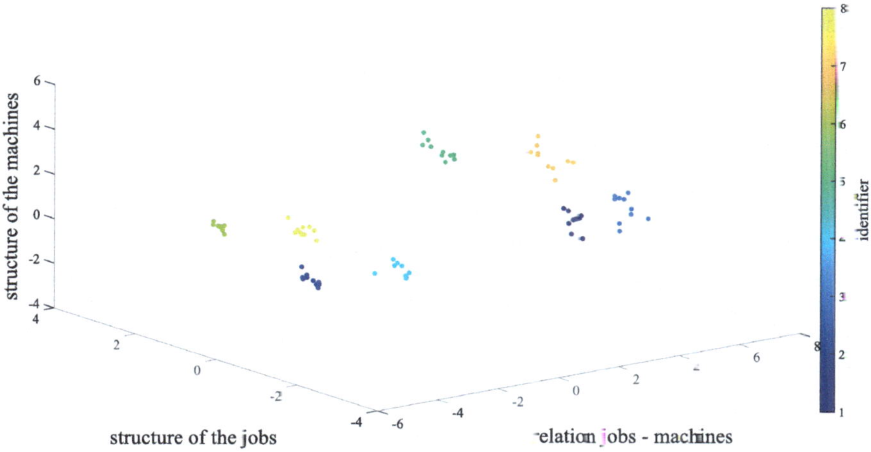


**Fig. 5** PCA of the twenty studied characteristics

are together, and the groups do not overlap. That is, the first three components are enough for discriminating among instances with different characteristics.

The first component, called relation job-machine, comprises indexes like the number of machines $m$ and the ration of the number of jobs by the number of machines ($\frac{n}{m}$) associated with the size of the problem. Furthermore it contemplates the form of the processing time distribution of the jobs (i.e., the uniformity of the jobs $uniformity(p_j)$) and variables that collect information of the machines that process fastest the jobs using the vector *diff_fastest* that includes the difference between the two fastest machines to process each job.

On the other hand, the second component, called job structure, contemplates characteristics of the instances related to the form (uniformity) and dispersion (range and coefficient of variation) of the processing time of the jobs.

Finally, the third component, machine structure, embraces information related to the form of the processing time distribution of the machines (i.e., uniformity of the machines $uniformity(p_i)$), the number of jobs that each machine could process in an extreme case $n$, and the frequency between the processing times required by the fastest machine to processes each job $multiplicity(lowest)$.

## 4.3 Phase 3: Study of Relations

The main goal of this phase is to explain the algorithm behavior of the heuristics under research, achieved by identifying how the properties immerse in the $R||C_{max}$ instances influence on its difficulty. During this stage, the information collected by the indexes is used to look for relations between the performance of the algorithms and the structure of the $R||C_{max}$ instances.

The Lineal Association Analysis was the first technique used to look for immersed relations among the structure of $R||C_{max}$ instances and the performances of the solution algorithms. Table 5 contains the results of the most outstanding indexes, those with a correlation greater or equal to 0.4 in at least one of the eleven construction heuristics (eleven of the twenty selected). Each cell indicates the linear relation among the solution algorithms (listed in the first column) and the eleven characteristics of $R||C_{max}$ instances (indicated in the remaining columns), highlighting in bold the correlations with an absolute value greater or equal than 0.4.

The correlation analysis results revealed that the heuristic Lowest exhibits a different behavior from the rest of the heuristics (see Table 5) and that its performance is mainly related to characteristics like $m$, *mach_corre*, $\frac{n}{m}$, *below* $\frac{n}{m}$, *max*(*diff_fastest*), *range*(*diff_fastest*), *uniformity*($p_j$), and *cv*(*n_fastest*). This behavior could be associated with the fact that as the number of machines $m$, the ratio of $\frac{n}{m}$, and the number of machines *below* $\frac{n}{m}$ grow, there is a greater likelihood that the shorter processing time of jobs will be better distributed among the machines, which can benefit the Lowest procedure. In contrast, its performance decreases by solving instances with high values of *mach_corre*, because this index indicates that machines present the same behavior for all jobs. Therefore, in the solutions generated by this heuristic, most jobs are assigned to the same machines (the fastest). Besides, characteristics of the property *diff_fastest* like *max* and *range* have a positive impact on the performance of Lowest. That is, as the difference between the two fastest machines to process each job decreases, its performance decreases, which suggests that when the shortest processing times of jobs are similar, the performances of Lowest decrease.

On the other hand, the correlation analysis results also revealed that the performance of the other ten heuristics (different to Lowest) is related to characteristics associated with the distribution of the processing times like $q$, $cv(min(p_j))$, and $cv(p_{ij})$, where $cv(p_{ij})$ stands out with the highest correlation (see Table 5). This behavior could be related to the fact that most of these heuristics use strategies to sort the jobs (like *lowest*, *highest*, *mean*, and *diff_fastest*) and the Min heuristic to assign them.

Next, in order to get an overview of the relations immersed among the performance of the heuristics and the difficulty of $R||C_{max}$ instances, we performed a tabular comparison of the quality of the solutions generated by each constructive heuristic for the 1400 test instances. For this goal, we analyzed the instances concerning three specific characteristics related to the general structure of the problem, including the number of jobs ($n$), the number of machines ($m$), and the criterion used to generate the processing times $p_{ij}$. Table 6 holds the results of this study, where each column represents a constructive heuristic, and every row indicates the variable used to group a subset of instances. In this way, Table 6 includes three different classifications of the instances. Each one with the 1400 instances distributed among different sets according to $n$, $m$, and the criterion used to generate the processing times $p_{ij}$, respectively. Hence, in the first block of rows of this table, the 1400 instances were categorized regarding the number of jobs $n$. Consequently, each cell indicates the average *RPD*, reached by each constructive heuristic, when the number of jobs to schedule is 100, 200, 500, and 1000, respectively. Similarly, in the second block of rows, the 1400 instances were distributed into five subsets concerning the number of

**Table 5** Characteristics with the stronger correlation to the solution algorithms performance. *Heuristic*: name of the constructive heuristic. *Index*: identifier of each feature

| Heuristic/ Index | $m$ | *mach_corre* | $cv(p_{ij})$ | $\frac{n}{m}$ | *below* $\frac{n}{m}$ | $q$ | *max* (*diff_fastest*) | *range* (*diff_fastest*) | $uniformity(p_j)$ | $cv(min(p_j))$ | $cv(n\_fastest)$ |
|---|---|---|---|---|---|---|---|---|---|---|---|
| Lowest | **0.49565** | **0.63081** | 0.17206 | **−0.47307** | **0.57414** | 0.08002 | **-0.55189** | **−0.48878** | **0.45302** | 0.08988 | **0.66946** |
| Lowest min | 0.14092 | 0.11283 | **0.78642** | −0.18623 | −0.00215 | **0.60611** | −0.11389 | 0.06834 | 0.1333 | **0.66662** | 0.13307 |
| Highest min | 0.08579 | 0.01603 | **0.77441** | −0.1363 | −0.06606 | **0.65734** | −0.05816 | 0.11622 | 0.0782 | **0.71158** | 0.02861 |
| Mean min | 0.0828 | −0.00769 | **0.76805** | −0.15573 | −0.10118 | **0.64232** | −0.05223 | 0.11872 | 0.08236 | **0.69829** | 0.01113 |
| Distance min | 0.06479 | 0.12713 | **0.81441** | −0.0966 | 0.03846 | **0.65376** | −0.03288 | 0.15742 | 0.05282 | **0.7224** | 0.12336 |
| Random | 0.09907 | 0.3001 | **0.86976** | −0.08664 | 0.16911 | **0.60137** | −0.0373 | 0.18823 | 0.10302 | **0.6869** | 0.30294 |
| Random min | 0.08881 | 0.0158 | **0.79914** | −0.13602 | −0.06882 | **0.67536** | −0.06748 | 0.1134 | 0.08165 | **0.73195** | 0.02658 |
| Random lowest bound min | 0.30376 | 0.19815 | **0.72139** | −0.30426 | 0.0486 | **0.62241** | −0.20382 | −0.03295 | 0.29377 | **0.62549** | 0.24325 |
| Lowest 4 g min | 0.04357 | 0.32031 | **0.70631** | −0.05399 | 0.20784 | 0.38979 | 0.06527 | 0.24521 | 0.03039 | **0.47214** | 0.32776 |
| Highest 4 g min | 0.04242 | 0.05106 | **0.70391** | −0.07398 | −0.04022 | **0.58254** | 0.07987 | 0.23659 | 0.03876 | **0.61567** | 0.07351 |
| Diff_fastest 4 g min | 0.06923 | 0.06116 | **0.80807** | -0.10843 | −0.02073 | **0.67034** | −0.04938 | 0.13648 | 0.06262 | **0.74294** | 0.06804 |

**Table 6** Tabular analysis of the eleven constructive heuristics and the 1400 instances grouped according to the number of jobs $n$, the number of machines $m$, and the range of the processing times ($p_{ij}$). *Heuristic*: name of the constructive heuristic. *Criterion*: parameter used to distribute the 1400 instances

| Criterion/Heuristic | | Lowest | Lowest min | Highest min | Mean min | Diff_fastest min | Random | Random min | Random lowest bound min | Lowest 4g min | Highest 4g min | Diff_fastest 4g min |
|---|---|---|---|---|---|---|---|---|---|---|---|---|
| $n$ | 100 | 0.549 | 0.186 | 0.172 | 0.184 | 0.176 | 0.493 | 0.171 | 0.337 | 0.132 | 0.148 | 0.167 |
| | 200 | 0.515 | 0.173 | 0.167 | 0.176 | 0.174 | 0.478 | 0.165 | 0.284 | 0.134 | 0.151 | 0.164 |
| | 500 | 0.45 | 0.142 | 0.138 | 0.143 | 0.15 | 0.464 | 0.139 | 0.221 | 0.12 | 0.132 | 0.141 |
| | 1000 | 0.41 | 0.133 | 0.13 | 0.134 | 0.144 | 0.456 | 0.131 | 0.187 | 0.12 | 0.127 | 0.134 |
| $m$ | 10 | 0.326 | 0.136 | 0.14 | 0.146 | 0.153 | 0.412 | 0.14 | 0.129 | 0.124 | 0.135 | 0.144 |
| | 20 | 0.44 | 0.158 | 0.155 | 0.162 | 0.164 | 0.462 | 0.154 | 0.225 | 0.131 | 0.145 | 0.155 |
| | 30 | 0.491 | 0.151 | 0.142 | 0.152 | 0.149 | 0.487 | 0.142 | 0.268 | 0.115 | 0.129 | 0.14 |
| | 40 | 0.549 | 0.164 | 0.149 | 0.16 | 0.158 | 0.5 | 0.15 | 0.312 | 0.122 | 0.135 | 0.15 |
| | 50 | 0.601 | 0.183 | 0.173 | 0.177 | 0.18 | 0.504 | 0.171 | 0.352 | 0.14 | 0.154 | 0.169 |
| $p_{ij}$ | $U(1, 100)$ | 0.406 | 0.3 | 0.307 | 0.326 | 0.313 | 0.919 | 0.303 | 0.647 | 0.196 | 0.299 | 0.3 |
| | $U(10, 100)$ | 0.38 | 0.243 | 0.223 | 0.244 | 0.238 | 0.801 | 0.222 | 0.425 | 0.202 | 0.219 | 0.217 |
| | $U(100, 120)$ | 0.512 | 0.061 | 0.06 | 0.063 | 0.061 | 0.114 | 0.06 | 0.054 | 0.06 | 0.06 | 0.06 |
| | $U(100, 200)$ | 0.377 | 0.139 | 0.136 | 0.142 | 0.138 | 0.373 | 0.136 | 0.136 | 0.134 | 0.136 | 0.134 |
| | $U(1000, 1100)$ | 0.387 | 0.034 | 0.035 | 0.035 | 0.034 | 0.059 | 0.034 | 0.023 | 0.034 | 0.034 | 0.034 |
| | Jobcorre | 0.541 | 0.148 | 0.148 | 0.15 | 0.152 | 0.337 | 0.153 | 0.138 | 0.076 | 0.076 | 0.153 |
| | Machcorre | 0.765 | 0.184 | 0.153 | 0.155 | 0.19 | 0.708 | 0.153 | 0.377 | 0.183 | 0.153 | 0.163 |

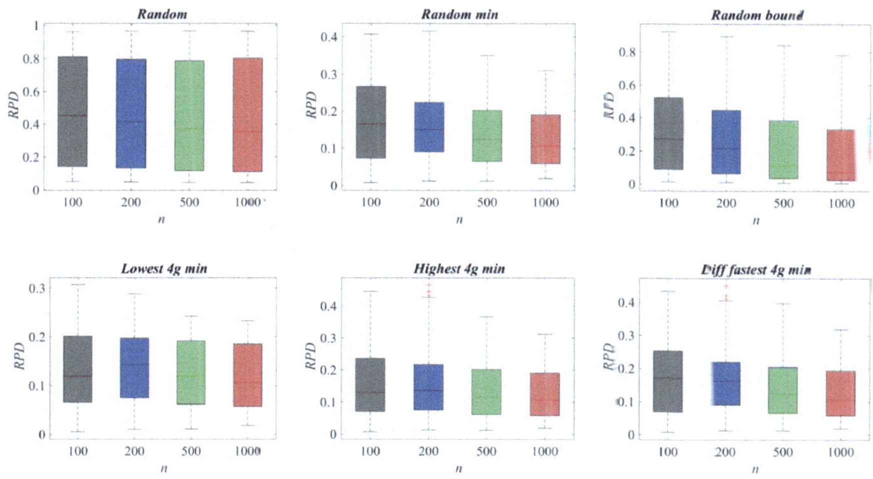


**Fig. 6** Performance of deterministic heuristics for instances grouped by number of jobs $n$

available machines they have. Recalling from Sect. 3.1, each instance can has either 10, 20, 30, 40, or 50 available machines. Therefore, every cell indicates the average *RPD* reached by each constructive heuristic when the available machines are 10, 20, 30, 40, and 50, respectively. Finally, the third block of rows compares the eleven constructive heuristics under research according to the criterion used to generate the processing times ($p_{ij}$). That is $U(1, 100)$, $U(10, 100)$, $U(100, 120)$, $U(100, 200)$, $U(1000, 1100)$, Machcorre, and Jobcorre, respectively.

Block one of Table 6 suggests that the number of jobs in an instance gives some information related to its difficulty. As can be seen in this block, most constructive heuristics reached a better PDR average in the instances with the greatest number of jobs (1000), and as the number of jobs decreases, the PDR average they reached grows.

As can be seen in Figs. 6 and 7, this behavior is presented by most deterministic and non-deterministic heuristics. Figure 6 includes five box plot graphs, one for each deterministic heuristic under research. In each graph, the x-axis indicates the number of jobs, and the y-axis the *RPD*. Similarly, Fig. 7 holds six box plot graphs, one for each non-deterministic heuristic under research. In each plot, the horizontal axis depicts the number of jobs, and the vertical axis indicates the *RPD*. From these plots can be observed that from the eleven heuristics studied, only the random heuristic not present the behavior mentioned above.

On the other hand, the second block of Table 6 suggests that the number of machines has a contrary impact on the difficulty of the instances. That is, most constructive heuristics get the best *RPD* average by solving instances with the lowest number of machines (10), and such average increases as the number of machines grows. Figures 8 and 9 show that most deterministic and non-deterministic heuristic have this behavior. Figure 8 holds five box plot graphs, one for each deterministic

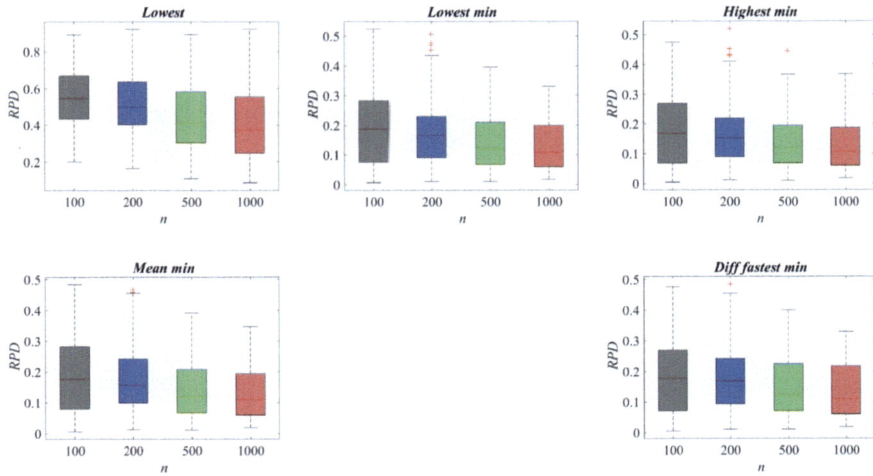


**Fig. 7** Performance of non-deterministic heuristics for instances grouped by number of jobs $n$

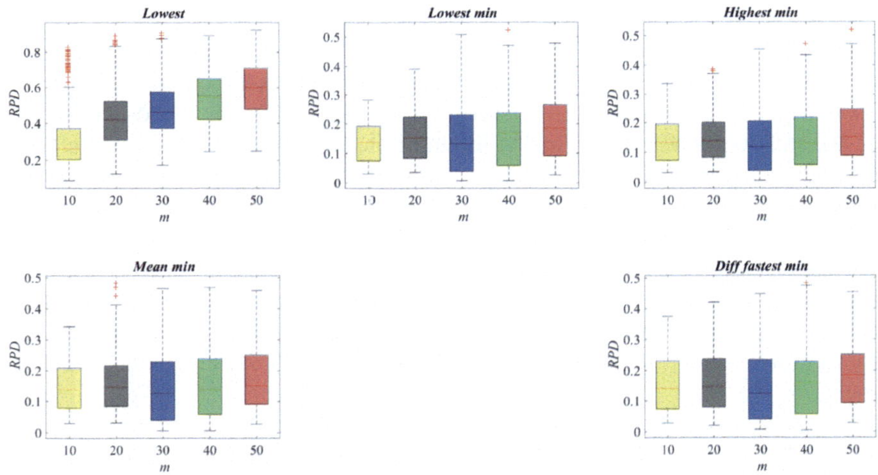


**Fig. 8** Performance of deterministic heuristics for instances grouped by number of machines $m$

heuristic under research. In each graph, the $x-$axis indicates the number of machines, and the $y-$axis the *RPD*. Similar, Fig. 9 holds six box plot graphs, one for each non-deterministic heuristic under research. In each plot, the horizontal axis depicts the number of machines, and the vertical axis indicates the *RPD*. From these plots it can be observed that the eleven heuristics studied present the behavior mentioned above. However, it is less marked in heuristics like Lowest 4 g min, Highest 4 g min, and Diff_fastest 4 g min.

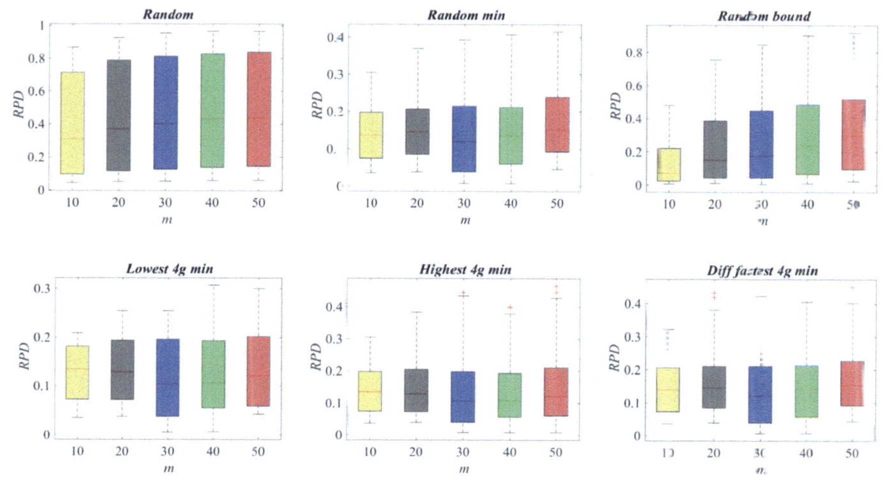


**Fig. 9** Performance of non-deterministic heuristics for instances grouped by number of machines $m$

Finally, from the third block of Table 6 can be inferred that the constructive heuristics generate solutions with a lower *RPD* when solving instances where the quotient $q$ of the maximum $max(p_{ij})$ by the minimum $min(p_{ij})$ processing time is less. Figure 10 indicates the average quotient of the instances of each set concerning the distribution of the processing times. As can be seen in this plot, the set $U(1000, 1100)$ contains the instances with the lowest $q$, while the ones with the highest $q$ are in the set $U(1, 100)$. Figures 11 and 12 allow observing how $q$ influences on the difficulty of the $R||C_{max}$ instances. Each figure contains a set of scatter plots where the $y-$axis indicates the performances of the algorithm according to the $q$ of the instances, indicated in the $x-$axis.

In Fig. 11 it can be seen that the deterministic constructive heuristics reach better results in instances where the processing time distribution is between 1000 and 1100 (i.e., where $q = 1.1$). A likely reason is that the processing times are quite similar since the highest processing time is only 1.1 times greater than the lowest one. Besides, these plots show how the performance of constructive heuristics decreases as the value of $q$ increases. That is, the heuristics reach the worst performance in instances where the processing time distribution is from 1 to 100. Similarly, Fig. 12 suggests that $q$ impacts on the performance of non-deterministic constructive heuristics, such that the higher the value of $q$, the greater the difficulty of the instance.

After the tabular analysis, we used graphic analysis techniques for analyzing the information collected by the final set of indexes. For this study, only the graphs of the heuristics Lowest and Lowest min will be shown since Lowest exhibits a different behavior from the rest of heuristics, and Lowest min shows a representative behavior of the ten heuristics that are different to Lowest. Figures 13 and 15 show the behavior of the most representative indexes. Figure 13 holds ten 3D scatter plots, in each of

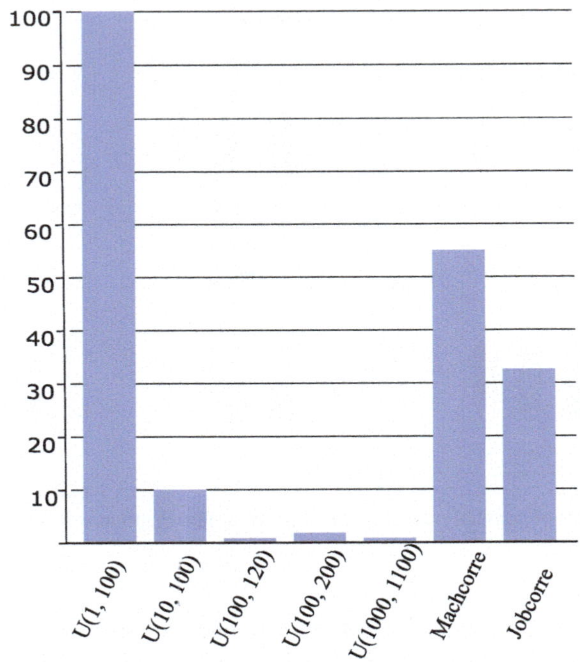


**Fig. 10** Average *quotient* of the instances in each set, grouped according to the criterion used to generate the values of $p_{ij}$

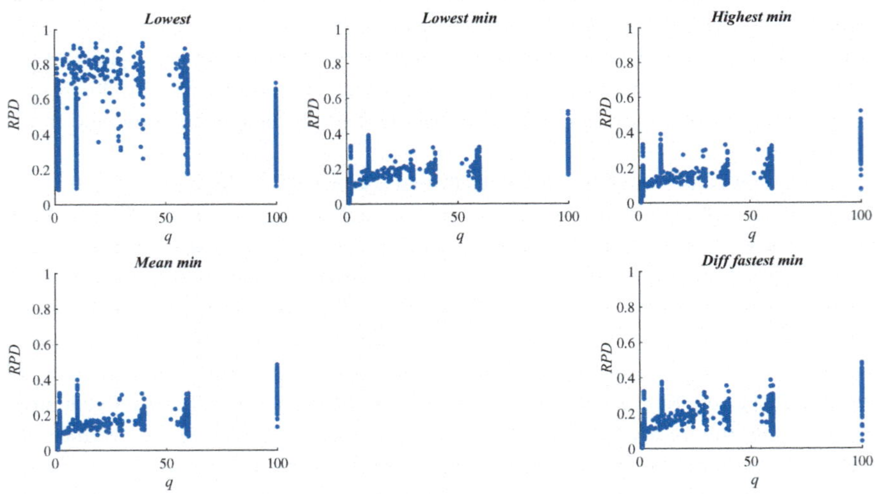


**Fig. 11** Performance of deterministic heuristics with references to the quotient of the maximum by the minimum processing time ($q$) of the instances

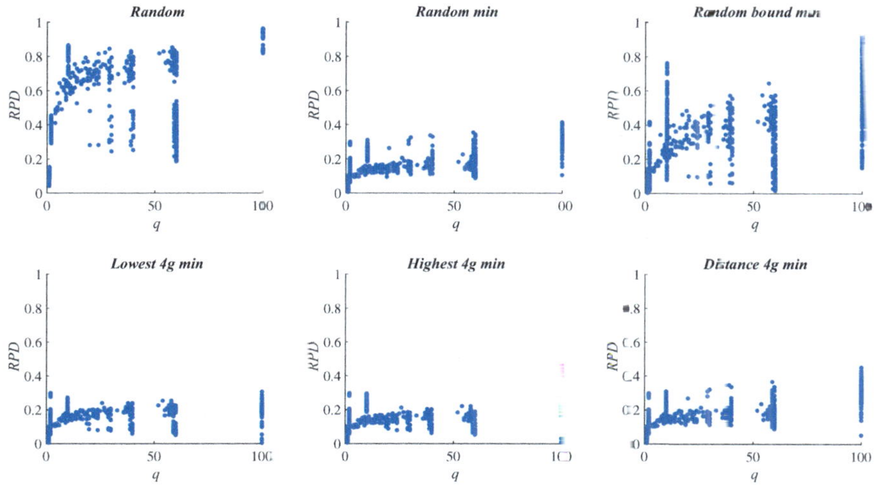


**Fig. 12** Performance of non-deterministic heuristics with references to the quotient of the maximum by the minimum processing time ($q$) of the instances

them the $x-$axis indicates the coefficient of variation of all the processing times in each instance $cv(p_{ij})$ since according to Table 5 it has the strongest relation with the performance of the constructive heuristics under research; the $y-$axis depicts one of the ten most representative indexes: $m$, $n$, $q$, $max(diff\_fastest)$, $range(diff\_fastest)$, $multiplicity(lowest)$, $job\_corre$, $mach\_corre$, $cv(min(p_j))$, and $uniformity(p_j)$; the $z-$axis indicates the error rate (*RPD*) of the solution generated by the Lowest heuristic for each instance; and the color bar shows the sub-set each instance belongs when considering the distribution used to generate its processing times $p_{ij}$ as grouping criterion. Recalling from Table 5, Lowest has the strongest correlations with the indexes $m$, $\frac{n}{m}$, $below\ \frac{n}{m}$, $max(diff\_fastest)$, $range(diff\_fastest)$, $uniformity(p_j)$, and $cv(n\_fastest)$. Similarly, Fig. 15 shows the impact of the ten most representative indexes in the constructive heuristic Lowest min, using the same structure of Fig. 13. That is, $x-$axis indicates the $cv(p_{ij})$, $y-$axis holds one of the ten representative indexes, $z-$axis includes the error rate of the Lowest min heuristic, and the color bar depicts the sub-set of instances. Recalling from Table 5, Lowest min and the remaining nine heuristics have the strongest correlations with the indexes $cv(p_{ij})$, $q$, and $cv(min(p_j))$.

Figures 13a, b allow observing how the indexes related to the size of the $R||C_{max}$ problem can impact on the performances of the constructive heuristic Lowest. In this sense, the scatter plot in Fig. 13a suggests that the performance of Lowest decreases when the size of $R||C_{max}$ instances grow concerning the number of machines $m$. Recalling from Table 5, the strongest linear relation of $m$ is 0.49, and it is with Lowest. In contrast, the scatter plot in Fig. 13b indicates that the performance of Lowest improves as the size of the problem increases regarding the number of jobs $n$, though this behavior is less marked, with a linear association of -0.27.

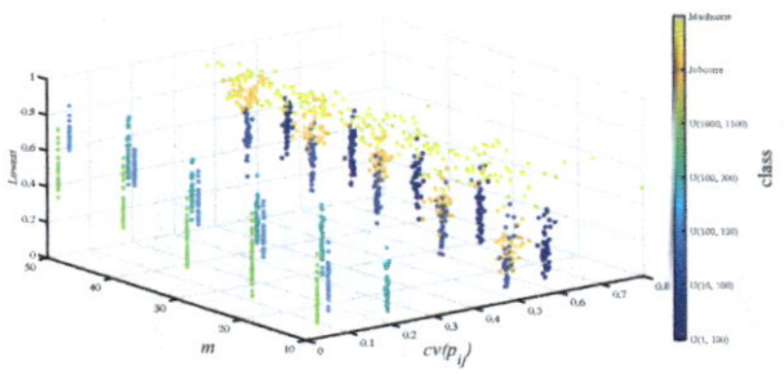


(a) Number of machines: $m$

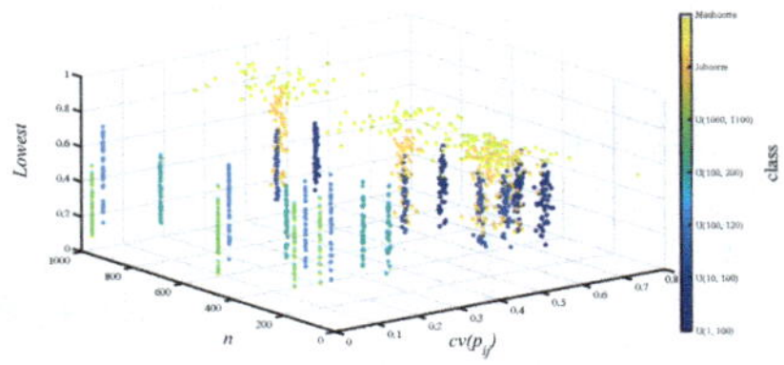


(b) Number of jobs: $n$

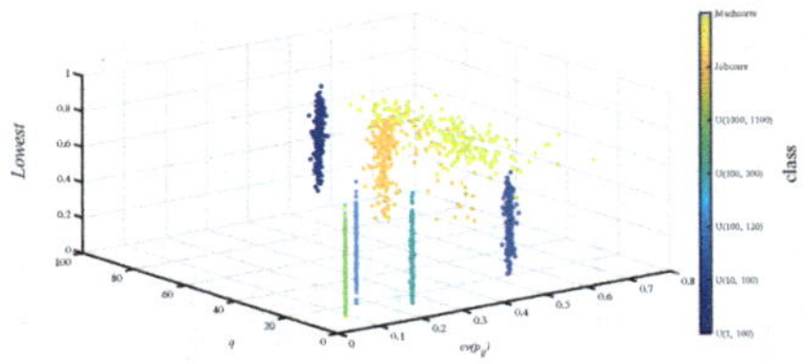


(c) *Quotient* between max and min processing time

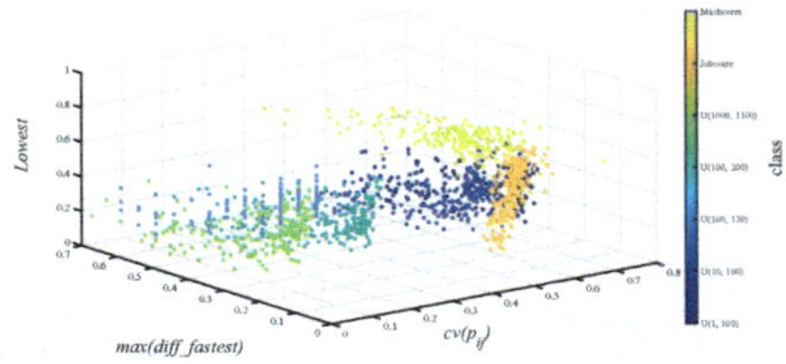


(d) Max difference of the two fastest machines

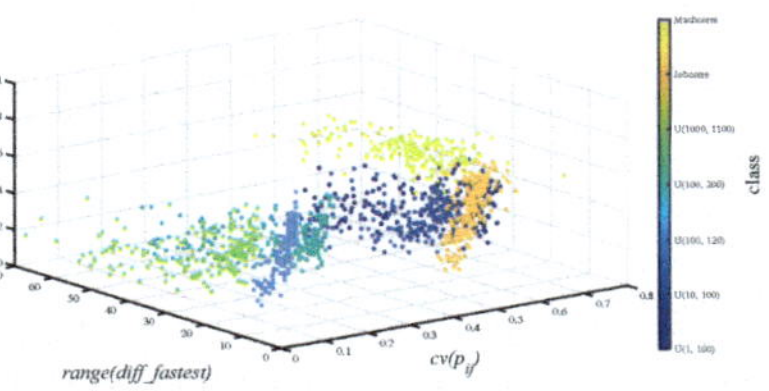


(e) Difference between the two fastest machines

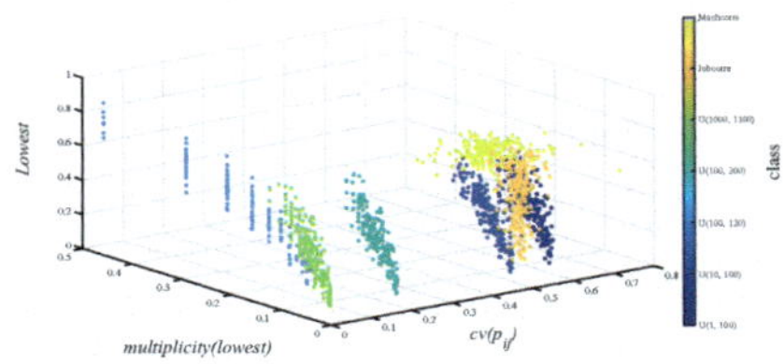


(f) *Multiplicity* of the lowest processing times

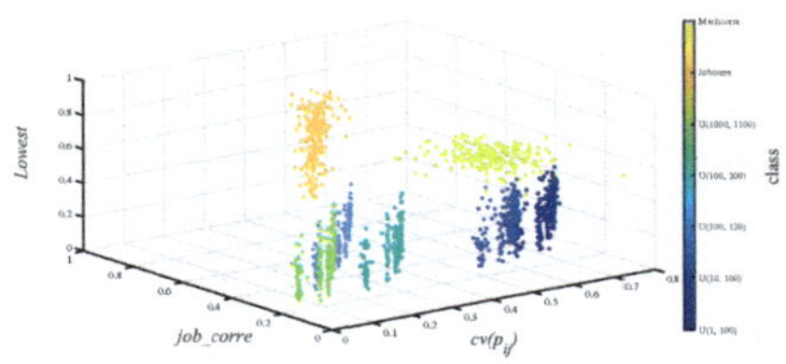


(g) Correlation among jobs: *job_corre*

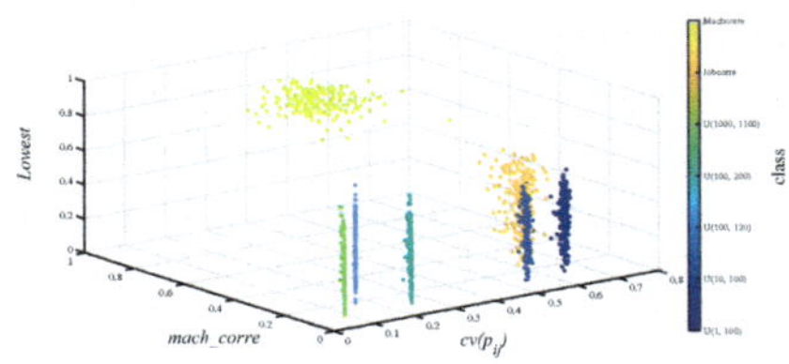


(h) Correlation among machines: *macs_corre*

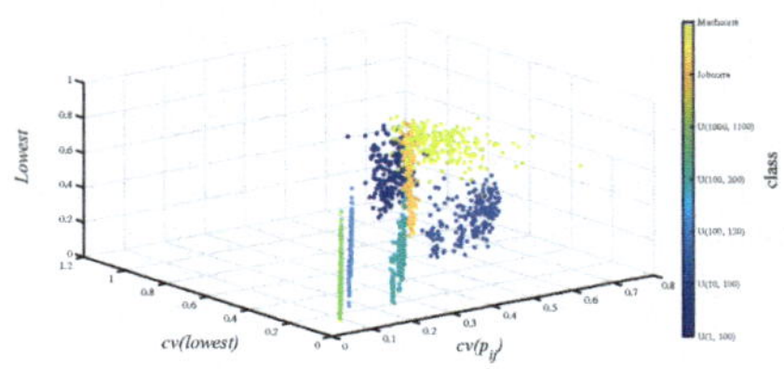


(i) Coefficient of variation of the lowest processing times

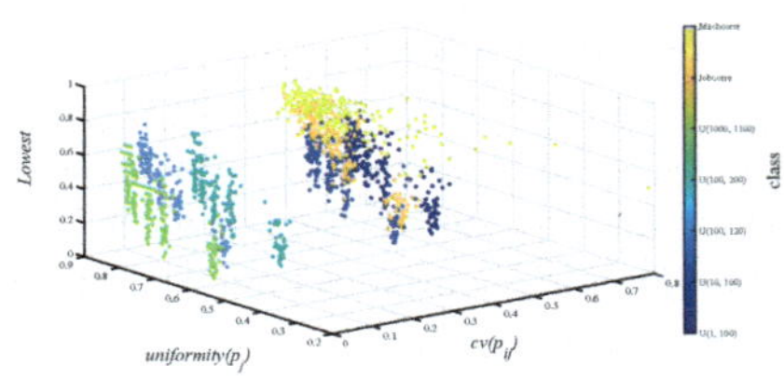


(j) Uniformity of the jobs

**Fig. 13** Influencing of the ten most representative characteristics of the structure of $R||C_{max}$ instances in the performances of the constructive heuristic Lowest

In the same spirit, different indexes that collect information about the processing time distribution were studied. A representative index is the quotient of the maximum and the minimum processing time ($q$), plotted in Fig. 13c. Reminding from Table 5, the index $q$ is strongly related to the performance of most constructive heuristics, except for Lowest only, which has a linear relation of 0.08. This behavior could be associated with its procedure since, unlike the other constructive heuristics, Lowest does not use any strategy to order jobs, and it assigns them to the machine that processes them faster without considering any criteria related to the distribution of the processing times. However, in this graph it can be seen that Lowest tends to get a worst performance in instances where $q$ is greater. That is the instances in the sub-set $U(1, 100)$, where $q = 100$.

As detailed in Fig. 4, another important aspect of the structure of $R||C_{max}$ instances is linked to the fastest machines to process each job. Among the most outstanding indexes that collect related information are *max(diff_fastest)* that holds the maximum difference between the two fastest machines to process each job and *range(diff_fastest)* that collects the difference of the greatest and the smallest *diff_fastest*. Evoking Table 5, *max(diff_fastest)* and *range(diff_fastest)* are strongly related to the performance of the constructive heuristic Lowest, with a linear association of $-0.55$ and $-0.48$, respectively. Figures 13d, e contains the scatter plots where it can be observed the impact of the values of *max(diff_fastest)* and *range(diff_fastest)* in the performance of Lowest, respectively. In these graphs it can be seen that the performance of Lowest tends to be better in instances where *max(diff_fastest)* and *range(diff_fastest)* are higher. Such behavior can be related to the fact that assigning jobs to the fastest machine without considering any other criteria can be beneficial, especially in instances with a significant difference among the lowest processing time of each job and the second-lowest. Therefore, as instances include jobs with smaller values of *max(diff_fastest)* and *range(diff_fastest)*, the performance of Lowest is worse. Graphs 13d, e also highlight that instances with correlated machines, marked in yellow, have the opposite behavior. That is, Lowest generates better solutions in instances where the values in *diff_fastest* are smaller. In this sub-set of instances, the machines are either fast or slow for all jobs; consequently, in some instances, one machine is the fastest for all jobs, which plays against the constructive heuristic Lowest. To analyze this phenomenon, Fig. 14 contains the instances of the extremes, that is the instance with the highest and lowest percentage of jobs that process fastest than the other ones of the 1400 cases. In each graph, the $x-$axis depicts the machines, and the $y-$axis indicates the percentage of jobs that each machine process fastest. Besides, in the upper right corner, each graph holds the general characteristics of the plotted instance, including the range of the values of $p_{ij}$, the number of machines $m$, the number of jobs $n$, and its *id*. As can be seen in these graphs, the benchmark considered in this work contains instances where the lowest processing times of jobs are distributed uniformly, such as the one shown in Fig. 14a. In contrast, Fig. 14b indicates that this benchmark also includes instances with the lowest processing times of jobs distributed poorly among machines, where even one machine is the fastest to process all the jobs, mainly in the set Machcorre.

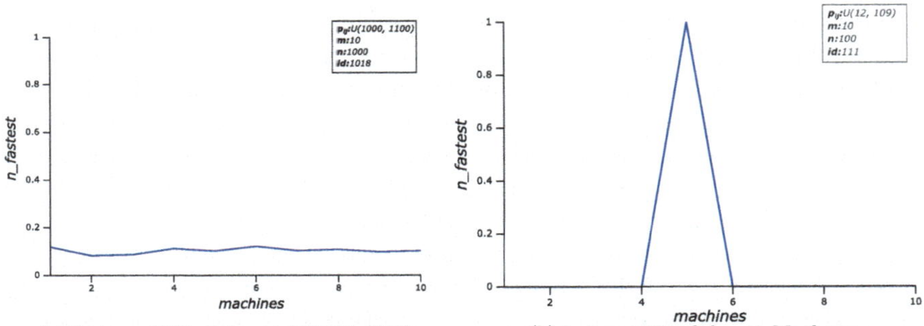


(a) Instance 1018 of the set $U(1000, 1100)$ (b) Instance 111 of the set *Machcorre*

**Fig. 14** Difference between the processing times of the two fastest machines to process each job (*diff_fastest*)

Given this phenomenon, the use of Lowest can be counterproductive when solving instances that belong to the said set.

Another significant aspect of the structure of the instances of $R||C_{max}$ is related to the form of the distribution of the processing times of the jobs, considering their minimum processing times only. Recalling from Fig. 3, in some instances, several jobs share the same lowest processing time while, in other ones, such frequency is almost zero. The scatter plot Fig. 13f shows how the *multiplicity*(*lowest*), a representative index that collects information related to Fig. 3, impacts in the algorithm behavior of Lowest. From this graph it can be seen that the quality of the solutions generated by Lowest decreases as the *multiplicity* of the number of jobs with the lowest processing time increases. This behavior is more marked in the instances of the set $U(100, 120)$, with processing times generated within the smallest range, so the *multiplicity*(*lowest*) increases considerably. Besides, this graph suggests that this behavior does not appear in instances with jobs- and machines-correlated since other features intervene.

As mentioned earlier, instances with jobs and machines correlated can have different properties to uncorrelated ones. Therefore, it is important to study these properties. Figures 13g, h holds the scatter plots of the indexes *job_corre* and *macs_corre*, respectively. Figure 13g shows that the index *job_corre* allows discriminating the instances with jobs correlated, marked in color orange. Furthermore, this plot allows seeing that the performance of Lowest is not related to the correlation of the jobs (*job_corre*), that indicates how similar are the processing times of jobs. The information collected by this index is important since recalling from Fig. 4, in some instances, the differences among processing a job in one machine or another vary little and even nothing, which can be useful when solving an instance of the problem.

Similarly, Fig. 13h indicates that the index *macs_corre* allows distinguishing the instances with machines correlated, marked in color yellow. Besides, this plot shows that the performance of Lowest increases as the value of *mach_corre* decreases, with a linear relation of 0.63. The correlation of machines implies that machines are either fast or slow for all the jobs, to such an extent that, in some instances, one machine is the fastest for most jobs, while other ones may even not process any job fastest. As can be inferred, this feature play against Lowest since in the solutions it generates, the fastest machines end overloaded, while the slow ones with few assigned jobs. For example, in the solution generated by the heuristic Lowest for the instance plotted in Fig. 4b, machine five has all the jobs assigned and the other machines none, because of machine five processes all the jobs fastest than the others. That is why, as can be seen in all the graphs in Fig. 13, Lowest obtained the worst performance in the instances with correlated machines. The last row of Table 6 shows this behavior, where it is possible to observe that Lowest has the worst average performance in the instances of the set *Machcorre*.

Finally, other important features of the structure of $R||C_{max}$ studied are the form and dispersion of the processing times of the jobs. To analyze this information, different indexes were studied, including the representative indexes $cv(min(p_{ij}))$ and $uniformity(p_j)$. Figures 13i and 13j hold scatter plots that indicates how $cv(min(p_{ij}))$ and $uniformity(p_j)$ impact the performance of Lowest, respectively. In Fig. 13i can be seen that, in some sub-set as $U(1, 100)$, $U(10, 100)$, $U(100, 120)$, and $U(100, 200)$, the performance of Lowest decreases as the $cv(min(p_j))$ decreases, although the behavior is not very pronounced by considering all the instances since its linear relation is only of 0.08. In contrast, in Fig. 13j is more clear how $uniformity(p_j)$ impact in the performances of Lowest since they have a linear relation of 0.45. Therefore, its performance worsens as the uniformity of processing times increases. This behavior could be related to the fact that as the distribution of the processing times of jobs has higher values of $uniformity(p_j)$, the likelihood that machines process fastest than the others a similar number of jobs grows.

Recalling from Table 5, the constructive heuristic Lowest has a behavior that differs from the others since its performance is strongly related to different indexes. To analyze the behavior that prevails in the remaining heuristics, Fig. 15 holds scatter plots of the ten most representative indexes and Lowest min, a constructive heuristic with typical behavior, with the same structure that Fig. 13; therefore, the exes and the color bar depict the same characteristics.

In Fig. 15a, b can be seen how $n$ and $m$, representative indexes of the size of the $R||C_{max}$ affect the performance of Lowest min. As can be seen in these plots, indexes related to the size of the problem affect the performance of the heuristic Lowest min similarly to the way they affect the performance of Lowest. That is, their performance decreases, as an instance has a greater number of machines $m$ and a smaller number of jobs $n$. However, these scatter plots also indicate that the performance of Lowest min is more stable since the error rate of the solution that generates oscillate between a smaller range. Besides, in these scatter plots can be seen that the quality of the solutions generated by Lowest min for the instances that belong to the set *Machcorre* are better than those generated by Lowest since Lowest has the worst performance in this set, while Lowest min generates better solutions for such set than for sets $U(1, 100)$ and $U(10, 100)$.

Concerning the quotient ($q$), when analyzing its impact on the performance of Lowest, it was not observed any relation. In contrast, Fig. 15c shows that $q$ is strongly related to the performances of Lowest min; therefore, it provides useful information

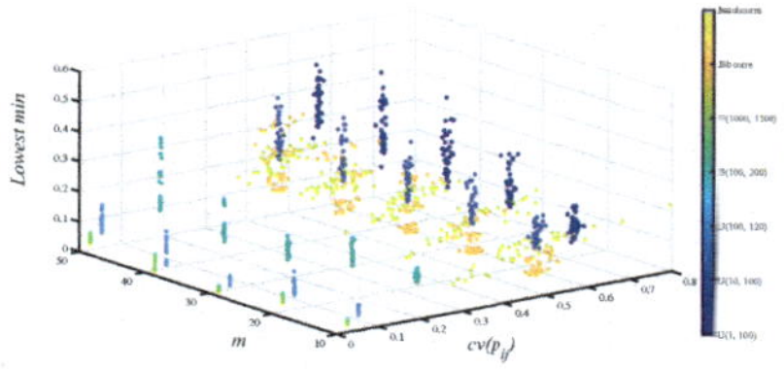


(a) Number of machines: $m$

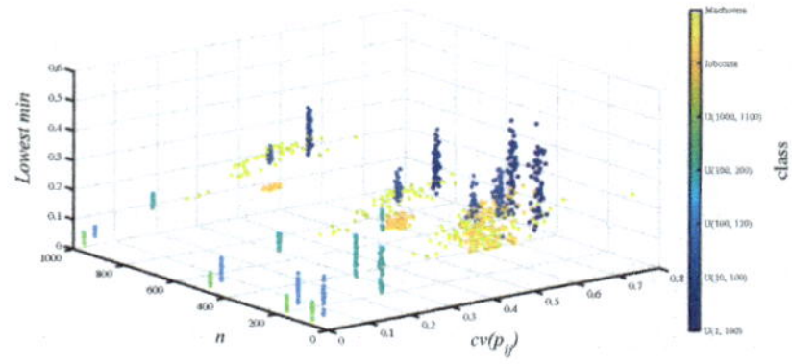


(b) Number of jobs: $n$

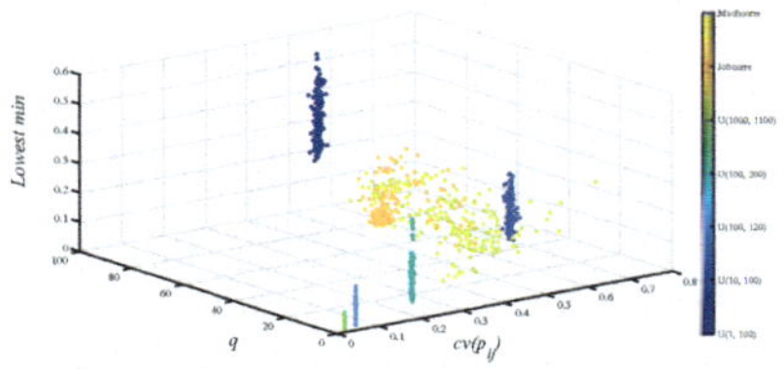


(c) *Quotient* between max and min processing time

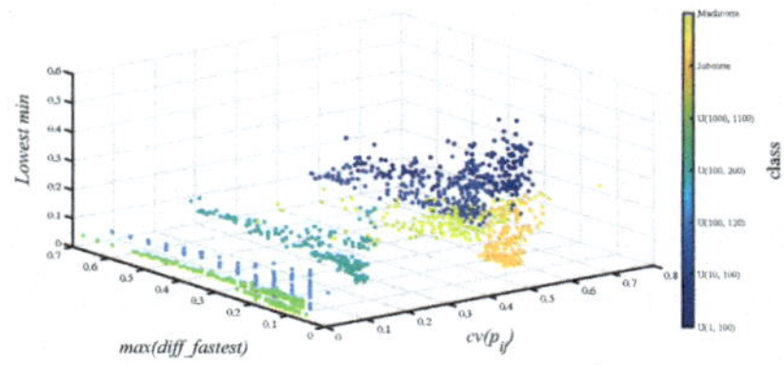


(d) Max difference of the two fastest machines

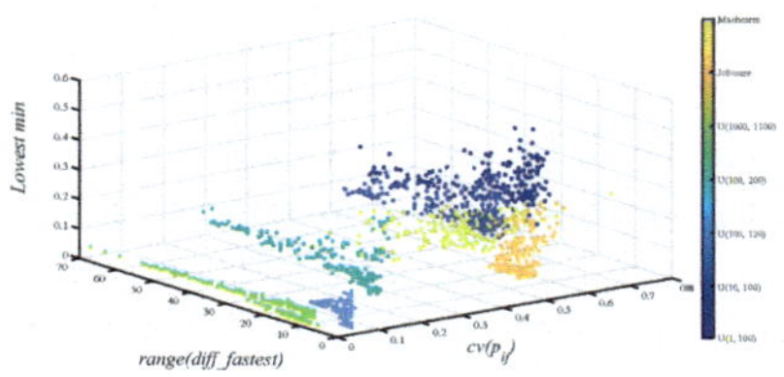


(e) Difference between the two fastest machines

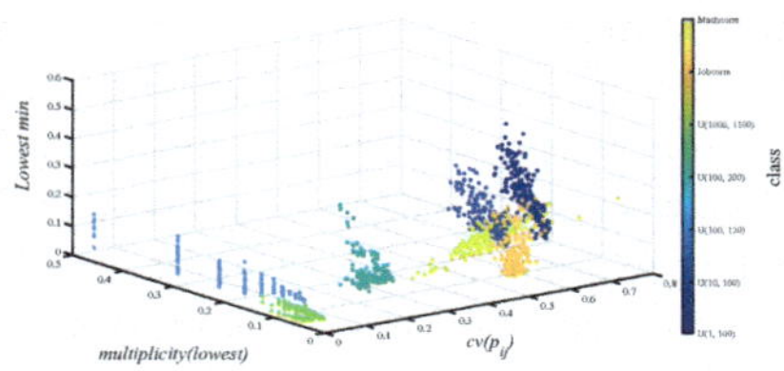


(f) *Multiplicity* of the lowest processing times

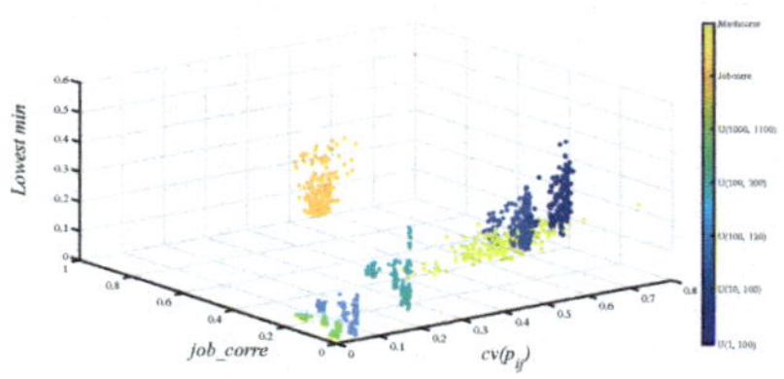


(g) Correlation among jobs: *job_corre*

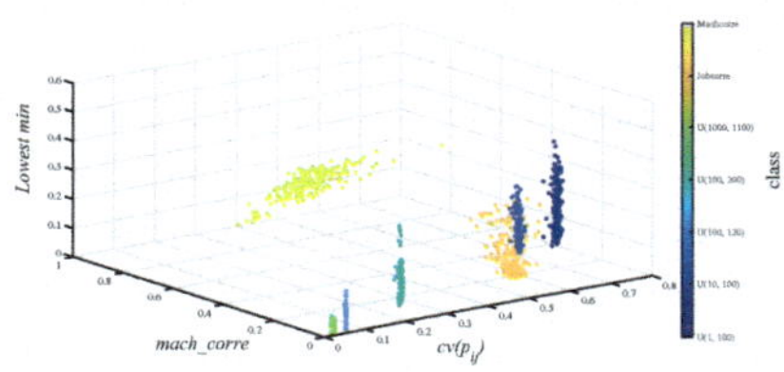


(h) Correlation among machines: *macs_ corre*

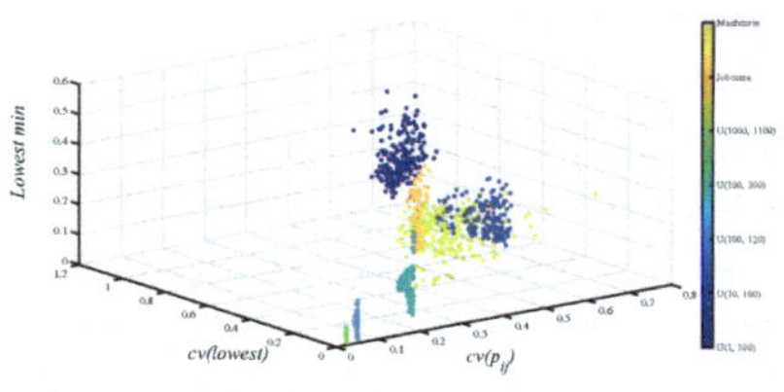


(i) Coefficient of variation of the lowest processing times

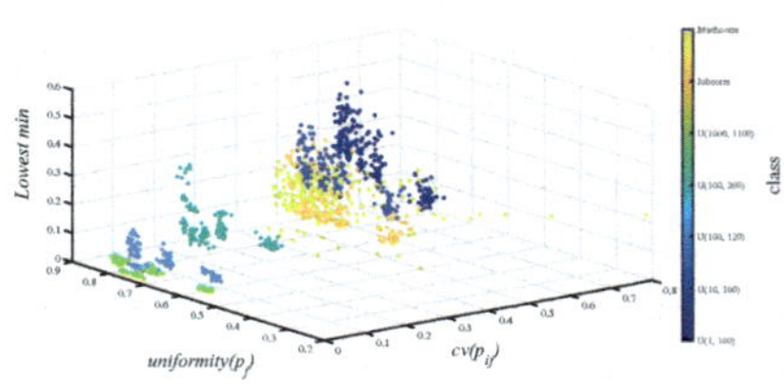


(j) Uniformity of the jobs

**Fig. 15** Influencing of the ten most representative characteristics of the structure of $R||C_{max}$ instances in the performances of the constructive heuristic Lowest min about the difficulty of the instances. Table 5 supports this fact since Lowest min and $q$ have a linear relation of 0.6. That is, as $q$ increases the performance of Lowest min decreases. In this way, the best solutions generated by Lowest min belong to the set $U(1000, 1100)$, where $q = 1.1$, and the worst ones for instances in the set $U(1, 100)$.

Figures 15d, e reinforces the fact that the information related to the difference between processing each job in the fastest and the second-fastest machine *diff_fastest* is related to the difficulty of the instances, mainly *max(diff_fastest)* and *range(diff_fastest)*. As can be seen in these scatter plots, as the maximum difference *max(diff_fastest)* and the range in which all the differences *range(diff_fastest)* of an instance are smaller, the Lowest min performance decreases.

Similarly, Fig. 15f shows that, like Lowest, as the *multiplicity(lowest)* of an instance grows, the performance of Lowest min decreases.

On the other hand, Fig. 15g.h indicate that indexes *job_corre* and *mach_corre* impact likewise Lowest and Lowest min. That is, the higher these values are, the lower the performance. However, from the scatter plots of Lowest min highlight that the solutions generated by this heuristic for the instances in the set *Machcorre* have better quality.

One of the indexes most related to the performance of Lowest min is $cv(min(p_{i,j}))$, with a linear relationship of 0.66. As can be seen in Fig. 15h, instances with high values of this characteristic represent a greater challenge for this constructive heuristic. In this way, the instances of the set $U(1, 100)$, marked in blue, have the higher values of $cv(min(p_{ij}))$, and they represent a greater challenge for Lowest min. On the contrary, the set $U(100, 120)$ holds the instances with the lowest $cv(min(p_{ij}))$, marked in green, in which Lowest min get the best performances.

Finally, Fig. 15i indicates that, similar to Lowest, as the $uniformity(p_j)$ of an instance increases, the performance of Lowest min decreases.

## 4.4 Phase 4: Explanations of the Algorithm Behavior

The results obtained from the application of data analysis techniques to the designed indexes indicate that the performance of the eleven constructive heuristics analyzed on solving $R||C_{max}$ instances have certain relation with the parameter used to generate the instances like the number of machines $m$, the number of jobs $n$, and its ratio $\frac{n}{m}$; measures extracted from all the processing times in the instances like the minimum $min(p_{ij})$, the maximum $max(p_{ij})$, and the average $mean(p_{ij})$; the form of the processing time distribution of the jobs $uniformity(p_j)$, the machines $uniformity(p_i)$, and the number of jobs that each machine process fastest than the others *uniformity(n_fastest)*; the maximum difference between the processing times of the two fastest machines to process each job *max(diff_fastest)*; the repetition of the shortest processing times of each job *multiplicity(lowest)*; and the machine correlation *mach_corre*. The results suggest that the difficulty the instances of this problem increases when $m$, $uniformity(p_j)$, $uniformity(p_i)$, *uniformity(n_fastest)*, $min(p_{ij})$, *max*($p_{ij}$), *multiplicity*(*lowest*), and *mach_corre* grow. In contrast, said difficulty also increases when $n$, $\frac{n}{m}$, *mean*($p_{ij}$), and *max*(*diff_fastest*) decreases.

On the other hand, the results obtained suggest that some indexes have a certain association with the performance of only ten heuristics. That is, all the heuristics analyzed, except for Lowest. Most of the indexes with this behavior collect information related to the dispersion of the processing times $p_{ij}$ of the $R||C_{max}$ instances, including the coefficient of variation *cv*($p_{ij}$), the difference *range*($p_{ij}$) and the quotient $q$ of the maximum and the minimum processing time. Additionally, this phenomenon is present in indexes like the coefficient of variation of the shortest processing time of jobs *cv*(*min*($p_{ij}$)), the central tendency of the difference between the processing times of the two fastest machines to process each job *mean*(*diff_fastest*), and the job correlation *job_corre*. The results suggest that the difficulty the $R||C_{max}$ instances for these ten heuristics increases when the values of $q$, *range*($p_{ij}$), *cv*($p_{ij}$) grows, as well as when the ones of *mean*(*diff_fastest*) and *job_corre* decrease. The low relationship between the performance of Lowest and these indices may be associated with the fact that Lowest no uses any sort strategy, and directly assigns each job to the fastest machine without considering any other criterion. Therefore, the indexes related to the dispersion of the processing times have not a great influence on its performance.

Finally, some characteristics of the structure of $R||C_{max}$ are associated with only a certain group of heuristics. For example, the difficulty these instances represent for the heuristics Lowest and Lowest 4 g min increases when the value of *below* $\frac{n}{m}$ (i.e., the number of machines processing faster a lower number of jobs that $\frac{n}{m}$) increases. Similarly, the difficulty of these instances for the heuristics Lowest and Random increases when the value of *cv*(*n_fastest*), i.e., the coefficient of variation of the number of jobs that each process fastest, increases.

## 5 Conclusions and Future Research

In this work, the characterization of the unrelated parallel machines scheduling problem ($R||C_{max}$) and the analysis of the algorithm behavior of eleven constructive heuristics, were addressed using data analysis techniques. In this sense, twenty indexes related to the instance structure were calculated, to later, studying how they affect the performances of the constructive heuristics.

The results generated by using different data analysis techniques suggest that the complexity of the case studies of $R||C_{max}$ are related to (1) the number of jobs and machines; (2) the general distribution of the possible processing times for each job; (3) properties extracted from the collection of processing times created with the minimum time each job can be processed; and (4) features generated from the compilation of the difference of the processing times of the two fastest machines to process each job. It is important to remark that the results of the study are limited to the set of instances and constructive heuristics used. Therefore, it is necessary to analyze a larger number of instances and algorithms for exploring more values of the properties that affect the difficulty of the instances and validating its impact.

As future work, we are planning to look for new and more robust indexes to study more deeply the structure of $R||C_{max}$ instances and identify other properties impacting in their difficulty. On the other hand, we will extend this study to analyze the impact of the indexes designed in more complex algorithms, including swarm intelligence metaheuristics and evolutionary algorithms. In this way, we expect to generate useful knowledge of the problem-domain to design high-performance metaheuristics that solve parallel machine scheduling problems and improve the performance of existing algorithms.

Finally, we expect that other parallel machine scheduling problems with similar structures can be studied, taking the methodology used in this work as a starting point.

## References


1. C. Blum, A. Roli, Metaheuristics in combinatorial optimization: overview and conceptual comparison. ACM Comput. Surv. (CSUR) **35**(3), 268–308 (2003)
2. M.R. Garey, D.S. Johnson, *Computers and Intractability*, vol. 29 (W. H, Freeman New York, 2002)
3. D. Hains, Structure in combinatorial optimization and its effect on heuristic performance. Ph.D. thesis, Colorado State University. Libraries (2013)
4. J. Kanda, A. de Carvalho, E. Hruschka, C. Soares, P. Brazdil, Meta-learning to select the best meta-heuristic for the traveling salesman problem: a comparison of meta-features. Neurocomputing **205**, 393–406 (2016)
5. M. Chiarandini, L. Paquete, M. Preuss, E. Ridge, Experiments on metaheuristics: methodological overview and open issues. Technical Report DMF-2007-03-003, The Danish Mathematical Society, Denmark (2007)
6. K. Sörensen, M. Sevaux, F. Glover, A history of metaheuristics, *Handbook of Heuristics* (2018) pp. 1–18
7. M. Quiroz-Castellanos, Caracterización del proceso de optimización de algoritmos heurísticos aplicados al problema de empacado de objetos en contenedores. Ph.D. thesis, Instituto Tecnológico de Tijuana (2014)
8. M. Quiroz, L. Cruz-Reyes, J. Torres-Jimenez, C.G. Santillán, H.J.F. Huacuja, P. Melin, Characterization of the optimization process, in *Recent Advances on Hybrid Approaches for Designing Intelligent Systems* (Springer, 2014), pp. 493–507
9. J.R. Rice, The algorithm selection problem, in *Advances in Computers*, vol. 15. (Elsevier, 1976), pp. 65–118
10. L. Kotthoff, Algorithm selection for combinatorial search problems: a survey, in *Data Mining and Constraint Programming* (Springer, 2016), pp. 149–190
11. P. Brazdil, C. Giraud-Carrier, Metalearning and algorithm selection: progress, state of the art and introduction to the 2018 special issue (2018)
12. K.A. Smith-Miles, Cross-disciplinary perspectives on meta-learning for algorithm selection. ACM Comput. Surv. (CSUR) **41**(1), 6 (2009)
13. K. Alyahya, J.E. Rowe, Landscape analysis of a class of np-hard binary packing problems Evol. Comput. **27**(1), 47–73 (2019)
14. H. Lu, R. Zhou, Z. Fei, C. Guan, Spatial-domain fitness landscape analysis for combinatorial optimization. Inf. Sci. **472**, 126–144 (2019)
15. E. Pitzer, M. Affenzeller, A comprehensive survey on fitness landscape analysis, in *Recent Advances in Intelligent Engineering Systems* (Springer, 2012), pp. 161–191

16. B. Chopard, M. Tomassini, Phase transitions in combinatorial optimization problems, in *An Introduction to Metaheuristics for Optimization* (Springer, 2018), pp. 171–189
17. Z. Wang, B. O'Gorman, T.T. Tran, E.G. Rieffel, J. Frank, M. Do, An investigation of phase transitions in single-machine scheduling problems, in *27th International Conference on Automated Planning and Scheduling* (2017)
18. T. Walsh, J. Slaney, Backbones in optimization and approximation, in *Proceedings of the Seventeenth International Joint Conference on Artificial Intelligence (IJCAI-01)* (2001)
19. K. Smith-Miles, L. Lopes, Measuring instance difficulty for combinatorial optimization problems. Comput. Operat. Res. **39**(5), 875–889 (2012)
20. E. Mokotoff, Parallel machine scheduling problems: a survey. Asia-Pac. J. Oper. Res. **18**(2), 193 (2001)
21. R.L. Graham, E.L. Lawler, J.K. Lenstra, A.R. Kan, Optimization and approximation in deterministic sequencing and scheduling: a survey, in *Annals of Discrete Mathematics*, vol. 5. (Elsevier, 1979), pp. 287–326
22. E.L. Lawler, J.K. Lenstra, A.R. Kan, Recent developments in deterministic sequencing and scheduling: a survey, in *Deterministic and Stochastic Scheduling* (Springer, 1982), pp. 35–73
23. L. Fanjul-Peyro, R. Ruiz, Iterated greedy local search methods for unrelated parallel machine scheduling. Eur. J. Oper. Res. **207**(1), 55–69 (2010)
24. M.R. Garey, A guide to the theory of np-completeness, in *Computers and Intractability* (1979)
25. J.K. Lenstra, D.B. Shmoys, E. Tardos, Approximation algorithms for scheduling unrelated parallel machines. Math. Program. **46**(1–3), 259–271 (1990)
26. S. Martello, F. Soumis, P. Toth, Exact and approximation algorithms for makespan minimization on unrelated parallel machines. Discrete Appl. Math. **75**(2), 169–188 (1997)
27. E. Mokotoff, P. Chrétienne, A cutting plane algorithm for the unrelated parallel machine scheduling problem. Eur. J. Oper. Res. **141**(3), 515–525 (2002)
28. E. Mokotoff, J. Jimeno, Heuristics based on partial enumeration for the unrelated parallel processor scheduling problem. Ann. Oper. Res. **117**(1–4), 133–150 (2002)
29. M. Ghirardi, C.N. Potts, Makespan minimization for scheduling unrelated parallel machines: A recovering beam search approach. Eur. J. Oper. Res. **165**(2), 457–467 (2005)
30. E. Davis, J.M. Jaffe, Algorithms for scheduling tasks on unrelated processors. J. ACM (JACM) **28**(4), 721–736 (1981)
31. P. De, T.E. Morton, Scheduling to minimize makespan on unequal parallel processors. Decis. Sci. **11**(4), 586–602 (1980)
32. E. Horowitz, S. Sahni, Exact and approximate algorithms for scheduling nonidentical processors. J. ACM (JACM) **23**(2), 317–327 (1976)
33. O.H. Ibarra, C.E. Kim, Heuristic algorithms for scheduling independent tasks on nonidentical processors. J. ACM (JACM) **24**(2), 280–289 (1977)
34. V. Sels, J. Coelho, A.M. Dias, M. Vanhoucke, Hybrid tabu search and a truncated branch-and-bound for the unrelated parallel machine scheduling problem. Comput. Operat. Res. **53**, 107–117 (2015)
35. K. Jansen, L. Porkolab, Improved approximation schemes for scheduling unrelated parallel machines. Math. Operat. Res. **26**(2), 324–338 (2001)
36. V. Kumar, M.V. Marathe, S. Parthasarathy, A. Srinivasan, A unified approach to scheduling on unrelated parallel machines. J. ACM (JACM) **56**(5), 28 (2009)
37. Y. Lin, M. Pfund, J. Fowler, Minimizing makespans for unrelated parallel machine scheduling problems, in *IEEE/INFORMS International Conference on Service Operations, Logistics and Informatics, 2009. SOLI'09* (IEEE, 2009), pp. 107–110
38. C.N. Potts, Analysis of a linear programming heuristic for scheduling unrelated parallel machines. Discrete Appl. Math. **10**(2), 155–164 (1985)
39. M. Serna, F. Xhafa, Approximating scheduling unrelated parallel machines in parallel. Comput. Optim. Appl. **21**(3), 325–338 (2002)
40. E.V. Shchepin, N. Vakhania, An optimal rounding gives a better approximation for scheduling unrelated machines. Operat. Res. Lett. **33**(2), 127–133 (2005)

41. D.B. Shmoys, É. Tardos, An approximation algorithm for the generalized assignment problem. Math. Program. **62**(1–3), 461–474 (1993)
42. C. Charalambous, K. Fleszar, K.S. Hindi, A hybrid searching method for the unrelated parallel machine scheduling problem, in *IFIP International Conference on Artificial Intelligence Applications and Innovations* (Springer, 2010), pp. 230–237
43. C. Glass, C. Potts, P. Shade, Unrelated parallel machine scheduling using local search. Math. Comput. Modell. **20**(2), 41–52 (1994)
44. Y. Guo, A. Lim, B. Rodrigues, L. Yang, Minimizing the makespan for unrelated parallel machines. Int. J. Artif. Intell. Tools **16**(03), 399–415 (2007)
45. N. Piersma, W. van Dijk, A local search heuristic for unrelated parallel machine scheduling with efficient neighborhood search. Math. Comput. Modell. **24**(9), 11–19 (1996)
46. P. Sivasankaran, T. Sornakumar, R. Panneerselvam, Efficient heuristic to minimize makespan in single machine scheduling problem with unrelated parallel machines. Intell. Inf. Manage. **2**(3), 188–198 (2010)
47. P. Sivasankaran, T. Sornakumar, R. Panneerselvam, Design and comparison of simulated annealing algorithm and grasp to minimize makespan in single machine scheduling with unrelated parallel machines. Intell. Inf. Manage. **2**(07), 406 (2010)
48. S.L. van de Velde, Duality-based algorithms for scheduling unrelated parallel machines. ORSA J. Comput. **5**(2), 192–205 (1993)
49. Y.K. Lin, M.E. Pfund, J.W. Fowler, Heuristics for minimizing regular performance measures in unrelated parallel machine scheduling problems. Comput. Operat. Res. **38**(6), 901–916 (2011)

# Comparison of Genetic Algorithm and Particle Swarm Optimization of Ensemble Neural Networks for Complex Time Series Prediction

**Martha Pulido and Patricia Melin**

**Abstract** In this paper the development of a method for the design of the general architecture of ensemble neural networks for time series prediction is presented. This method consists on the optimization of the ensemble neural network with response integration based on type-1 and type-2 fuzzy logic. The optimization algorithms that are used to illustrate the proposed method are the genetic algorithm and the particle swarm optimization algorithm and the purpose of applying these methods is to find the best possible architecture of the neural network ensemble with the lowest prediction error. Simulation results with several complex times series are presented to show the advantage of the proposed method, also comparisons are made with other time series, such as Mexican Stock Exchange, and Dow Jones Stock Exchange.

**Keywords** Ensemble neural networks · Genetic algorithms · Optimization · Particle swarm · Time series prediction · Fuzzy logic

## 1 Introduction

Time series predictions are very important because based on their results we can examine past events to know the possible behavior of futures events and thus we can be able to take preventive or corrective decisions to help avoid unwanted circumstances [1–23].

The selection and implementation of an appropriate method for prediction has been usually a major concern for corporations that seek to ensure the benefit and survival of business. The predictions give the company the ability to make decisions in the medium and long term, and due to the accuracy or inaccuracy of data this could mean predicted growth or profits or financial losses [24–48].

M. Pulido (✉) · P. Melin
Tijuana Institute of Technology, Tijuana, Mexico
e-mail: marthapulido_84@hotmail.com

P. Melin
e-mail: pmelin@tectijuana.mx



P. Melin et al. (eds.), *Recent Advances of Hybrid Intelligent Systems Based on Soft Computing*, Studies in Computational Intelligence 915,
https://doi.org/10.1007/978-3-030-58728-4_3

The main contribution of this paper is the proposed approach for prediction based on ensemble neural networks with fuzzy integration of responses and optimization with genetic algorithms and particle swarm optimization.

This paper is organized as follows: In Sect. 2 the problem statement and the proposed method are presented, Sect. 3 shows the simulation results of the proposed method, Sect. 4 presents the comparison of results and Sect. 5 offers the conclusions.

## 2 Problem Statement and Proposed Method

The main goal of this paper is to present a model that allows us to predict historical data of chaotic time series. The series that is used to illustrate the approach is the US Dollar/MX Peso exchange rate, as we can also mention that this method can be applied to any problem expressed as a time series. In Sect. 4 we show results of other series that were also considered using ensemble neural networks and their optimization by means of bio-inspired methods, as well as the response integration with type-1 and type-2 fuzzy systems. Genetic algorithms and particle swarm optimization algorithms are also used to optimize the fuzzy systems. Figure 1 illustrates the general architecture of the method, where the historical data of each time series for prediction is indicated, and then the data is provided to the modules that are optimized with a genetic algorithm or the particle swarm optimization for the ensemble network. Then results of these modules are integrated with an integration method based on type-1 and type-2 fuzzy logic, which is also optimized with a genetic algorithm or the particle swarm optimization algorithm.

Data of the US Dollar/MX Peso time series: The data consist of 800 points that correspond to a period from 07/04/08 to 09/05/11 (as shown in Fig. 2).

The proposed method consists in the following steps:

### 2.1 *Analyze the Time Series*

Plot and analyze the time series to observe the behavior, trend, seasonality and irregular changes (US/Dollar MX time series).

### 2.2 *Creation of the Ensemble Neural Network*

The parameters used for the ensemble neural network are: 5 modules, number of layers from 1 to 3, neurons number of 1 to 30, the network parameters used to train the neural network ensemble for each of the ensembles are 100 epochs, learning rate of = 0.01, the error goal is 0.01, and training method of Levenberg-Marquardt (trainlm).

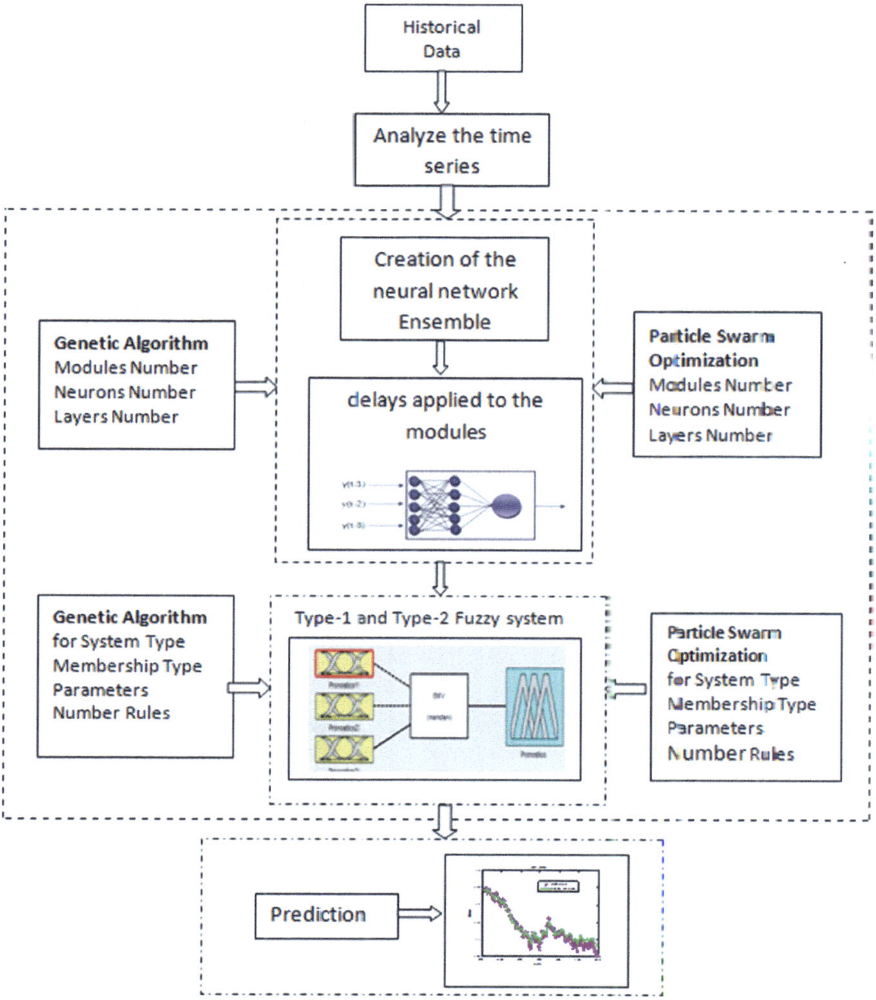


**Fig. 1** General architecture of the proposed method

These parameters are used in this work because previous tests were performed and we managed to obtain a good prediction error with these parameter values.

The architecture for each of the modules is based on three delays of the time series, as shown in Fig. 3.

The net input $\bar{x}_j$ of a node is defined by the weighted sum of the incoming signals plus a bias term. For instance, the net input and the output of the node $j$ are given as follows.

$$\bar{x}_j = \sum_i \omega_{ij} + \omega_j,$$

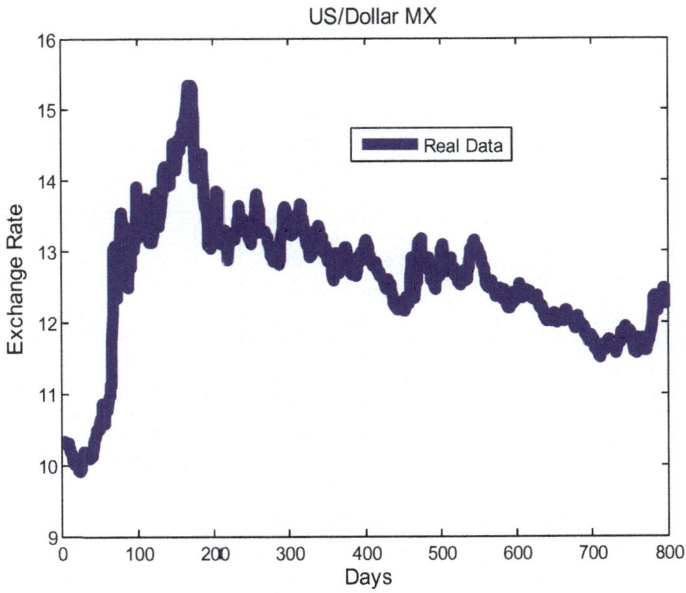


**Fig. 2** Series data of the US Dollar/MX time series

**Fig. 3** Shows that a module is a neural network and calculations are performed as follows

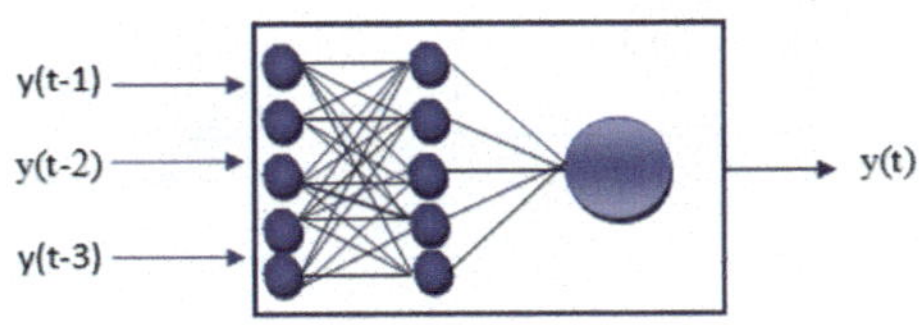


$$x = f\left(\bar{x}_j\right) = \frac{1}{1 + \exp\left(-\bar{x}_j\right)}, \tag{1}$$

where $x_i$ is the output of node *i* located in any of the previous layers, $\omega_{ij}$ is the weight associated with the link connecting nodes *i* and *j*, and $\omega_j$ is the bias of node *j*. Since the weights $\omega_{ij}$ are actually internal parameters associated with each node *j* changing the weights of the node will alter behavior of the node and in turn alter the behavior of the whole backpropagation Multilayer Perceptron.

A squared error measure for the *p*th input-output pair is defined as:

$$E_P = \sum (d_k - x_k)^2, \tag{2}$$

where $d_k$ is the desired output for the node, $x_k$ is the actual output for the node when the input part of the *p*th data pair is presented.

### 2.2.1 GA Applied for the Optimization of the Ensemble Neural Network

For the optimization of the structure of the ensemble neural network a genetic algorithm can be used. The chromosome of the genetic algorithm represents the structure of the ensemble. The objective of using these functions is because in previous studies, tests were performed with this function and better prediction error was obtained.

The objective function is defined to minimize the prediction error:

$$EM = \left( \sum_{i=1}^{D} |a_i - x_i| \right) / \mathrm{D} \tag{3}$$

where $a$, corresponds to the predicted data depending on the output of the network modules, $X$ represents real data, and $D$ the Number of Data points and $EM$ is the total prediction error.

The corresponding chromosome structure is shown in Fig. 4.

The chromosome shows the proposed structure to optimize the architecture of the ensemble neural network with a genetic algorithm. First, the number of modules is represented in which a limit of 5 modules is considered, and then the number of layers with a maximum of 3 layers and the number of neurons with a maximum of 30 neurons are shown. The selection method is the Roulette, the percentage of crossover and mutation are changed at random, as suggested in current works, like in [24]. The crossover rate is changed in the range of [0.5, 1], the mutation rate is changed in

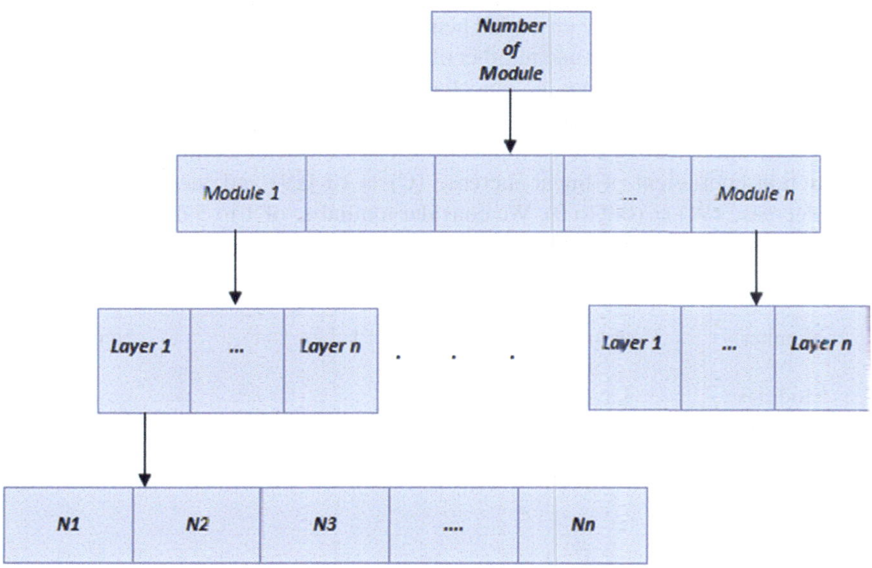


**Fig. 4** Chromosome structure to optimize the neural network

the range of [0, 0.1] and 100 individuals and 100 generations are used. Previous experiments were performed with three modules, and up to five modules consider that when using 6 modules the improvement in the prediction was not significant and were performed with three layers and 30 neurons in the ensemble neural networks giving better results of the prediction.

### 2.2.2 PSO Applied for the Optimization of the Ensemble Neural Network

For the optimization of the structure of the ensemble neural network a particle swarm optimization algorithm is also used. The structure of the particle represents the structure of the ensemble.

The objective function is defined to minimize the prediction error as follows:

$$EM = \left(\sum_{i=1}^{D} |a_i - x_i|\right)/\text{D} \tag{4}$$

where $a$, corresponds to the predicted data depending on the output of the network modules, ***X*** represents the real data. *D* is the Number of Data points and *EM* is the total average prediction error.

The corresponding particle structure is shown in Fig. 5.

Figure 5 represents the Particle Structure to optimize the ensemble neural network, where the parameters that are optimized are the number of modules, number of layers, and number of neurons of the ensemble neural network. PSO determines the number of modules, number of layers and number of neurons per layer that the neural network ensemble should have, to meet the objective of achieving a better Prediction error.

The parameters for the particle swarm optimization algorithm are: 100 Particles, 100 iterations, Cognitive Component (C1) = 2, Social Component (C2) = 2, Constriction coefficient of linear increase (C) = (0–0.9) and Inertia weight with linear decrease (W) = (0.9 to 0). We consider a number of 1 to 5 modules, number of layers of 1 to 3 and neurons number from 1 to 30.

| | | | | |
|---|---|---|---|---|
| Number of Modules | Number of Layers | Neurons 1 | ... | Neurons n |

**Fig. 5** Particle structure to optimize the ensemble neural network

## 2.3 Type-1 and Type-2 Fuzzy System Integration

The aggregation of the responses of the optimized ensemble neural network is performed with type-1 and type-2 fuzzy systems. In this work the fuzzy system consisting of 5 inputs depending on the number of modules of the neural network ensemble and one output is used. Each input and output linguistic variable of the fuzzy system uses 2 Gaussian membership functions. The performance of the type-2 fuzzy aggregators is analyzed under different levels of uncertainty to find out the best design of the membership functions for the 32 rules of the fuzzy system. Previous tests have been performed only with a three input fuzzy system and the fuzzy system changes according to the responses of the neural network to give us better prediction error. In the type-2 fuzzy system we also change the levels of uncertainty to obtain the best prediction error.

Figure 6 shows a fuzzy system consisting of 5 inputs depending on the number of modules of the neural network ensemble and one output. Each input and output linguistic variable of the fuzzy system uses 2 Gaussian membership functions. The performance of the type-2 fuzzy aggregators is analyzed under different levels of uncertainty to find out the best design of the membership functions for the 32 rules of the fuzzy system. Previous experiments were performed with triangular, Gaussian and the Gaussian and given the better results of the prediction.

Figure 7 represents the 32 possible rules of the fuzzy system; we have 5 inputs in the fuzzy system with 2 membership functions, and the outputs with 2 membership functions. These fuzzy rules are used for both the type-1 and type-2 fuzzy systems. In

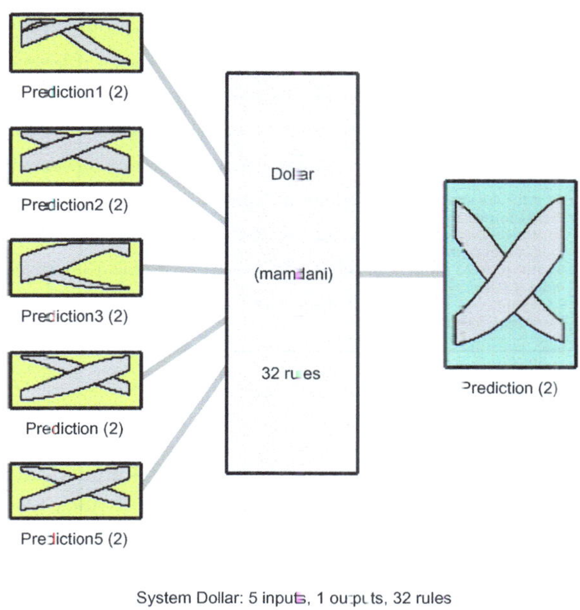


**Fig. 6** Fuzzy inference system for integration of the ensemble neural network

1. If (Prediction1 is Pron1Bajo) and (Prediction2 is Pron2Bajo) and (Prediction3 is Pron3Bajo) and (Prediction4 is Pron4Bajo) and (Prediction5 is Pron5Bajo) then (Prediction is Bajo)
2. If (Prediction1 is Pron1Alto) and (Prediction2 is Pron2Alto) and (Prediction3 is Pron3Alto) and (Prediction4 is Pron4Alto) and (Prediction5 is Pron5Alto) then (Prediction is Alto)
3. If (Prediction1 is Pron1Bajo) and (Prediction2 is Pron2Bajo) and (Prediction3 is Pron3Bajo) and (Prediction4 is Pron4Bajo) and (Prediction5 is Pron5Alto) then (Prediction is Bajo)
4. If (Prediction1 is Pron1Alto) and (Prediction2 is Pron2Alto) and (Prediction3 is Pron3Alto) and (Prediction4 is Pron4Alto) and (Prediction5 is Pron5Bajo) then (Prediction is Alto)
5. If (Prediction1 is Pron1Bajo) and (Prediction2 is Pron2Bajo) and (Prediction3 is Pron3Bajo) and (Prediction4 is Pron4Alto) and (Prediction5 is Pron5Alto) then (Prediction is Bajo)
6. If (Prediction1 is Pron1Alto) and (Prediction2 is Pron2Alto) and (Prediction3 is Pron3Alto) and (Prediction4 is Pron4Bajo) and (Prediction5 is Pron5Bajo) then (Prediction is Alto)
7. If (Prediction1 is Pron1Bajo) and (Prediction2 is Pron2Bajo) and (Prediction3 is Pron3Alto) and (Prediction4 is Pron4Alto) and (Prediction5 is Pron5Alto) then (Prediction is Alto)
8. If (Prediction1 is Pron1Alto) and (Prediction2 is Pron2Alto) and (Prediction3 is Pron3Bajo) and (Prediction4 is Pron4Bajo) and (Prediction5 is Pron5Bajo) then (Prediction is Bajo)
9. If (Prediction1 is Pron1Bajo) and (Prediction2 is Pron2Alto) and (Prediction3 is Pron3Alto) and (Prediction4 is Pron4Alto) and (Prediction5 is Pron5Alto) then (Prediction is Alto)
10. If (Prediction1 is Pron1Alto) and (Prediction2 is Pron2Bajo) and (Prediction3 is Pron3Bajo) and (Prediction4 is Pron4Bajo) and (Prediction5 is Pron5Bajo) then (Prediction is Bajo)
11. If (Prediction1 is Pron1Bajo) and (Prediction2 is Pron2Alto) and (Prediction3 is Pron3Bajo) and (Prediction4 is Pron4Alto) and (Prediction5 is Pron5Bajo) then (Prediction is Bajo)
12. If (Prediction1 is Pron1Alto) and (Prediction2 is Pron2Bajo) and (Prediction3 is Pron3Alto) and (Prediction4 is Pron4Bajo) and (Prediction5 is Pron5Alto) then (Prediction is Alto)
13. If (Prediction1 is Pron1Bajo) and (Prediction2 is Pron2Alto) and (Prediction3 is Pron3Bajo) and (Prediction4 is Pron4Bajo) and (Prediction5 is Pron5Bajo) then (Prediction is Bajo)
14. If (Prediction1 is Pron1Alto) and (Prediction2 is Pron2Bajo) and (Prediction3 is Pron3Alto) and (Prediction4 is Pron4Alto) and (Prediction5 is Pron5Alto) then (Prediction is Alto)
15. If (Prediction1 is Pron1Bajo) and (Prediction2 is Pron2Bajo) and (Prediction3 is Pron3Alto) and (Prediction4 is Pron4Bajo) and (Prediction5 is Pron5Bajo) then (Prediction is Bajo)
16. If (Prediction1 is Pron1Alto) and (Prediction2 is Pron2Alto) and (Prediction3 is Pron3Bajo) and (Prediction4 is Pron4Alto) and (Prediction5 is Pron5Alto) then (Prediction is Alto)
17. If (Prediction1 is Pron1Bajo) and (Prediction2 is Pron2Bajo) and (Prediction3 is Pron3Bajo) and (Prediction4 is Pron4Alto) and (Prediction5 is Pron5Bajo) then (Prediction is Bajo)
18. If (Prediction1 is Pron1Alto) and (Prediction2 is Pron2Alto) and (Prediction3 is Pron3Alto) and (Prediction4 is Pron4Bajo) and (Prediction5 is Pron5Alto) then (Prediction is Alto)
19. If (Prediction1 is Pron1Bajo) and (Prediction2 is Pron2Bajo) and (Prediction3 is Pron3Alto) and (Prediction4 is Pron4Alto) and (Prediction5 is Pron5Bajo) then (Prediction is Bajo)
20. If (Prediction1 is Pron1Alto) and (Prediction2 is Pron2Alto) and (Prediction3 is Pron3Bajo) and (Prediction4 is Pron4Bajo) and (Prediction5 is Pron5Alto) then (Prediction is Alto)
21. If (Prediction1 is Pron1Bajo) and (Prediction2 is Pron2Alto) and (Prediction3 is Pron3Alto) and (Prediction4 is Pron4Bajo) and (Prediction5 is Pron5Bajo) then (Prediction is Bajo)
22. If (Prediction1 is Pron1Alto) and (Prediction2 is Pron2Bajo) and (Prediction3 is Pron3Bajo) and (Prediction4 is Pron4Alto) and (Prediction5 is Pron5Alto) then (Prediction is Alto)
23. If (Prediction1 is Pron1Bajo) and (Prediction2 is Pron2Bajo) and (Prediction3 is Pron3Alto) and (Prediction4 is Pron4Alto) and (Prediction5 is Pron5Bajo) then (Prediction is Bajo)
24. If (Prediction1 is Pron1Alto) and (Prediction2 is Pron2Alto) and (Prediction3 is Pron3Bajo) and (Prediction4 is Pron4Bajo) and (Prediction5 is Pron5Alto) then (Prediction is Alto)
25. If (Prediction1 is Pron1Bajo) and (Prediction2 is Pron2Alto) and (Prediction3 is Pron3Alto) and (Prediction4 is Pron4Bajo) and (Prediction5 is Pron5Alto) then (Prediction is Alto)
26. If (Prediction1 is Pron1Alto) and (Prediction2 is Pron2Bajo) and (Prediction3 is Pron3Bajo) and (Prediction4 is Pron4Alto) and (Prediction5 is Pron5Alto) then (Prediction is Bajo)
27. If (Prediction1 is Pron1Bajo) and (Prediction2 is Pron2Alto) and (Prediction3 is Pron3Bajo) and (Prediction4 is Pron4Alto) and (Prediction5 is Pron5Bajo) then (Prediction is Bajo)
28. If (Prediction1 is Pron1Alto) and (Prediction2 is Pron2Bajo) and (Prediction3 is Pron3Alto) and (Prediction4 is Pron4Bajo) and (Prediction5 is Pron5Alto) then (Prediction is Alto)
29. If (Prediction1 is Pron1Bajo) and (Prediction2 is Pron2Alto) and (Prediction3 is Pron3Alto) and (Prediction4 is Pron4Alto) and (Prediction5 is Pron5Alto) then (Prediction is Bajo)
30. If (Prediction1 is Pron1Alto) and (Prediction2 is Pron2Bajo) and (Prediction3 is Pron3Bajo) and (Prediction4 is Pron4Alto) and (Prediction5 is Pron5Alto) then (Prediction is Bajo)
31. If (Prediction1 is Pron1Bajo) and (Prediction2 is Pron2Alto) and (Prediction3 is Pron3Bajo) and (Prediction4 is Pron4Alto) and (Prediction5 is Pron5Alto) then (Prediction is Alto)
32. If (Prediction1 is Pron1Alto) and (Prediction2 is Pron2Bajo) and (Prediction3 is Pron3Alto) and (Prediction4 is Pron4Bajo) and (Prediction5 is Pron5Bajoo) then (Prediction is Bajo)

**Fig. 7** Rules of the type-2 fuzzy system

previous work tests were performed with 3 inputs, and the prediction error obtained was significantly higher and the number of rules was greater, and this is why we changed to 2 inputs.

## 2.4 GA Applied for the Optimization of the Fuzzy System

### 2.4.1 Optimization of the Type-1 Fuzzy System with a GA

The chromosome to optimize the type-1 fuzzy system is shown in Fig. 8 and is composed of 180 genes, which allow us to optimize the structure of the fuzzy integrator. The first gene (binary) determines the type of system (Mamdani or Sugeno), the second gene (real data) determines the type of membership functions (Gaussian, generalized bell, triangular and trapezoidal). Genes 3–147 (real numbers) allow us to manage the parameters of the membership functions, for the inputs and output (depending on the type of system and type of functions we use only part of the chromosome). Genes 179–180 (binary) allow us to reduce the number of fuzzy rules, activating or deactivating them.

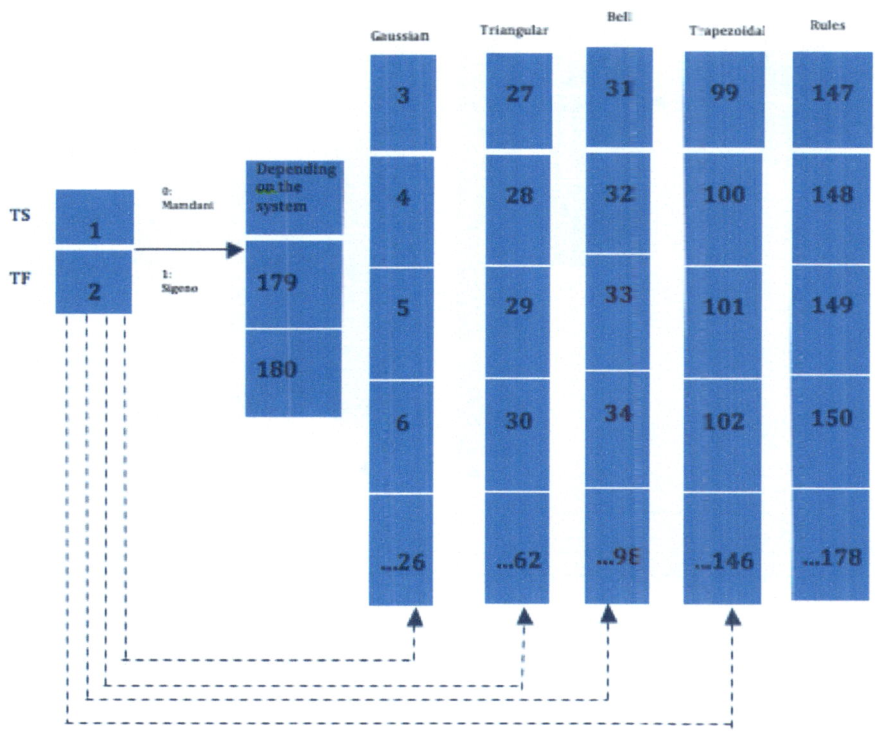


Fig. 8 Chromosome structure to optimize the type-1 fuzzy system

### 2.4.2 Optimization of the Type-2 Fuzzy System with the GA

The chromosome to optimize the type-2 system is shown in Fig. 9 and is composed of 296 genes, which allow us to optimize the structure of the fuzzy integrator. The first (binary) gene determines the type of system (Mamdani or Sugeno), the second gene determines the type of membership functions (Gaussian, generalized bell, triangular and trapezoidal). Genes 3-260 (real numbers) allow us to manage the parameters of the membership functions for the inputs and output (depending on the type of system and type of functions we use only part of the chromosome). Genes 261–294 (binary) allow us to reduce the number of fuzzy rules, activating or deactivating them.

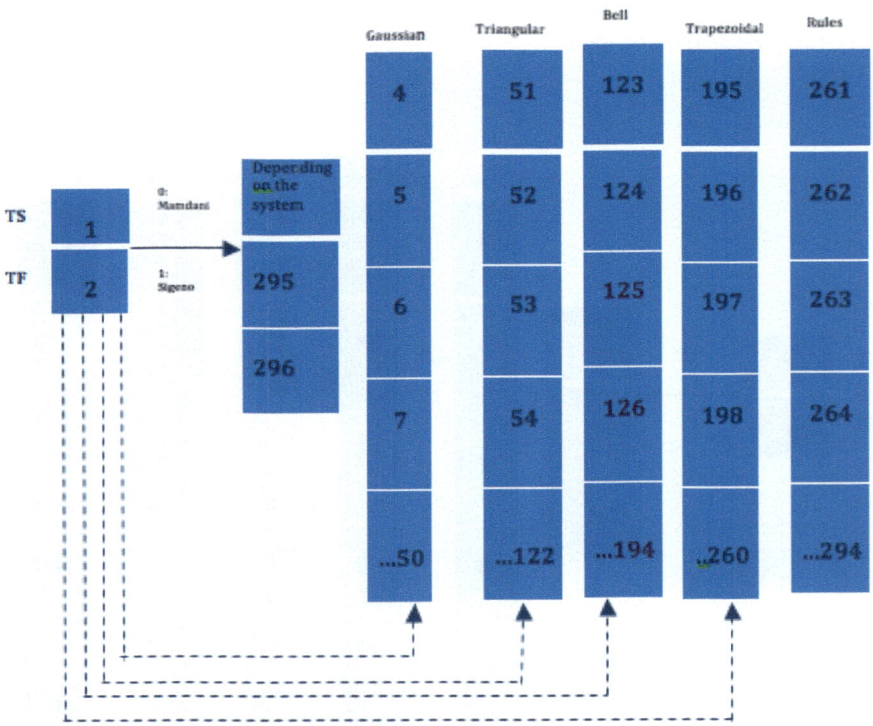


Fig. 9 Chromosome structure to optimize the type-2 fuzzy system

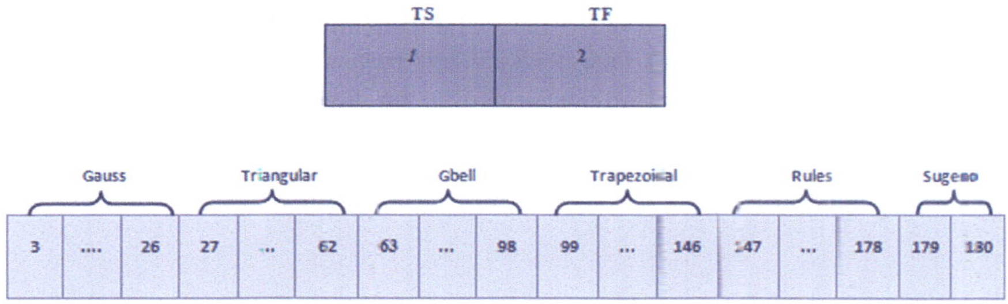


**Fig. 10** Particle structure to optimize the type-1 fuzzy integration

## 2.5 *Optimization of the Fuzzy Systems with PSO*

### 2.5.1 Optimization of Type-1 Fuzzy System with PSO

The parameters for the particle swarm optimization method are defined as follows, $C_1$ and $C_2$ have a value of 2, and the maximum speed is 1. Figure 10 represents the Particle Structure to optimize the Type-1 Fuzzy System, and is composed of 180 parameters, which allow us to optimize the structure of the fuzzy integrator. Parameter 1 determines the type of system (Mamdani or Sugeno), parameter 2 determines the type of membership functions (Gaussian, generalized bell, triangular and trapezoidal). Parameters 3–26 (real numbers) allow us to manage the parameters of the membership functions for the inputs and output (depending on the type of system and type of functions we use only part of the chromosome). Parameters 147–178 allow us to reduce the number of fuzzy rules, activating or deactivating them.

### 2.5.2 Optimization of Type-2 Fuzzy Systems with PSO

Figure 11 represents the Particle Structure to optimize the Type-2 Fuzzy System and it is composed of 296 parameters, which allow us to optimize the structure of the fuzzy integrator. Parameter 1 determines the type of system (Mamdani or Sugeno),

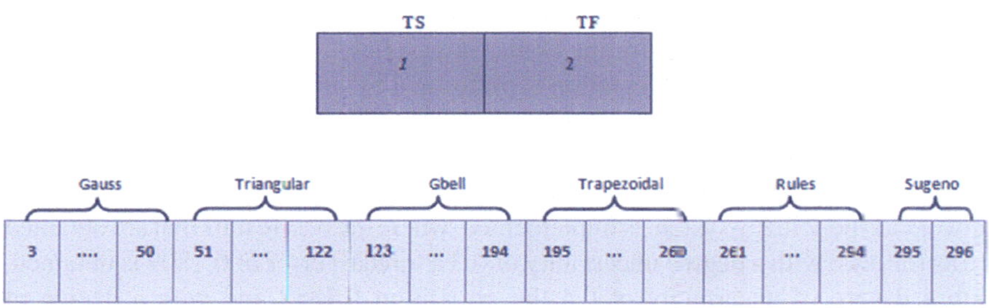


**Fig. 11** Particle structure to optimize the type-2 fuzzy integration

and parameter 2 determines the type of membership functions (Gaussian, generalized bell, triangular and trapezoidal). Parameters 3–260 (real numbers) allow us to manage the parameters of the membership functions for the inputs and output (depending on the type of system and type of functions we use only part of the chromosome). Parameters 261–294 allow us to reduce the number of fuzzy rules, activating or deactivating them.

## 3 Simulation Results

This section shows results of the neural network ensemble optimization with the genetic algorithm and PSO technique, as well as results for the type-1 and type-2 fuzzy integration and the optimization of these systems with GA and PSO. The main goal is finding the best network architecture for each of the time series: In this case the method is illustrated with the Dollar Time Series.

### *3.1 US Dollar/MX Time Series Prediction Based on the GA*

The genetic algorithm is applied to optimize the structure of the ensemble neural network, having considered 5 modules at the most, and the best achieved architecture is shown in Fig. 12.

In this architecture we have two layers in each module. In module 1 in the first layer we have 21 neurons and 24 neurons in the second layer. In module 2 we use 12 neurons in the first layer and 33 neurons in the second and in module 3 we use 11 neurons in the first layer and 23 neurons in the second. In module 4 we use 6 neurons in the first layer and 21 neurons in the second and in module 5 we use 21 neurons in the first layer and 10 neurons in the second. The Levenberg-Marquardt (LM) training method is used; 3 delays for the network are considered.

When using a genetic algorithm to optimize the structure of the ensemble neural network and considering 5 modules at the most, the best achieved architecture for this time series is shown in Table 1, in row number 7.

Figure 13 shows the plot of the real data against the predicted data generated by the ensemble neural network for the Dollar time series.

In the first phase, fuzzy integration is performed by implementing a type-1 fuzzy system, where the best result is achieved in experiment number 8 with an error: 0.2631, which is shown in Table 2.

As a second phase, to integrate the results of the optimized ensemble neural network a type-2 fuzzy system is implemented, where the best results that are obtained are as follows: with a degree uncertainty of 0.3 a forecast error of 0.1809 is obtained, with a degree of uncertainty of 0.4 the error is of 0.1842 and with a degree of uncertainty of 0.5 the error is of 0.3028, as shown in Table 3.

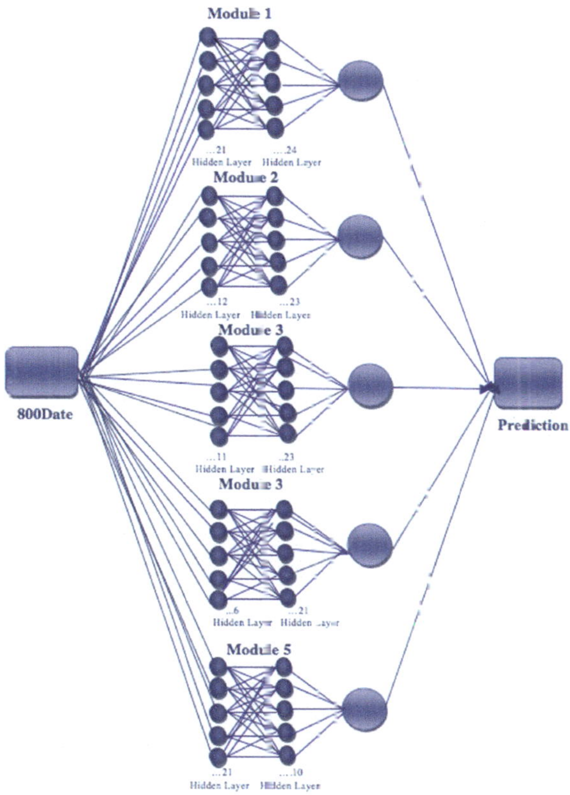


**Fig. 12** Ensemble neural network architecture for the Dollar time series

Fuzzy integration is performed initially by implementing a type-1 fuzzy system in which the best result is in the experiment of row number 7 of Table 4 with an error of: 0.00502056.

Fuzzy integration is also performed by implementing a type-2 fuzzy system in which the best result is in the experiment of row number 8 of Table 5 with an error of: 0.00389.

## 3.2 *US Dollar/MX Time Series Prediction for the PSO*

Particle Swarm is also applied to optimize the structure of the ensemble neural network and having considered 5 modules at the most, the best achieved architecture is shown in Fig. 14.

In this architecture we have two layers in each module. In module 1 in the first layer we have 20 neurons and 26 neurons in the second layer. In module 2 we use 23

**Table 1** Results of the ensemble neural network for the Dollar time series

| No. | Gen. | Ind. | Pm | Pc | Num. modules | Num. layers | Num. neurons | Duration | Prediction error |
|---|---|---|---|---|---|---|---|---|---|
| **1** | 100 | 100 | 0.09 | 0.5 | 5 | 3 | 19,7,20 27,17,14<br>16, 1,29<br>15,30,15<br>20,14, 9 | 29:11:12 | 0.0030949 |
| **2** | 100 | 100 | 0.09 | 0.5 | 5 | 3 | 14,18,17 1,25,18<br>16,27,20 7,21,30<br>11,14,13 | 32:22:45 | 0.0024102 |
| **3** | 100 | 20 | 0.01 | 0.6 | 4 | 3 | 25,12,13 6,10,22<br>3,13,18 2,14.23 26,17,22 | 13:24:39 | 0.0030126 |
| **4** | 100 | 20 | 0.5 | 0.08 | 5 | 3 | 6,23,8 24,20,18<br>6,14,21 21,15 | 05:12:34 | 0.0031568 |
| **5** | 100 | 100 | 0.07 | 0.5 | 5 | 3 | 17,18,20 1,6,22<br>21,22,14 8,3,30<br>4,13,9 | 32:29:09 | 0.0025886 |
| **6** | 100 | 30 | 0.09 | 0.7 | 5 | 3 | 11,14,25 15,20,16<br>9,3,21 11,11,2<br>4,13,9 | 24:23:38 | 0.0026431 |
| **7** | 100 | 30 | 0.02 | 0.9 | 5 | 2 | 23,24, 12,11<br>23 6 23,6<br>21,10 | 26:12:32 | 0.0021415 |

(continued)

**Table 1** (continued)

| No. | Gen. | Ind. | Pm | Pc | Num. modules | Num. layers | Num. neurons | Duration | Prediction error |
|---|---|---|---|---|---|---|---|---|---|
| **8** | 100 | 30 | 0.08 | 0.5 | 5 | 3 | 5,17,22 27,30,9<br>8,21,22 6,10,12<br>9,22,19 | 26:08:22 | 0.0028733 |
| **9** | 100 | 100 | 0.1 | 0.1 | 5 | 3 | 18,19,7 22,23,19<br>22,23,16 12,18,11 24,12.12 | 23:22:18 | 0.0031 |
| **10** | 100 | 30 | 0.7 | 0.7 | 5 | 3 | 22,2,28 3,12,18<br>8,2,1 23,10,22 11,16,13 | 25:56:31 | 0.0031886 |

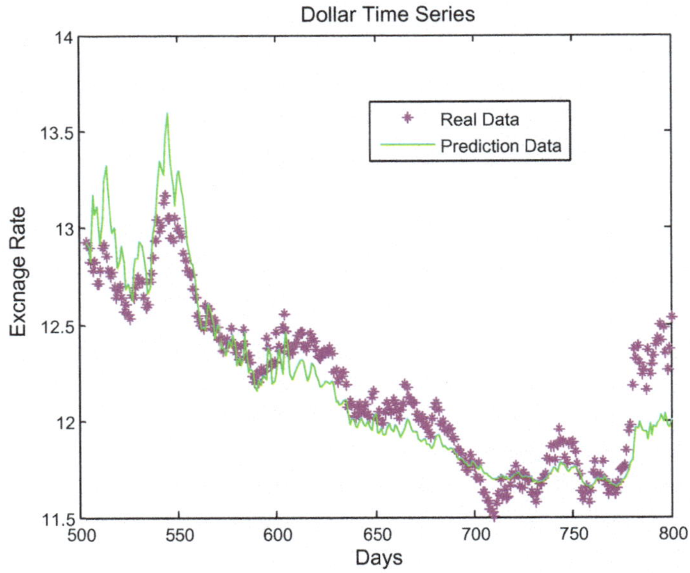


**Fig. 13** Prediction with the optimized ensemble neural network of the Dollar

**Table 2** Results type-1 fuzzy integration of the Dollar time series

| Experiments | Integration prediction error type-1 fuzzy |
|---|---|
| Experiment 1 | 0.3713 |
| Experiment 2 | 0.2787 |
| Experiment 3 | 0.2688 |
| Experiment 4 | 0.2912 |
| Experiment 5 | 0.2631 |
| Experiment 6 | 0.3072 |
| Experiment 7 | 0.2926 |
| Experiment 8 | 0.2954 |
| Experiment 9 | 0.2666 |
| Experiment 10 | 0.4248 |

neurons in the first layer and 22 neurons in the second. The Levenberg-Marquardt (LM) training method is used; 3 delays for the network are considered.

When using a particle swarm algorithm to optimize the structure of the ensemble neural network and considering 5 modules at the most, the best achieved architecture for the time series is shown in Table 6, in row number four.

Figure 15 shows the plot of the real data against the predicted data generated by the ensemble neural network for the Dollar time series.

**Table 3** Results of type-2 fuzzy integration of the Dollar time series

| Experiments | Prediction error with footprint uncertainty 0.3 | Prediction error with footprint uncertainty 0.4 | Prediction error with foot print uncertainty 0.5 |
|---|---|---|---|
| Experiment 1 | 0.2217 | 0.2117 | 0.3882 |
| Experiment 2 | 0.2231 | 0.2054 | 0.3809 |
| Experiment 3 | 0.2278 | 0.2116 | 0.3775 |
| Experiment 4 | 0.2227 | 0.1842 | 0.3976 |
| Experiment 5 | 0.2559 | 0.2204 | 0.3757 |
| Experiment 6 | 0.2756 | 0.2592 | 0.3028 |
| Experiment 7 | 0.1809 | 0.2228 | 0.3765 |
| Experiment 8 | 0.1904 | 0.2151 | 0.3759 |
| Experiment 9 | 0.2018 | 0.2373 | 0.3560 |
| Experiment 10 | 0.2076 | 0.2003 | 0.3770 |

**Table 4** Results GA type-1 fuzzy integration of the Dollar time series

| No. | Ind. | Gen. | Pm | Pc | Pm rules | TS | TF | Num rules | Duration | GA error | Prediction error |
|---|---|---|---|---|---|---|---|---|---|---|---|
| 1 | 100 | 100 | 0.09 | 0.5 | 0.02 | M | Tri | 6 | 00:51:29 | 0.0058071 | 0.0058071 |
| 2 | 100 | 100 | 0.05 | 0.8 | 0.02 | M | Gau | 8 | 00:55:22 | 0.0058134 | 0.0048134 |
| 3 | 100 | 100 | 0.05 | 0.5 | 0.02 | M | Gau | 7 | 00:50:19 | 0.006885 | 0.005885 |
| 4 | 100 | 100 | 0.02 | 0.2 | 0.02 | S | Tri | 23 | 00:48:12 | 0.0055768 | 0.0045768 |
| 5 | 100 | 100 | 0.06 | 0.9 | 0.02 | S | Tri | 30 | 00:56:15 | 0.0070661 | 0.0060661 |
| 6 | 100 | 100 | 0.6 | 0.08 | 0.02 | M | Tri | 18 | 00:43:22 | 0.0062172 | 0.0052172 |
| 7 | 100 | 100 | 0.9 | 0.07 | 0.02 | M | Tri | 12 | 00:41:36 | 0.00602056 | 0.00502056 |
| 8 | 100 | 100 | 0.9 | 0.07 | 0.02 | M | Tri | 21 | 00:45:21 | 0.0067298 | 0.0057298 |
| 9 | 100 | 100 | 0.05 | 0.8 | 0.02 | M | Tri | 18 | 00:53:18 | 0.0052831 | 0.0058231 |
| 10 | 100 | 100 | 0.06 | 0.5 | 0.02 | S | Gau | 32 | 00:51:33 | 0.0031067 | 0.0071607 |

Fuzzy integration is performed initially by implementing a type-1 fuzzy system in which the best result is in experiment of row number 8 of Table 7 with an error of: 0.4251.

As a second phase, to integrate the results of the optimized ensemble neural network a type-2 fuzzy system is implemented, where the best results that are obtained are as follows: with a degree uncertainty of 0.3 a forecast error of 0.4040 is obtained, with a degree of uncertainty of 0.4 the error is of 0.4088 and with a degree of uncertainty of 0.5 the error is of 0.04039, as shown in Table 8.

Optimization of the type-1 fuzzy system with the Particle Swarm algorithm of the best experiment of the ensemble neural network optimization in which the best result is in the experiment of row number 3 of Table 9 with an error of: 0.001187.

**Table 5** Results GA type-2 fuzzy integration of the Dollar time series

| No. | Ind. | Gen. | Pm | Pc | Pm rules | TS | TF | Num rules | Duration | GA error | Prediction error |
|---|---|---|---|---|---|---|---|---|---|---|---|
| 1 | 100 | 100 | 0.09 | 0.5 | 0.02 | M | Tri | 12 | 06:55:29 | 0.00556 | 0.00456 |
| 2 | 100 | 100 | 0.05 | 0.8 | 0.02 | S | Tra | 12 | 06:53:22 | 0.006073 | 0.005073 |
| 3 | 100 | 100 | 0.05 | 0.5 | 0.02 | S | Gbell | 15 | 05:55:13 | 0.006102 | 0.005102 |
| 4 | 100 | 100 | 0.02 | 0.2 | 0.02 | M | Gau | 22 | 06:22:23 | 0.005283 | 0.004283 |
| 5 | 100 | 100 | 0.06 | 0.9 | 0.02 | S | Gau | 30 | 06:32:12 | 0.005241 | 0.004241 |
| 6 | 100 | 100 | 0.6 | 0.08 | 0.02 | M | Gau | 31 | 07:10:20 | 0.005283 | 0.004283 |
| 7 | 100 | 100 | 0.9 | 0.07 | 0.02 | S | Tra | 12 | 06:23:36 | 0.005087 | 0.004087 |
| 8 | 100 | 100 | 0.9 | 0.07 | 0.02 | M | Gau | 21 | 07:05:23 | 0.00489 | 0.00389 |
| 9 | 100 | 100 | 0.05 | 0.8 | 0.02 | S | Tra | 18 | 06:57:38 | 0.005224 | 0.004224 |
| 10 | 100 | 100 | 0.06 | 0.5 | 0.02 | M | Gbell | 19 | 06:49:33 | 0.006299 | 0.005299 |

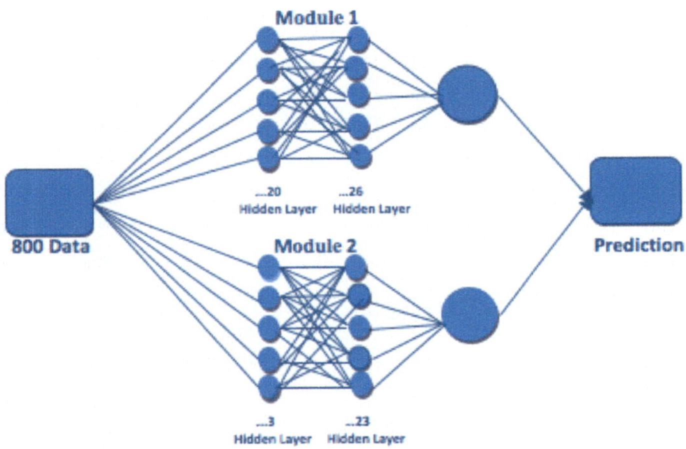


**Fig. 14** Ensemble neural network architecture with PSO for the Dollar time series

Optimization of the type-2 the fuzzy system with the Particle Swarm algorithm of the best experiment of ensemble neural network optimization in which the best result is in an experiment of row number 8 of Table 10 with an error of: 0.001084.

## 4 Comparison of Results

In this section we are presenting the comparison of the results obtained for each time series that was considered in the study: Mackey-Glass, Dow jones, Mexican Stock Exchange, and US Dollar Time series.

**Table 6** Particle Swarm algorithm results for the Ensemble Neural Network of the Dollar time series

| No. | Iterations | Particles | Number of neurons | Number of layers | Number of neurons | Duration | Prediction error |
|---|---|---|---|---|---|---|---|
| 1 | 100 | 100 | 2 | 3 | 27,13,15<br>24,15,25 | 02:01:52 | 0.0028523 |
| 2 | 100 | 100 | 2 | 2 | 16,27<br>14,24 | 01:04:08 | 0.0027317 |
| 3 | 100 | 100 | 2 | 3 | 7,27,11<br>19,13,15 | 01:27:07 | 0.0018746 |
| 4 | 100 | 100 | 2 | 2 | 20,26<br>3,23 | 01:54:33 | 0.0015855 |
| 5 | 100 | 100 | 2 | 2 | 19,12<br>18,14 | 01:40:33 | 0.0019656 |
| 6 | 100 | 100 | 3 | 1 | 11<br>13<br>5 | 01:17:07 | 0.0038019 |
| 7 | 100 | 100 | 2 | 2 | 16,19<br>3,12 | 01:28:19 | 0.0042207 |
| 8 | 100 | 100 | 3 | 2 | 24,18<br>15,23<br>15,27 | 02:20:03 | 0.0031941 |
| 9 | 100 | 100 | 2 | 3 | 7,12.19<br>21,18,22 | 01:45:00 | 0.0029591 |
| 10 | 100 | 100 | 2 | 3 | 20,23,23<br>17,10,11 | 02:05:09 | 0.0027134 |

## 4.1 *Statistical T Student Test for the Ensemble Neural Network*

In Table 10 the results with the t student test for all the time series are presented. The number of samples used for the Genetic Algorithm is 30 and for the Particle Swarm method are also 30 samples. The t values obtained for the Mackey-Glass with $\tau = 17, 34$ time series, Dow Jones series 1 and series 2, Mexican Stock Exchange series 1 and series 2 and US Dollar Time series with a confidence interval of 99% are presented in Table 11.

For the Mackey-Glass time series with $\tau = 17$, it can be noticed that there is no statistical evidence of significant difference in the results between the GA and PSO. In other words, it can be said that for this time series, prediction is only slightly better for the GA than with the PSO for the ensemble neural network.

For the Mackey-Glass with $\tau = 34$, it can be noticed that there is statistical evidence of significant difference in the results between the GA and PSO. In other words, it can be said that for this time series, the prediction is better with PSO for the ensemble neural network.

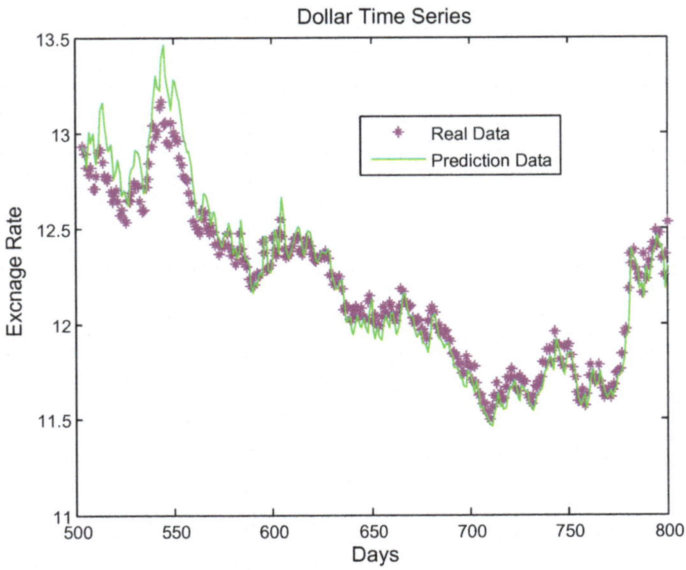


**Fig. 15** Prediction with the optimized PSO ensemble neural network of the Dollar time series

**Table 7** Results PSO for type-1 fuzzy integration of the Dollar time series

| Experiment | Prediction error with fuzzy integration type-1 |
|---|---|
| Experiment 1 | 0.4860 |
| Experiment 2 | 0.1724 |
| Experiment 3 | 0.7310 |
| Experiment 4 | 1.8360 |
| Experiment 5 | 0.9306 |
| Experiment 6 | 1.7777 |
| Experiment 7 | 0.5433 |
| Experiment 8 | 0.4251 |
| Experiment 9 | 2.3293 |
| Experiment 10 | 0.4653 |

For the Dow jones time series 1, it can be noticed that there is statistical evidence of significant difference in the results between the GA and PSO. In other words, it can be said that for this time series, prediction is better with PSO for the ensemble neural network. For the Dow Jones time series 2, it can be noticed that there is statistical evidence of significant difference in the results between the GA and PSO. In other words, it can be said that for this time series prediction is better with PSO for the ensemble neural network.

For the MSE of series 1, it can be noticed that there is statistical evidence of significant difference in the results between the GA and PSO. In other words, it can

**Table 8** Results of PSO for type-2 fuzzy integration of the Dollar series

| Experiment | Prediction error 0.3 uncertainty | Prediction error 0.4 uncertainty | Prediction error 0.5 uncertainty |
|---|---|---|---|
| Experiment 1 | 1.2346 | 0.7402 | 0.9029 |
| Experiment 2 | 0.4627 | 0.4756 | 0.4157 |
| Experiment 3 | 0.5313 | 0.6192 | 0.6717 |
| Experiment 4 | 1.3986 | 1.3056 | 1.2210 |
| Experiment 5 | 0.5394 | 0.4194 | 0.4269 |
| Experiment 6 | 0.7996 | 1.2783 | 1.2804 |
| Experiment 7 | 0.5866 | 0.5346 | 0.4039 |
| Experiment 8 | 0.4040 | 0.4088 | 0.4073 |
| Experiment 9 | 4.125[illegible] | 5.4394 | 5.5476 |
| Experiment 10 | 1.8488 | 1.8799 | 1.9070 |

**Table 9** Results of PSO type-1 fuzzy integration of series for the Dollar series

| No. | Interations | Particles | TS | TF | Núm. rules | Duration | PSO error | Prediction error |
|---|---|---|---|---|---|---|---|---|
| 1 | 100 | 100 | M | Gau | 8 | 00:53:13 | 0.0035 | 0.0025 |
| 2 | 100 | 100 | M | Gau | 8 | 00:54:26 | 0.00623 | 0.00518 |
| 3 | 100 | 100 | S | Gau | 7 | 01:15:34 | 0.002137 | 0.001187 |
| 4 | 100 | 100 | M | Gau | 6 | 01:20:02 | 0.002562 | 0.001562 |
| 5 | 100 | 100 | M | Tri | 8 | 01:22:10 | 0.004578 | 0.003578 |
| 6 | 100 | 100 | S | Gau | 6 | 01:15:15 | 0.002836 | 0.001836 |
| 7 | 100 | 100 | M | Tri | 8 | 01:15:15 | 0.004715 | 0.003715 |
| 8 | 100 | 100 | S | Gau | 7 | 01:17:15 | 0.003699 | 0.002699 |
| 9 | 100 | 100 | M | Gau | 8 | 00:51:22 | 0.003555 | 0.002555 |
| 10 | 100 | 100 | M | Gau | 6 | 00:49:54 | 0.002866 | 0.001366 |

be said that for this time series, prediction is better with PSO for the ensemble neural network.

For the MSE of series 2, it can be noticed that there is statistical evidence of significant difference in the results between the GA and PSO. In other words, it can be said that for this time series, prediction is better with PSO for the ensemble neural network.

For the US Dollar Time series, it can be noticed that there is no statistical evidence of significant difference in the results between the GA and PSO. In other words, it can be said that for this time series, prediction is slightly better for GA and PSO for the ensemble neural network.

**Table 10** Results PSO type-2 fuzzy integration for the Dollar time series

| No. | Integrations | Particles | TS | TF | Num. Rules | Duration | PSO error | Prediction error |
|---|---|---|---|---|---|---|---|---|
| 1 | 100 | 100 | M | Gbell | 7 | 01:52:28 | 0.00325 | 0.00225 |
| 2 | 100 | 100 | S | Tri | 8 | 01:56:26 | 0.00439 | 0.00339 |
| 3 | 100 | 100 | M | Gau | 7 | 01:51:24 | 0.002277 | 0.001277 |
| 4 | 100 | 100 | M | Gau | 6 | 01:42:02 | 0.002525 | 0.001525 |
| 5 | 100 | 100 | M | Gbell | 7 | 01:56:22 | 0.003214 | 0.002214 |
| 6 | 100 | 100 | S | Gau | 6 | 01:54:25 | 0.004596 | 0.003596 |
| 7 | 100 | 100 | S | Gbell | 8 | 01:55:25 | 0.003651 | 0.002651 |
| 8 | 100 | 100 | S | Gau | 7 | 01:52:04 | 0.002084 | 0.001084 |
| 9 | 100 | 100 | M | Gbell | 7 | 01:52:34 | 0.003665 | 0.002665 |
| 10 | 100 | 100 | S | Gau | 6 | 01:53:24 | 0.002976 | 0.001976 |

**Table 11** Results of *t* student test for the time series

| Time series | N (GA) | N (PSO) | T value | *P* value |
|---|---|---|---|---|
| Mackey glass $\tau = 17$ | 30 | 30 | 0.93 | 0.58 |
| Mackey glass $\tau = 34$ | 30 | 30 | 27.24 | 0.0001 |
| Dow Jones serie1 | 30 | 30 | −6.70 | 0.0001 |
| Dow Jones series 2 | 30 | 30 | −4.97 | 0.0001 |
| MSE serie 1 | 30 | 30 | −8.62 | 0.0001 |
| MSE serie 2 | 30 | 30 | −6.49 | 0.0001 |
| US Dollar Time series | 30 | 30 | 1.79 | 0.079 |

## 4.2 *The Best Architecture for the Ensemble Neural Network*

Table 12 shows the results of 30 experiments of the neural network ensemble, the best architecture obtained is shown with the optimization algorithms and the average time obtained for each time series. This table shows the best architectures obtained for each of the time series for each of the optimization algorithms, and we can observe that the PSO optimization algorithm, gave us the best architecture for the presented time series, as well as also the execution time was lower for the time series and we also added the best prediction error for each of architectures obtain of the ensemble neural network.

Table 13 shows results of the Prediction Error for the Ensemble Neural Network, Prediction Error with Type-2 Fuzzy Integration and Prediction Error with a 0.3 level of Uncertainty for the time series 1.

Table 14 shows results of the Prediction Error for the Ensemble Neural Network, Prediction Error with Type-2 Fuzzy Integration and Prediction Error with 0.3 level of Uncertainty for the time series 1.

**Table 12** The best architecture for the tests of the time series

| Time series | Optimization algorithm | Number of modules | Number of layers | Number of neurons | Duration average | Prediction error |
|---|---|---|---|---|---|---|
| Mackey glass $\tau = 17$ | GA | 5 | 2 | 1,4 6,30<br>16,30<br>6,1 19,30 | 04:43:03 | 0.0010989 |
| Mackey glass $\tau = 17$ | PSO | 2 | 3 | 17, 5, 18<br>6, 25, 24 | 01:47:22 | 0.0018838 |
| Mackey glass $\tau = 34$ | GA | 2 | 2 | 19,30<br>9,21 | 01:49:10 | 0.00489 |
| Mackey glass $\tau = 34$ | PSO | 2 | 3 | 17, 5, 18<br>6, 25, 24 | 02:13:05 | 0.002068 |
| Dow Jones series 1 | GA | 4 | 3 | 2,29,28<br>8,15,2<br>2,13,1 | 08 43:08 | 0.0020627 |
| Dow Jones series 1 | PSO | 2 | 1 | 3<br>5 | 06:43:58 | 0.0037691 |
| Dow Jones series 2 | GA | 4 | 2 | 15,21<br>9,10<br>16,18<br>15,26 | 01:56:35 | 0.0022606 |
| Dow Jones series 2 | PSO | 2 | 2 | 8,14<br>5,4 | 01:32:11 | 0.0037229 |
| MSE series 1 | GA | 3 | 3 | 6,8,28<br>8,14,8<br>22,8,24 | 2:13:26 | 0.0057659 |
| MSE series 2 | PSO | 2 | 2 | 11,24<br>5,10 | 02:33:22 | 0.0097645 |
| MSE series 2 | GA | 3 | 3 | 6,8,28<br>8,14,8<br>22,8,24 | 02:05 17 | 0.0037 |
| MSE series 2 | PSO | 2 | 2 | 12,11<br>6,18 | 02:22:20 | 0.0079875 |
| US Dollar Time series | GA | 5 | 5 | 23,24 12,23<br>11,23<br>11, 23 6,21<br>21,10 | 13:04:31 | 0.0021415 |
| US Dollar Time series | PSO | 2 | 2 | 20,26<br>3,23 | 06:31:52 | 0.0015855 |

**Table 13** Results of MSE for time series 1

| Experiment | Prediction error for ensemble neural network | Prediction error with type-1 fuzzy integration | Prediction error with type-2 fuzzy integration |
|---|---|---|---|
| Experiment 1 | 0.305 | 0.3713 | 0.2217 |
| Experiment 2 | 0.6763 | 0.2787 | 0.2231 |
| Experiment 3 | 0.7333 | 0.2688 | 0.2278 |
| Experiment 4 | 0.8024 | 0.2912 | 0.2227 |
| Experiment 5 | 0.7938 | 0.2631 | 0.2559 |
| Experiment 6 | 0.7880 | 0.3072 | 0.2756 |
| Experiment 7 | 0.4447 | 0.2926 | 0.1809 |
| Experiment 8 | 0.3.824 | 0.2954 | 0.1904 |
| Experiment 9 | 0.2563 | 0.2666 | 0.2018 |
| Experiment 10 | 0.1756 | 0.4248 | 0.2076 |

**Table 14** Results of MSE for time series 1

| Experiment | Prediction error for ensemble neural network | Prediction error with fuzzy integration type-1 | Prediction error with fuzzy integration type-2 |
|---|---|---|---|
| Experiment 1 | 0.405 | 0.4860 | 1.2346 |
| Experiment 2 | 0.5763 | 0.1724 | 0.4627 |
| Experiment 3 | 0.3333 | 0.7310 | 0.5313 |
| Experiment 4 | 1.8024 | 1.8360 | 1.3986 |
| Experiment 5 | 1.7938 | 0.9306 | 0.5394 |
| Experiment 6 | 1.7880 | 1.7777 | 0.7996 |
| Experiment 7 | 0.3337 | 0.5433 | 0.5866 |
| Experiment 8 | 0.2724 | 0.4251 | 0.4040 |
| Experiment 9 | 1.121 | 2.3293 | 4.1251 |
| Experiment 10 | 0.2916 | 0.4653 | 1.8488 |

In these tables it is clearly noted that using the response integration results are significantly improved with respect to only using the results of the modules in the ensemble for each method GA and PSO.

## 5 Conclusions

Based on the results of this paper the effectiveness of the ensemble neural network optimization for forecasting time series has been proven as it allows the problem to be solved with expert modules and thus to obtain better results. As is the case in this work, the results obtained in each of the time series can be considered good because the margin of error that is obtained is very small. Another important part of this work is the training method that is used, which is the Levenberg Marquardt algorithm. This is a good method as is able to solve problems faster and enables easier forecasts, and as it guarantees a high learning speed. It is clear that for obtaining the optimal outcome this also depends on the amount of data that we have, the complexity of the time series, the manner in which the problem is divided as well as the design of the neural network architecture for the necessary training. The testing methods are different in order to find an algorithm that in addition to providing quickness it also ensures stability in the results.

An important part of this work was finding a suitable architecture of the ensemble neural network and obtaining a good result to be reported. To make the integration of the ensemble neural network, type-1 and type-2 fuzzy systems were used and tests were performed with different time series in order to minimize the prediction error. Another objective is to have a good integration with the type-1 and type-2 fuzzy systems, which is based on the need to apply an optimization method to find the parameter values and the membership functions and rules. The optimization of the neural network ensemble and type-1 and type-2 fuzzy systems is performed with genetic algorithms and particle swarm optimization. A comparison between the time series for the neural network architectures with each of the proposed methods is also performed. As future work, this method can be applied in other series in another area of application, as well as other methods can also optimization as such Ant Colony Optimization, the optimization algorithm based on the chemical reactions, and bee colony optimization. We can also consider other types of applications, like in [49–52].

**Acknowledgements** We would like to express our gratitude to the CONACYT, Tijuana Institute of Technology for the facilities and resources granted for the development of this research.

## References

1. P.D. Brockwell, A.D. Richard, *Introduction to Time Series and Forecasting* (Springer, New York, 2002), pp. 1–219
2. O. Castillo, P. Melin, Hybrid intelligent systems for time series prediction using neural networks, fuzzy logic, and fractal theory. Neural Netw. IEEE Trans. **13**(6), 1395–1408 (2002)
3. O. Castillo, P. Melin, Simulation and forecasting complex economic time series using neural networks and fuzzy logic. Proc. Int. Neural Netw. **3**, 1805–1810 (2001)
4. O. Castillo, P. Melin, Simulation and forecasting complex financial time series sing neural networks and fuzzy logic. Proc. IEEE Int. Conf. Syst. Man Cybern. **4** 2664–2669 (2001)
5. Cherkauer K.J., Human expert level performance on a scientific image analysis task by a system using combined artificial neural networks, in *Proceedings of AAAI-96 Workshop on Integrating Multiple Learned Models for Improving and Scaling Machine Learning Algorithms* ed by P. Chan, S. Stolfo, D. Wolpert (AAAI Press, Portland, OR, Menlo Park, CA, 1996), pp. 15–21
6. P. Cunningham, J. Carney, S. Jacob, Stability problems with artificial neural networks and the ensemble solution. Artif. Intell. Med. **20**(3), 217–225 (2000)
7. N. Davey, S. Hunt, R. Frank, *Time Series Prediction and Neural Networks* (University of Hertfordshire, Hatfield, UK, 1999)

8. M. Dorigo, T. Stutzle, Ant Colony Optimization "Ant Colony Optimization Theory" (A Bradford Book The Milt Press, London England, Massachusetts Institute of Technology, 2004) pp. 121–151
9. H. Drucker, R. Schapire, P. Simard, Improving performance in neural networks using a boosting algorithm, in *Advances in Neural Information Processing Systems*, vol. 5, ed. by S.J. Hanson, J.D. Cowan, C.L. Giles (Morgan Kaufmann, Denver, CO, San Mateo, CA, 1993), pp. 42–49
10. Dow Jones Company. http://www.dowjones.com. 10 Jan 2010
11. Dow Jones Indexes. http://www.djindexes.com. 5 Sep 2010
12. Engelbrech. P., in Fundamentals of Computational of Swarm Intelligence, Basic Particle Swarm Optimization, (Wiley y Sons, 2005), pp. 93–129
13. R.R. Yager et al., in Fuzzy Sets and Applications: Selected Papers, ed. by L.A. Zadeh (John Wiley, New York, 1987)
14. D. Goldberg, Genetic Algorithms in Search, Optimization and Machine Learning. Ed. (Addison Wesley, 1989)
15. S. Gutta, H. Wechsler, Face recognition using hybrid classifier systems, in *Proceedings of ICNN-96* (IEEE Computer Society Press, Washington, DC, Los Alamitos, CA, 1996), pp. 1017–1022
16. J. Hampshire, A. Waibel, A novel objective function for improved phoneme recognition using time- delay neural networks. IEEE Trans. Neural Netw. **1**(2), 216–228 (1990)
17. L.K. Hansen, P. Salomon, Neural network ensembles. IEEE Trans. Pattern Anal. Mach. Intell. **12**(10), 993–1001 (1990)
18. L.K. Hansen, P. Salomon, Ensemble methods for hand written digit recognition, in *Proceedings of IEEE Workshop on Neural Networks for Signal Processing* (IEEE Press, Helsingoer, Denmark, Piscataway, NJ, 199)2, pp. 333–342
19. J. Holland, Adaptation in Natural and Artificial Systems (University of Michigan Press, 1975)
20. F.J. Huang, Z. Huang. H.-J. Zhang, T.H. Chen, Pose invariant face recognition, in *Proceedings of 4th IEEE International Conference on Automatic Face and Gesture Recognition* (IEEE Computer Society Press, Grenoble, France, Los Alamitos, CA, 2000), pp. 245–250
21. N. Karnik, J.M. Mendel, Applications of type-2 fuzzy logic systems to forecasting of time-series. **120**(1–4), 89–111 (1999)
22. A. Kehagias, V. Petridis, Predictive modular neural networks for time series classification. Neural Netw. **10**(1), 31–49 (1997)
23. A. Krogh, J. Vedels, Neural network ensembles, cross validation, and active learning, in *Advances in Neural Information Processing Systems 7*, ed. by G. Tesauro, D. Touretzky, T. Leen (MIT Press, CO, Denver, Cambridge, MA, 1995), pp. 231–238
24. M.C. Mackey, Adventures in Poland: having fun and doing research with Andrzej Lasota. Mat. Stosow, 5–32 (2007)
25. M.C. Mackey, L. Glass, Oscillation and chaos in physiological control systems. Science **197**, 287–289 (1997)
26. Maclin R., Shavlik J.W., Combining the predictions of multiple classifiers: using competitive learning to initialize neural networks, in *Proceedings of IJCAI-95* (Montreal, Canada, Morgan Kaufmann, San Mateo, CA, 1995), pp. 524–530
27. K. Man, K. Tang, S. Kwong, in *Genetic Algorithms and Designs, "Introduction, Background and Biological Background Ed.* (Springer, London Limited, 1998), pp. 1–62
28. L.P. Maguire, B. Roche, T.M. McGinnity, J. McDaidL, Predicting a chaotic time series using a fuzzy neural network. Inf. Sci. **112**(1–4), 125–136 (1998)
29. J. Mao, A case study on bagging, boosting and basic ensembles of neural networks for OCR, in *Proceedings of IJCNN-98*, vol. 3. (IEEE Computer Society Press, Anchorage, AK, Los Alamitos, CA, 1998), pp. 1828–1833
30. MATLAB 2007, Historical Mackey Glass data
31. Mexico Bank Database, http://www.banxico.org.mx. 30 Aug 2011
32. P. Melin, O. Castillo, S. Gonzalez, J. Cota, W. Trujillo, P. Osuna, *Design of Modular Neural Networks with Fuzzy Integration Applied to Time Series Prediction*, vol. 41 (Springer, Berlin/Heidelberg, 2007), pp. 265–273

33. I.M. Multaba, M.A. Hussain, *Application of Neural Networks and Other Learning. Technologies in Process Engineering* (Imperial Collage Press, London, 2001)
34. J.S.R. Jang, C.T. Sun, E. Mizutani, *Neuro-Fuzzy and Soft Computing* (Prentice Hall, Englewood Cliffs NJ, 1996)
35. D.W. Opitz, J.W. Shavlik, Generating accurate and diverse members of a neural network ensemble, in *Advances in Neural Information Processing Systems 8*, ed. by D.S. Touretzky, M.C. Mozer, M.E. Hasselmo (MIT Press, Cambridge, MA, Denver, CO, 1996), pp. 535–541
36. E.A. Plummer, in *Time Series Forecasting With Feed-Forward Neural Networks: Guidelines and Limitations*. University of Wyoming (2000)
37. C. Reeves, J. Row, *Genetic Algorithms: Principles and Perspectives, "Optimization"* (Kluwer Academic Publishers, New York, Boston, Dordrecht, London, 2002), pp. 4–8
38. A. Sharkey, *One combining Artificial of Neural Nets* (Department of Computer Science University of Sheffield, U.K., 1996)
39. A. Sharkey, *Combining Artificial Neural Nets: Ensemble and Modular Multi-Net Systems* (Springer, London, 1999)
40. P. Sollich, A. Krogh, Learning with ensembles: how over fitting can be useful, in *Advances in Neural Information Processing Systems*, vol. 8, ed. by D.S. Touretzky, M.C. Mozer, M.E. Hasselmo (MIT Press, Cambridge, MA, Denver, CO, 1996), pp. 190–196
41. S. Soltani, On the use of the wavelet decomposition for time series prediction. Neurocomputing **48**(1–4), 267–277 (2002)
42. M. Štěpnička, P. Cortez, J.P. Donate, L. Štěpničková, Forecasting seasonal time series with computational intelligence: On recent methods and the potential of their combinations. Expert Syst. Appl. **40**(6), 1981–1992 (2013)
43. W.W.S. Wei, *Time Series Analysis: Univariate and Multivariate Methods*, vol. 1 (Addison-Wesley, Reading, MA, 1994), pp. 40–100
44. R.N. Yadav, P.K. Kalra, J. John, Time series prediction with single multiplicative neuron model. Soft Comput. Time Ser. Prediction Appl. Soft Comput. **7**(4), 1157–1163 (2007)
45. X. Yao, Y. Liu, Making use of population in formation in evolutionary artificial neural networks. IEEE Trans. Syst. Man Cybern. Part B Cybern. **28**(3), 417–425 (1998)
46. J. Yen, R. Langari, *Fuzzy Logic: Intelligence, Control and Information Ed.* (Prentice Hall, Englewood Cliffs NJ, 1999)
47. L. Zhao, Y. Yang, PSO-based single multiplicative neuron model for time series prediction. Expert Syst. Appl. **36**(2), 2805–2812 (2009). (Part 2)
48. Z.-H. Zhou, Y. Jiang, Y.-B. Yang, S.-F. Chen, Lung cancer cell identification based on artificial neural network ensembles. Artif. Intell. Med. **24**(1), 25–36 (2002)
49. M.A. Sanchez, O. Castillo, J.R. Castro, P. Melin, Fuzzy granular gravitational clustering algorithm for multivariate data. Inf. Sci. **279**, 498–511 (2014)
50. D. Sánchez, P. Melin, Optimization of modular granular neural networks using hierarchical genetic algorithms for human recognition using the ear biometric measure. Eng. Appl. Artif. Intell. **27**, 41–56 (2014)
51. M.A. Sanchez, J.R. Castro, O. Castillo, O. Mendoza, A. Rodriguez-Diaz, P. Melin, Fuzzy higher type information granules from an uncertainty measurement. Granular Comput. **2**(2), 95–103 (2017)
52. F. Gaxiola, P. Melin, F. Valdez, J.R. Castro, O. Castillo, Optimization of type-2 fuzzy weights in backpropagation learning for neural networks using GAs and PSO. Appl. Soft Comput. **38**, 860–871 (2016)

# Path Planning by Search Algorithms in Graph-Represented Workspaces

**Ivan Dario Vanegas Perez, Oscar Montiel, and Ulises Orozco-Rosas**

**Abstract** Path planning is an essential task in autonomous mobile robotics that demands to navigate following a minimum-cost path, which involves partitioning the landscape in nodes and the use of combinatorial optimization methods to find the optimal sequence of nodes to follow. Traditional algorithms such as the A* and Dijkstra are computationally efficient in landscapes with a reduced number of nodes. Most of the practical applications require to use a significantly large number of nodes up to the point that the problem might be computationally explosive. This work contributes to state-of-the-art with two heuristics for the A* algorithm that allows finding the optimal path in landscapes with a large number of nodes. The heuristics used the Euclidean and Manhattan distance in the estimation function. We present a comparative analysis of our proposal against the Dijkstra's and A* algorithms. All the experiments were achieved using a simulation-platform specially designed for testing important algorithm features, such as the grid size, benchmark problems, the design of custom-made test sceneries, and others. Relevant results are drawn to continue working in this line.

**Keywords** Path planning · Knowledge representation · Graph traversal · Algorithms · Simulation

---

I. D. V. Perez · O. Montiel (✉)
CITEDI-IPN, Instituto Politécnico Nacional, Av. Instituto Politécnico Nacional, no. 1310, Nueva Tijuana, C.P. 22435 Tijuana, Baja California, Mexico
e-mail: oross@ipn.mx

I. D. V. Perez
e-mail: ivanegas@citedi.mx

U. Orozco-Rosas
Centro de Innovación y Diseño (CEID), CETYS Universidad, Av. CETYS Universidad. no 4., Fracc. El Lago, C.P 22210 Tijuana, Mexico
e-mail: ulises.orozco@cetys.mx


P. Melin et al. (eds.), *Recent Advances of Hybrid Intelligent Systems Based on Soft Computing*, Studies in Computational Intelligence 915,
https://doi.org/10.1007/978-3-030-58728-4_4

# 1 Introduction

The path planning problem is part of a highly general community interest for achieving the development of robots that possess the capacity of autonomous navigation, according to authors like Latombe [1] and LaValle [2]. It consists of four elemental stages: sensing the environment, planning the trajectory, generating the sequence of action data, and then, following the returned trajectory. The main concern in path planning is to find a free-collisions route from a starting point noted as A to a destination point B [3], independently of the technique employed, the optimal results are measured in terms of paths distance, execution time, and other different metrics according to the necessity of the research [4].

For more than five decades, ever since the 60's decade [5], formal heuristics have been established for modeling and simulating the path planning problem with obstacles avoidance, i.e., for finding free-collision trajectories on a given workspace. In mobile robotics [6], path planning is yet an open field of study full of opportunity niches for researching [7]. Most of the known solution until now, demand significantly high computational cost. Search algorithms by graph traversal such as Dijkstra's [8] and A* [9] are some of the many heuristics that can be found on state of the art and, likewise, are highly trusted by computational scientists for solving the discussed problem of path planning.

There are several scenarios where path planning is demanded in navigation with minimum cost paths; we can list, for example, roads between cities, arms trajectories, deliver sequences, network routing, circuits design, autonomous robots, drug design, and videogames, among others. As for the scope of this work we mainly care about the knowledge representation of the problem for a mobile robot requiring to move from a starting to a target position on a 2D plane, that means, the computational data structures necessary for handling the information of the workspaces, their available locations, and the transition costs associated.

Workspace representation is essential and a nontrivial task to execute. According to seminal geometric representations of the agent's environment [5], it is observable that, making an appropriate abstraction of environment data into relevant information for the planner, implies optimal results providing the agent the capability to know which positions are either obstructed or free and therefore available for being occupied according to a parametrized configuration of the mobile robot, i.e., location and orientation angle [10].

It is possible to make use of different strategies for solving the path planning problem, some of them are probabilistic planning with roadmap sampling, which is a technique designed to generate and evaluate random points in the space and if the partial path seems to be approaching to the goal, then it is accepted. One more strategy that we can find under this category is the construction of trees for generating the paths as the rapid-exploring random trees RRT [11]. Geometric models as the Voronoi diagrams and visibility graphs presented on the seminal work of Lozano-Perez [5] that allow working with polyhedral representations of obstacles have been as important as the probabilistic methods as the previously discussed.

Lastly, we can also find mathematical models as potential fields [12]; inspired on physical laws of attraction and repulsion that simulate the displacement of a fluid drop on a vector space. More sophisticated algorithms, now make use of computational intelligence, using inspirations on nature's behaviors as well as mixing different metaheuristics, some of them are ant colony optimization [13], artificial bee colonies [14] and membrane evolutionary artificial potential fields [15], among others.

The paper is organized as follows. Section 2 presents a description of the methodology boarded for achieving the implementation of Dijkstra's and two-heuristics A* algorithms according to a novel designed scheme of the workspace's representation. Section 3 shows the performance of the compared heuristics; the results with workspace illustrations and quantitative graphical charts are also included. Finally, Sect. 4 holds the conclusions and future work for this research.

## 2 Methodology

Employing careful programming practices, we could achieve efficient simulations performance of the designed experiments. On the different scenarios that were tested, the dimensions of the implicit workspace grids representation were of varying sizes, embracing magnitudes from 10 × 10 cells up to 100 × 100 in the reported results; that means a workspace size of 10,000 units of information that are translated to the nodes of the built representation graph.

Workspace definition, according to [3], is established considering a set of obstacles that live, in this case, statically on a two-dimensional grid of cells. The size of the cell is depreciated, and scenario maps are drawn over an 800 by 800 pixels screen. In the beginning, the default size of the workspaces is 10 × 10. To provide the system with a variable resolution trajectory, we add the possibility of rescaling the resolution of the scene increasing the number of cells division proportionally by five and ten times, i.e., workspaces grids of 50 × 50 and 100 × 100 cells; increasing this way the total size of the problem but conserving the relative size of obstacles.

The planner agent is a holonomic robot represented as a punctual particle, i.e., it does not have any physical displacement restrictions either on orientation nor direction, and also, the occupancy of the agent was set to centrically encompass the size of one cell at all times. Working with this type of robot's model makes easy the relative location of the agent on the space because the connectivity between the discrete cells of the grids is entirely stored in the representation graph of the workspaces.

It is important to take into consideration that the given workspaces might have different forms of representations, e.g., matrix forms or adjacency list forms. Taking advantage of the sophisticated data structures called graphs [8], we can represent the knowledge of the system, i.e., the sets of obstacles and free spaces, as well as the initial and ending position and configuration of the mobile agent with the help of a Python dictionary. Most commonly, the reader can find in the state of the art continuous and discrete representations for the motion planning problem; there also exist other kinds of representation, such as topological maps [2, 16].

After choosing the appropriate configuration space representation, it is possible to start planning with the best suiting algorithms. As we mentioned in the introduction, different approaches such as probabilistic, bio-inspired, and geometrical algorithms already exist in the field, each one of them, with its advantages inherent to the problem nature. As for the scope of this work, we only present the description of the two most common algorithms for static information space searching by graph traversal implementation, i.e., Dijkstra and A*.

## *2.1 Algorithms Description and Analysis*

While Dijkstra makes use of Breath First Search queue, as can be reviewed on [8], A* utilizes one open and one closed list, in which it stores partial costs from source to current visited node and, if 'revisits' any of them, then updates their value for the best one. The former, illustrated in Fig. 1, has transition cost restrains; that is, it assumes that all costs are positives. The latter one has different possibilities because it subordinates a search heuristic that guides the solution towards the objective. The heuristic is defined by a cost function of, for example, the estimated distance from the current cell to the target one. In this work, we made use of Euclidean and Manhattan distance, according to the literature [17], it is not supposed that the value of the function overestimates the real value.

For the A* algorithm, two heuristics were implemented; for achieving optimal search guidance, the estimation of the distance value to the source from every current node labeled as 'h' was obtained using the Euclidean and Manhattan distance in each case. The main idea is to choose the minimum 'f' value from all the possibilities, so we used 'f' as the sum of the real partial distance 'g' plus the estimated 'h'. In every scenario, the user had the possibility of choosing which one of the heuristics wanted to execute, the description of this algorithm can be visualized for better understanding in Fig. 2.

As we have mentioned, two search heuristics were implemented for the A* algorithm in this work, the Euclidean and Manhattan distance between the target and currently visited nodes as are defined in Eqs. (1) and (2) respectively; Eq. (1) calculates $h_e$ as typical difference vectors norm, whereas Eq. (2) calculates $h_m$ as the total distance vertical and horizontally between two points. These values are calculated according to the x and y coordinates of the cells on the grid map of the workspaces; then they are assigned to the estimated value 'h' of the nodes.

$$h_e = \sqrt{(x_2 - x_1)^2 + (y_2 - y_1)^2} \tag{1}$$

$$h_m = |x_2 - x_1| + |y_2 - y_1| \tag{2}$$

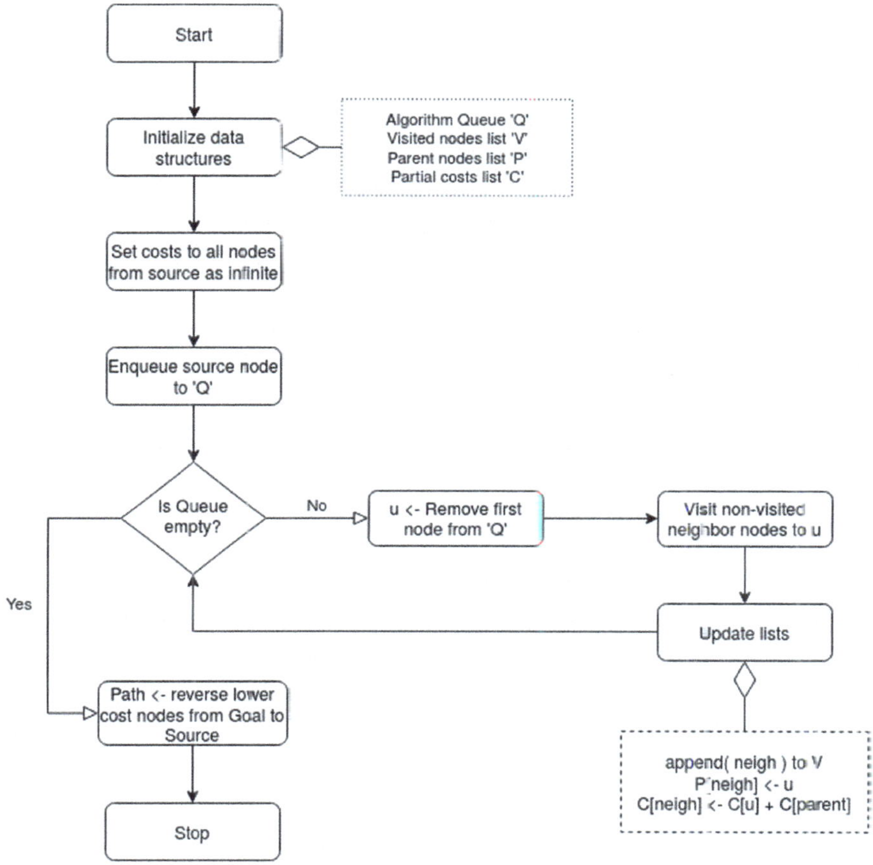


**Fig. 1** Diagram description of Dijkstra's algorithm

## 2.2 *Workspace Design*

Three elemental configurations were implemented considering rectangle polygonal tessellated obstacles, as can be seen in Fig. 3. Each one of them is represented by reading a plain text file that allocates the vertex points that define the corners of the obstacles. All the scenarios have the possibility of varying the resolution of the scenarios by scaling the cells magnitudes by five and ten times; this is to provide the possibility of working the simulations with more details. The Python dictionary stored each cell with a unique index as the key counting from left to right and from the top to the bottom of the grid. The elements of the data structure held the available neighbors of each cell and the respective transition cost considering the possibility of moving up and down, sidewise, and to the four diagonals as well.

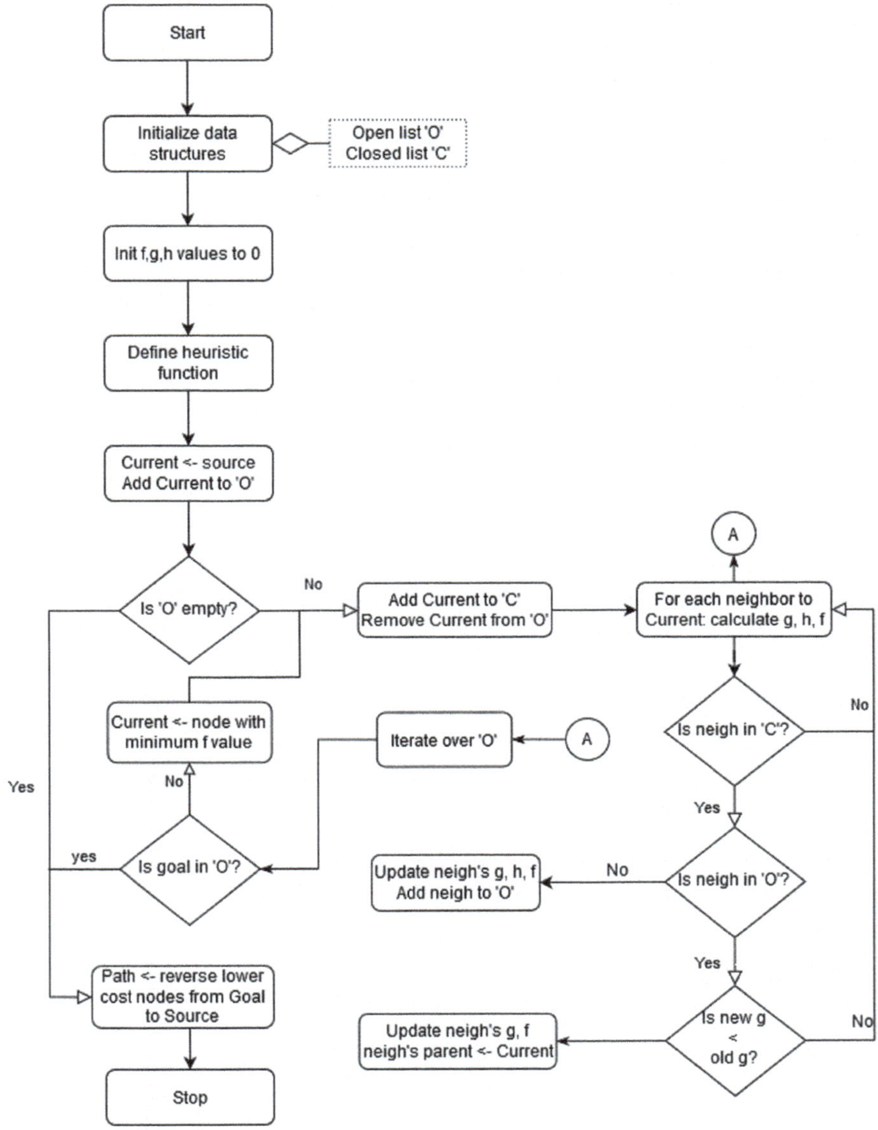


**Fig. 2** Diagram description of A* algorithm

## 2.3 *Experiments*

Simulations were executed in a computer with the following hardware specifications:

$$\text{Processor} : \text{Intel Core I7930@2.80 GHz}$$

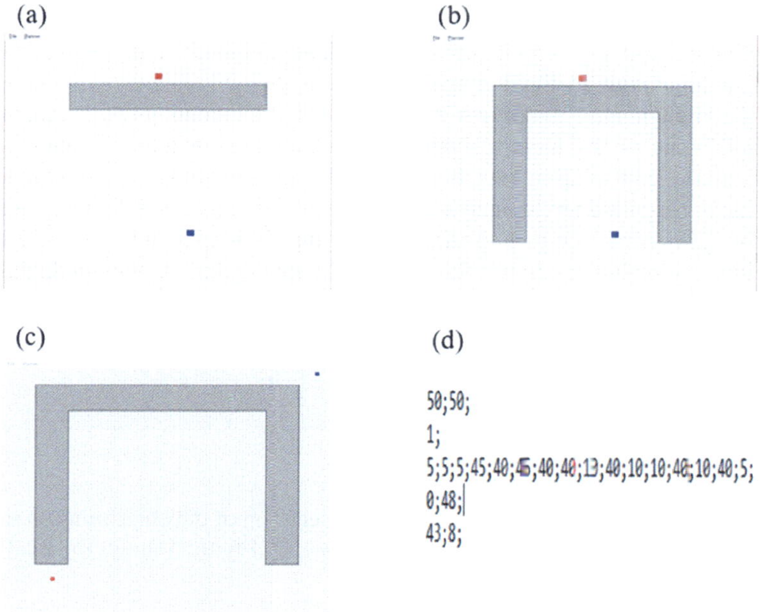


**Fig. 3** Test scenery 1: One single obstacle located between the source and goal points on 50 × 50 workspace, **b** Test scenery 2: One single obstacle between source and goal, the source is surrounded by an obstacle, and the best solution needs to move away from the desired location to find a feasible trajectory in a 50 × 50 workspace, **c** Test scenery 3: One single obstacle between source and goal, on this scenario, the source, and the goal location were outside of the surrounded area by the single obstacle; the best solutions can be obtained straight forward in a 100 × 100 workspace, **d** plain text preview for space information of sample workspace #3 abstraction

$$\text{Memory} : 2 \times 2048\ \text{Mb RAM} @ 1066\ \text{MHz}$$

For each one of the three workspaces, the three algorithms were executed under the three different resolutions of 10 × 10, 50 × 50 and, 100 × 100. For gathering the average results, each simulation was run several times; it was not necessary to make more than thirty attempts because of the determinism of the planners. In total, there were nine tested scenes or cases simulated. In Fig. 3 are shown different workspace configurations with their mission details such as starting and target points.

The first scenario, in Fig. 3a, is a single obstacle located between the mission's source and target nodes, the next workspace in Fig. 3b, corresponds to a trap in which the starting node is inside the obstacle's bounds, and finally, the third map in Fig. 3c has the same topology than the previous one but, with a difference in the initial configuration of the mission.

The scheme codification of the file, observable in Fig. 3d is simple, the first line indicates the resolution of the system, then, the next line is the number 'n' of obstacle's sets, and the following 'n' lines indicate the mentioned vertex points sets. Once the location of the obstacles was determined, the corresponding nodes were discarded from the representation graph. The mission for the planner is declared with two tuples of x and y coordinates that are matched to the starting node (colored in red) and a target node (colored in blue).

All the implemented algorithms were executed in each one of the three different scenarios. Some of the most relevant metrics for evaluation are execution time, memory space, and solution cost. For this work, time performance and search sub-space size or magnitude were analyzed and compared for all cases. While, regarding path cost, it is guaranteed that all these algorithms, Dijkstra's, A*, and its different search heuristic variances, return the minimum distance trajectory, even though, the number of nodes of the paths is not equal, because of the size of the cells under the different scales, but the cost is calculated by integrating the parts values.

## 3 Results and Discussion

For the obtained trajectories, Fig. 4 shows the execution of the algorithms on sample workspace #1. Figure 5 corresponds to workspace #2. Figure 6 shows the final paths

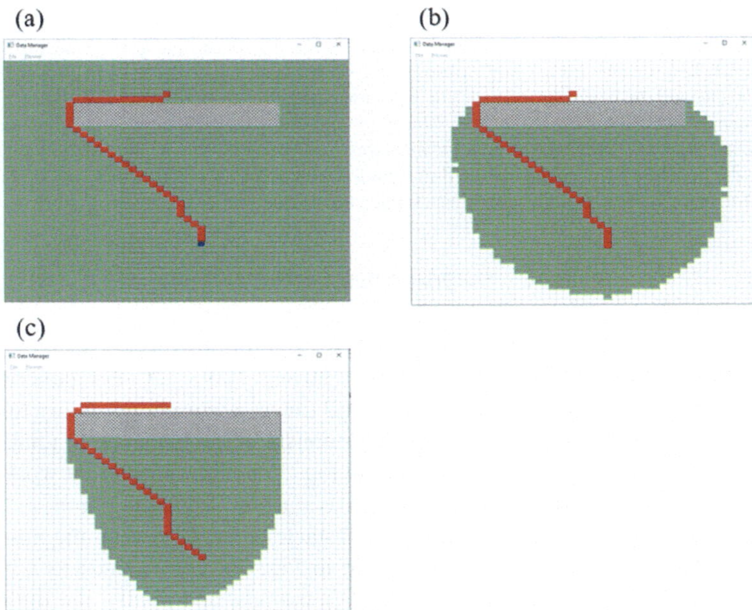

**Fig. 4** Test scenery 1. Simulations on the workspace shown in Fig. 3a. **a** The solution path by Dijkstra's algorithm (in red color) and explored sub-space (gray color cells with green contour), **b** The solution path by A* algorithm with Euclidean heuristic (in red color) and explored sub-space (gray color cells with green contour), **c** The solution path by A* algorithm with Manhattan heuristic (in red color) and explored sub-space (gray color cells with green contour)

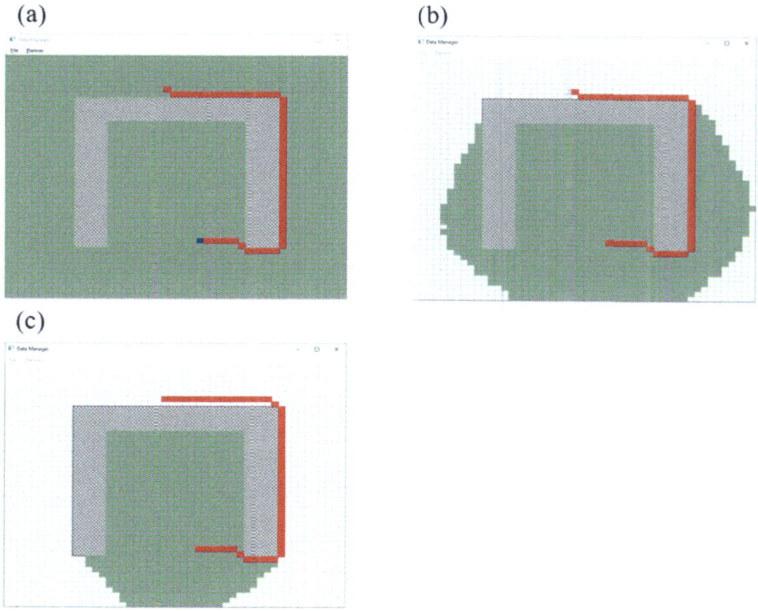


**Fig. 5** Test scenery 2. Simulations on the workspace shown in Fig. 3b. **a** The solution path by Dijkstra's algorithm (in red color) and explored sub-space (gray color cells with green contour), **b** The solution path by A* algorithm with Euclidean heuristic (in red color) and explored sub-space (gray color cells with green contour) **c** The solution path by A* algorithm with Manhattan heuristic (in red color) and explored sub-space (gray color cells with green contour)

for workspace #3. It is evident that the final solution (in red color) is the same in terms of the shortest distance for Dijkstra's and A* algorithms; however, the important difference is highlighted as the minimization of search space (colored with gray cells and green contour) as can be observed in the mentioned figures.

The change of resolution of the workspaces into different scales helps to soften and smooth the final trajectory but increases the computational complexity in execution time. In Figs. 7, 9, and 11 is shown that the increase in time is proportional to the analytic complexity of the implemented algorithm in terms of the polynomial order. For the $O(n^2)$ the growth of the execution time by the space size, i.e., the amount of cells or units of information on the workspace, grows quadratically; as well as for A*, although in best cases, its complexity might be given linearly by $O(|E| + |V|)$, the execution time also grows linearly; but that is entirely dependent on the problem's nature.

The previous chart shows that for small sceneries, the execution time is low, and the performance seems reasonable, but more than one-third of a second for larger workspaces is a warning for real-time systems that demand to make decisions in the minimum time possible.

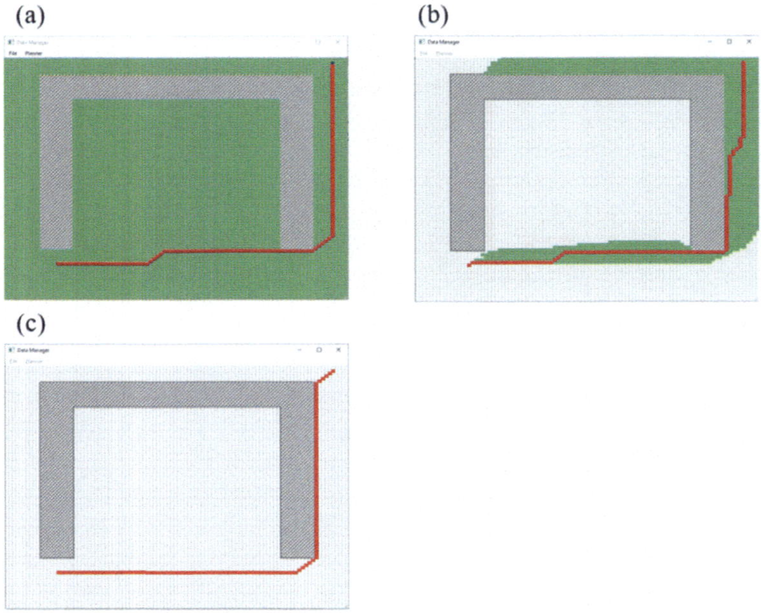


**Fig. 6** Test scenery 3. Simulations on the workspace shown in Fig. 3c. **a** The solution path by Dijkstra's algorithm (in red color) and explored sub-space (gray color cells with green contour), **b** The solution path by A* algorithm with Euclidean heuristic (in red color) and explored sub-space (gray color cells with green contour), **c** The solution path by A* algorithm with Manhattan heuristic (in red color) and explored sub-space (gray color cells with green contour)

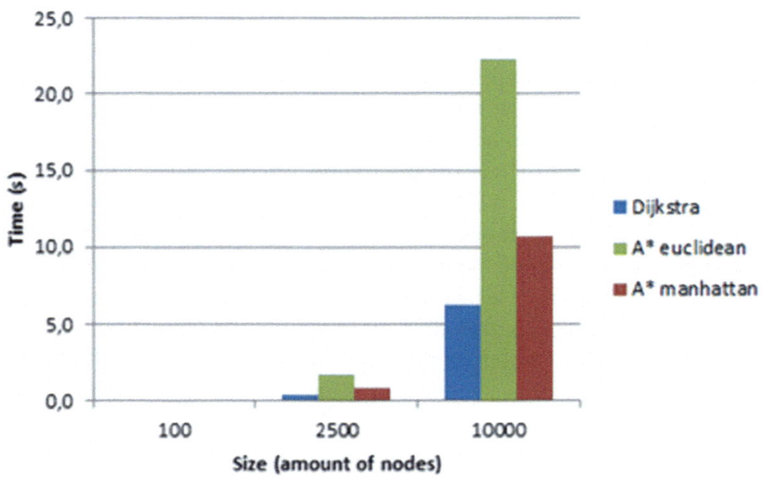


**Fig. 7** The execution time of the Dijkstra algorithm for test scenery 1 was not outperformed by A*

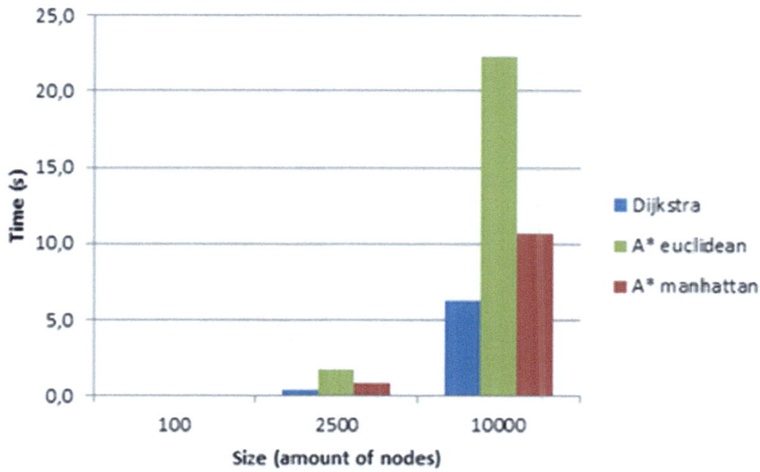


**Fig. 8** Space reduction of A* algorithm was more than 50%

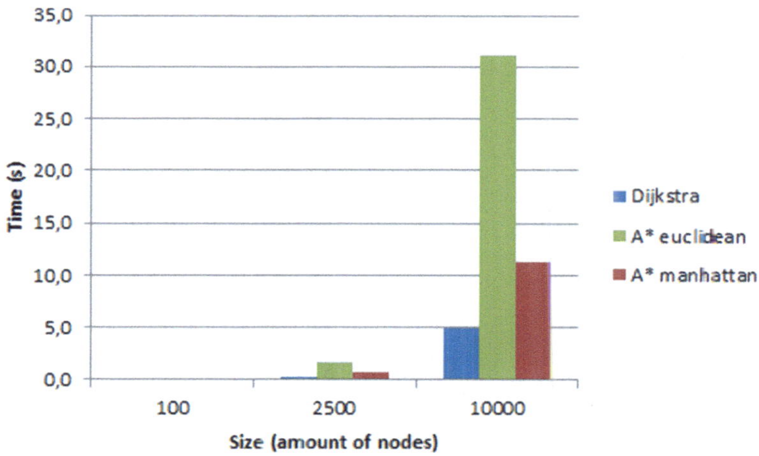


**Fig. 9** The execution time of the Dijkstra algorithm for test scenery 2 was not outperformed by A*

In Figs. 8, 10, and 12 is observable that space reduction reaches more than fifty percent of the total amount of nodes of the workspace graph. With this kind of performance, the possibility of implementing a yet more sophisticated metaheuristic that could improve the overall metrics of the implemented algorithms is high and brings hope to the researcher for pursuing it.

For worst-case situations as in sceneries 1 and 2, which exhibit difficult obstacles to avoid intuitively for a robot, the execution time was higher for both of the proposed heuristics; however, the space reduction was around fifty to sixty percent with Euclidean heuristic and seventy percent with Manhattan's heuristic. Whereas

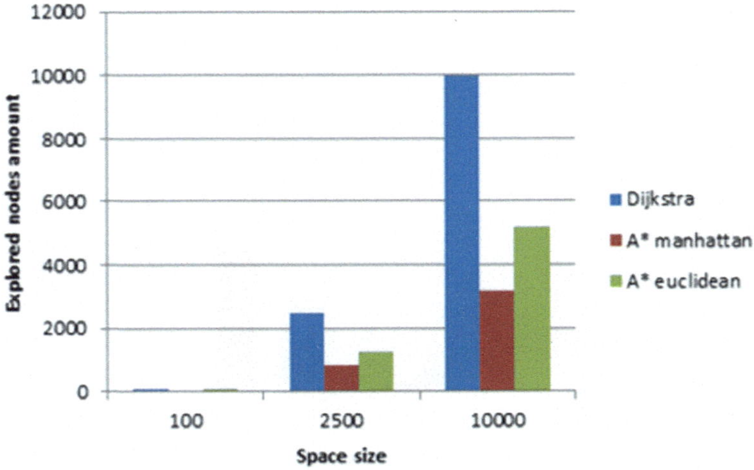


**Fig. 10** Space reduction of A* algorithm was more than 50%

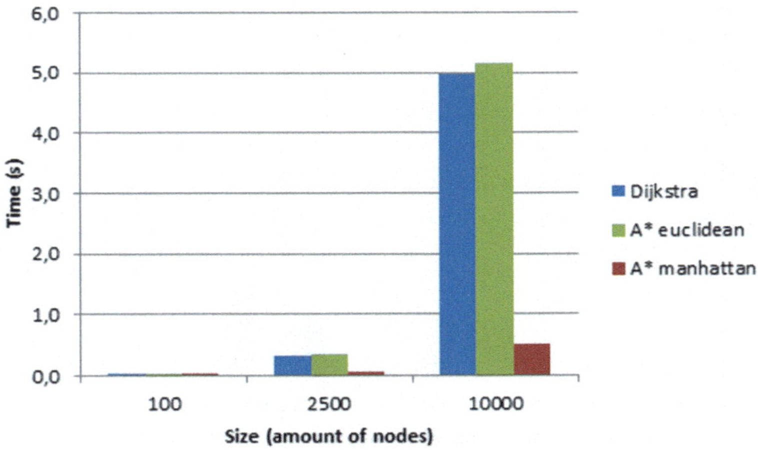


**Fig. 11** The execution time of the Dijkstra algorithm for test scenery 3 was highly significantly outperformed by A*

for best-case situations as in scenery 3, which seems easily solvable, space reduction was eighty and ninety percent by Euclidean and Manhattan's heuristic, respectively, the latter one, also reducing the execution time significantly.

With these results, it is observable that although the achieved implementations of all A* were not always faster than Dijkstra's, the Manhattan distance heuristic overcomes the other algorithms in terms of search space reduction; it is essential because it means computing memory saving. For future works, it is expected to find

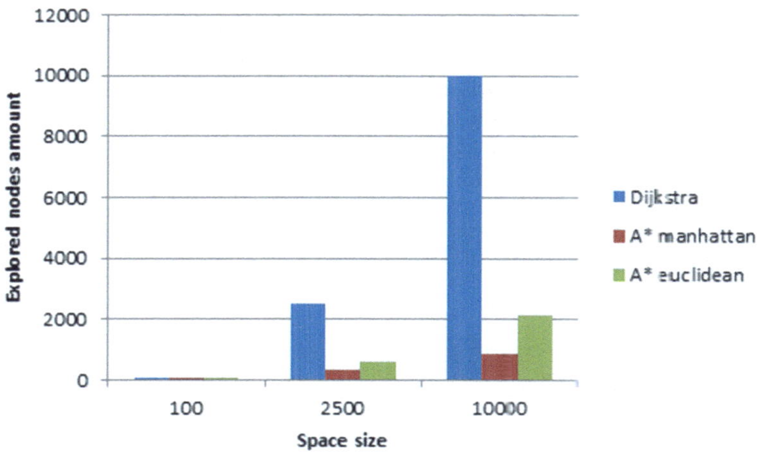


**Fig. 12** Space reduction of A* algorithm was above 80% for both heuristics

a technique that best suits all general problems and offers an optimal balance between time and space complexity, reducing that way the trade-off.

## 4 Conclusions and Future Work

In this work, we discussed essential topics in the field of search algorithms in graph-represented workspaces. Also, we presented the methodology used for testing the different path planning algorithms that are included in this research, particularly the A* and Dijkstra's algorithms. We contribute to state-of-the-art with the implementation of two heuristics that exhibit how the search space is significantly reduced, i.e., optimized in comparison with strategies that stick on their locally optimal values.

For performing the experiments, we developed a generic graphics simulation software platform of the path planning algorithms on different workspaces. One of the main characteristics of this evaluation platform is the scalability of the discretized workspaces to different grid sizes, ranging from small scenarios to big ones above 100 by 100 cell grids.

The adequate codification of workspaces allows fast modification and creation of new ones in a highly scalable way, at the same time as an adaptation of the scenarios to different GUI as their representation consists mainly of the most relevant system's elements configuration and regions of interest determined all of them by problem's nature.

Proper implementation of A* algorithm achieves optimal guided solution as well as a search subspace reduction and, hence, achieves better performing efficiency than other classical algorithms in static scenarios. Implementations evidence that high-performance computing is mandatory for large scale scenarios, i.e., more than 100

× 100 nodes on regular computer architectures. Execution time and space reduction are also important and should be considered for improving resource cost-saving, e.g., process memory. The search space reduction is a range of sixty to ninety percent of total graph size, which implies a nearly direct guided solution.

**Acknowledgements** This work was supported by Instituto Politécnico Nacional (institutional project SAPI (2020053)), and by the Mexican National Council of Science and Technology (CONACYT, México). We additionally thank to the Coordinación Institucional de Investigación of CETYS Universidad for their collaboration with the work.

## References

1. J.-C. Latombe, *Robot Motion Planning* (Kluwer Academic Publishers, Boston, 1991)
2. S.M. LaValle (ed.), *Planning Algorithms* (Cambridge University Press, Illinois, 2006)
3. C. Behring, M. Bracho, M. Castro, J.A. Moreno, An algorithm for robot path planning with cellular automata, in *Theory and Practical Issues on Cellular Automata* (Springer, London, 2001), pp. 11–191
4. U. Orozco-Rosas, K. Picos, O. Montiel, Hybrid path planning algorithm based on membrane pseudo-bacterial potential field for autonomous mobile robots. IEEE Access **7**(1), 156787—156803 (2019)
5. T. Lozano-Pérez, M.A. Wesley, An algorithm for planning collision-free paths among polyhedral obstacles, Watson Research Center. Commun. ACM **22**(10), 560–570 (1979)
6. P. Jiménez, Understanding motion planning better: a comparative review of principles of robot motion: theory, algorithms, and implementations, by H. Choset et al. Comput. Sci. Rev. **1**(1), 67–72 (2007)
7. K. Chu, M. Lee, M. Sunwoo, Local path planning for off-road autonomous driving with avoidance of static obstacles. IEEE Trans. Intell. Transp. Syst. **13**(4), 1599–1616 (2012)
8. T.H. Cormen, C.E. Leiserson, R.L. Rivest, C. Stein, (eds.), in *Introduction to Algorithms*. (MIT Press and McGraw-Hill, 2001)
9. P.E. Hart, N.J. Nilsson, B. Raphael, A formal basis for the heuristic determination of minimum cost paths. Syst. Sci. Cybern. IEEE Trans. **4**(2), 100–107 (1968)
10. O. Montiel, R. Sepúlveda, U. Orozco-Rosas, Optimal path planning generation for mobile robots using parallel evolutionary artificial potential field. J. Intell. Robot. Syst. **79**(2), 237–257 (2015)
11. L.E. Kavraki, P. Svestka, J.-C. Latombe, M.H. Overmars, Probabilistic roadmaps for path planning in high-dimensional configuration spaces. Robot. Autom. IEEE Trans. **12**(4), 566–580 (1996)
12. O. Montiel-Ross, U. Orozco-Rosas, R. Sepúlveda, Path planning for mobile robots using bacterial potential field for avoiding static and dynamic obstacles. Expert Syst. Appl. **42**, 5177–5191 (2015)
13. O. Montiel-Ross, R. Sepúlveda, O. Castillo, P. Melin, Ant colony test center for planning autonomous mobile robot navigation. Comput. Appl. Eng. Edu. **21**, 214–229 (2013)
14. M.A. Contreras-Cruz, V. Ayala-Ramirez, U.H. Hernandez-Belmonte, Mobile robot path planning using artificial bee colony and evolutionary programming. Appl. Soft Comput. **30**, 319–328 (2015)
15. U. Orozco-Rosas, O. Montiel, R. Sepúlveda, Mobile robot path planning using membrane evolutionary artificial potential field. Appl. Soft Comput. **77**, 236–251 (2019)
16. U. Orozco-Rosas, O. Montiel, R. Sepúlveda, Pseudo-bacterial potential field based path planner for autonomous mobiel robot navigation. Int. J. Adv. Robot. Syst. **12**(7), 1–14 (2015)

17. M. Marzolla, Parallel implementations of cellular automata for traffic models, in *Cellular Automata. ACRI 2018, Lecture Notes in Computer Science*, vol 11115. (2018), pp. 503–512

# Evaluation of Deep Learning Algorithms for Traffic Sign Detection to Implement on Embedded Systems

**Miguel Lopez-Montiel, Ulises Orozco-Rosas, Moisés Sánchez-Adame, Kenia Picos, and Oscar Montiel**

**Abstract** Nowadays, machine learning algorithms are trendy and are used to solve different problems of autonomous vehicles obtaining good results. Among these algorithms, deep learning has emerged as an excellent alternative to improve the results of the state-of-the-art in machine vision applications. An essential task in autonomous vehicles is the detection of traffic signs. Some metrics used for these detectors focus on assessing precision and recall. However, it is necessary to consider other factors, such as the implementation of these models on an embedded system. In this work, we implement deep learning algorithms on an embedded system to evaluate two different detection algorithms: Faster R-CNN and Single Shot Multibox Detector (SSD) with two feature extractors, ResNet V1 101 and MobileNet V1 to determine the location of traffic signs within the observed scenario. The contribution of this work focuses on evaluating the implementation of traffic sign detection systems based on deep learning algorithms on embedded systems. The experiments were achieved on the experimental embedded system board Nvidia Jetson Nano. The inference time and memory consumption of these detection systems were evaluated; they delivered good performance (81–98%) measure by average precision for each superclass (prohibitory, warning, and mandatory).

M. Lopez-Montiel · M. Sánchez-Adame · O. Montiel (✉)
Instituto Politécnico Nacional, CITEDI-IPN Av, Instituto Politécnico Nacional, No. 1310, Nueva Tijuana, C.P. 22435 Tijuana, Baja California, Mexico
e-mail: oross@ipn.mx

M. Lopez-Montiel
e-mail: milopez@citedi.mx

M. Sánchez-Adame
e-mail: msanchez@citedi.mx

U. Orozco-Rosas · K. Picos
Centro de Innovación y Diseño (CEID), CETYS Universidad, Av. No 4., Fracc. El Lago, C.P 22210 Tijuana, Baja California, Mexico
e-mail: ulises.orozco@cetys.mx

K. Picos
e-mail: kenia.picos@cetys.mx



P. Melin et al. (eds.), *Recent Advances of Hybrid Intelligent Systems Based on Soft Computing*, Studies in Computational Intelligence 915,
https://doi.org/10.1007/978-3-030-58728-4_5

**Keywords** Traffic sign detection · Deep learning · Computer vision · Autonomous vehicles · Object detection · Embedded systems

## 1 Introduction

In this chapter, we continue with the work [36] that uses Arcos research [23] as a basis where different combinations for the detection systems in the traffic sign detection problem were evaluated. This work proposes a methodology design for the evaluation of traffic sign detection systems based on deep learning on embedded systems for soft real-time applications that use Nvidia fast graphics processing units (GPUs). The factors of memory consumption and inference time (understanding by inference time the time it takes for the deep learning algorithm to process the input and deliver a result) are evaluated in this methodology. These factors impact the selection of meta-architectures and feature extractors to its implementation on embedded systems. The Jetson Nano embedded system board of Nvidia was used to perform the experiments, as well as the evaluations and validations of the results of the detection systems. The LISA database [19] was selected for the experiments.

The contribution of this work focuses on a proposal to evaluate the memory consumption and inference time of these systems when TSD (Traffic Sign Detection) is implemented on the mentioned embedded system of Nvidia.

This chapter is divided as follows. In Sect. 2, the meta-architectures and feature extractors used in the two traffic sign detection systems are presented. In Sect. 3, the proposed methodology design for the evaluation of the traffic sign detection system on an embedded system is shown. In Sect. 4, the experiments and results obtained from TSD systems are shown, where a subset of examples was selected from the LISA database that had occlusion, blur, and illumination cases. In Sect. 5, the analysis of memory consumption, inference time, and the precision-sensitivity curve of the evaluated systems are shown. In Sect. 6, the conclusions and future work are presented.

## 2 Background and Motivation

In recent years, the trends of emerging technologies that involve Autonomous Driving Systems are significant to develop better solutions for vehicles and transportation systems. Over the years, there have been different stages of safety; the relevance of some of them are still changing. They evolved from a set of features to provide advanced driver assistance to a reasonable degree of autonomy. The Society of Automotive Engineers (SAE) proposed levels of autonomy to classify the needs of each level. Some of these systems are ADAS (Advanced driver-assistance system), which are based on different sensors such as LiDAR, sonars, and cameras to automate different functionalities [10, 35]. The investment in research and development of autonomous vehicles today is calculated in billions of dollars. It is predicted that in the year 2025, approximately 8 million autonomous or semi-autonomous vehicles will be deployed with SAE levels of 3, 4, or 5 [3]. These autonomy levels measure the intelligence and automation capabilities of vehicles from 0 to 5 levels based on their abilities.

The SAE represents these levels in the following way: level 0 is manual driving, which involves vehicle control by people. Level 1 is driver assistance with only one automated aspect, such as direction, speed, or braking control. Level 2 is partial driving automation that includes the capabilities of an ADAS steering and acceleration/deceleration system. Level 3 is conditional driving automation, that can detect the environment to make decisions for themselves, such as overtaking slower-moving vehicles. Level 4 is high driving automation; the vehicles can intervene if something goes wrong with the system. However, a human still has the option to override manually. Levels 5 is full driving automation, that requires no human intervention, there are no typical driving controls such as steering wheels, acceleration, and brake pedals, and there is no driver. They can also perform driving tasks like an experienced human would [5, 28].

The company Gartner in the report called *Hype Cycle for Emerging Technologies* 2019 reveals that some technologies concerning Autonomous Driving Levels 4 and 5, in these levels, the automation is almost complete [8]. The National Highway Traffic Safety Administration established the evolution of automated vehicles that will bring security to society through machine-aided driver assistance systems for improving the automation of vehicles [6].

An essential functionality in these systems is the Traffic Sign Recognition System (TSR), which provides information regarding signs in the environment to execute different tasks and get a level of autonomy. These systems are divided into Traffic Sign Detection (TSD) and Traffic Sign Classification (TSC). Different works reach both tasks [12, 13, 29]. Recent works are oriented on TSD [24, 44] due to the relevance to detect the signs in different environmental conditions such as urban, rural, and highway, which are more challenging.

Some commons problems in TSD are related to the scene or with the method used. These problems can be separated into internal and external variables. The external variables, such as illumination, occlusion, blur, and weather conditions, involve the scene. On the other hand, the internal variables such as the Frames Per Second (FPS), execution time as well as memory consumption affect the method and system used. These conditions suggest the necessity of developing new methods and systems to improve the performance, not only considering the precision of the system but also its implementation in a real-time system [15, 31, 34].

The artificial intelligence approach to solving the problem of TSD has emerged with deep learning algorithms [30, 41], where the Convolutional Neural Networks (CNN) have obtained good performance in the precision of the models [9, 27, 38, 43], that is one of the more important metrics to consider in this problematic. The hardware implementation of these systems presents an area of opportunity for reducing memory consumption and the inference time of the models. These two issues are very relevant to the development of these systems. Therefore, it is necessary to optimize them for the implementation of this kind of system on embedded systems.

## *2.1 Deep Learning Method for Traffic Sign Detection*

The interest in deep feedforward networks was revived around 2006 by a group of researchers brought together by the Canadian Institute for Advanced Research (CIFAR) [1, 17, 26, 27]. The researchers introduced unsupervised learning procedures that could create layers of feature detectors without requiring labeled data. The first significant application of this pre-training approach was in speech recognition, and it was made possible by the advent of GPUs that were convenient to program and allowed researchers to train networks 10- or 20-times faster [18].

During the period when neural networks were obsolete, and deep learning was rehabilitated, a particular type of feed-forward network that was much easier to train and generalized much better than other networks with full connectivity between layers emerged. We refer to the convolutional neural network (ConvNet), which achieved many practical successes and was adopted by the computer community [22].

### 2.1.1 Convolutional Neural Networks

ConvNets are designed to process data that come in the form of multiple arrays, for example, a color image composed of three 2D arrays containing pixel intensities in the three-color channels. Many data modalities are in the form of multiple arrays: 1D for signals and sequences, including languages; 2D for images or audio spectrograms; and 3D for video or volumetric images. Since the early 2000s, ConvNets have been applied with great success to the detection, segmentation, and recognition of objects and regions in images. These images can be labeled at the pixel level, allowing the training of learning-based algorithms. These algorithms make it possible to develop applications for different technologies, such as autonomous mobile robots and autonomous automobiles. Companies such as Mobileye and NVIDIA are using such ConvNet-based methods in their upcoming vision systems cars [20].

The architecture of a typical ConvNet [50] is structured as a series of stages see Fig. 1. The first few stages are composed of two types of layers: convolutional layers and pooling layers. Units in a convolutional layer are organized in feature maps, withing which each unit is connected to local patches in the feature maps of the previous layer through a set of weights called a filter bank.

The essential components of these architectures consist of three types of layers, namely convolutional, pooling, and fully-connected layers. The convolutional layer aims to learn feature representations of the inputs. The pooling layer aims to achieve shift-invariance by reducing the resolution of the feature maps. After several convolutional and pooling layers, there may be one or more fully-connected layers that

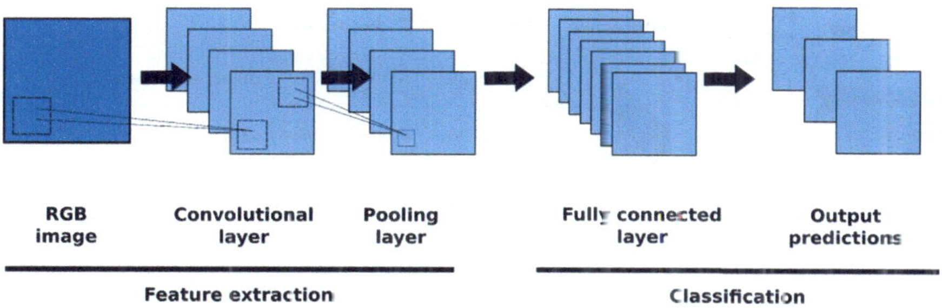


**Fig. 1** General architecture of convolutional neural network

aim to perform high-level reasoning. They take all neurons in the previous layer and connect them to every single neuron of the current layer to generate global semantic information. The last layer of CNNs is an output layer [45, 50].

Equation (1) represents the convolution operation [45].

$$x^{l-1} * w^{l} \tag{1}$$

Between an input feature map $l-1$, convolved with a 2D convolutional kernel (or filter or weights) $w^l$. This convolution appears over a sequence of layers, subject to the nonlinear operation σ, as seen in Eq. (2).

$$x_j^l = \sigma\left(\sum_{i=1}^{N^{l-1}} x_i^{l-1} * w_{i,j}^l + b_j^l\right) \tag{2}$$

With a convolution now between the $N^{l-1}$ input feature maps $x_i^{l-1}$ and the corresponding kernel $w_{i,j}^l$, plus a bias term $b_j^l$. The elementwise nonlinear function σ(.) is typically a rectified linear unit (ReLU) for each element, see Eq. (3).

$$\sigma(x) = \max(x, 0). \tag{3}$$

Finally, pooling corresponds to the downsampling/upsampling of feature maps. Most layers of a CNN consist of several feature maps, within each feature maps each pixel acts as a neuron. Each neuron in a convolutional layer is connected to feature maps of the previous layer through a set of weights $w_{i,j}$ (essentially a set of 2D filters) [45].

### 2.1.2 Faster R-CNN

In [33], the entire system is presented as a unique and unified network for object detection, see Fig. 2. Using the recently popular terminology of neural networks

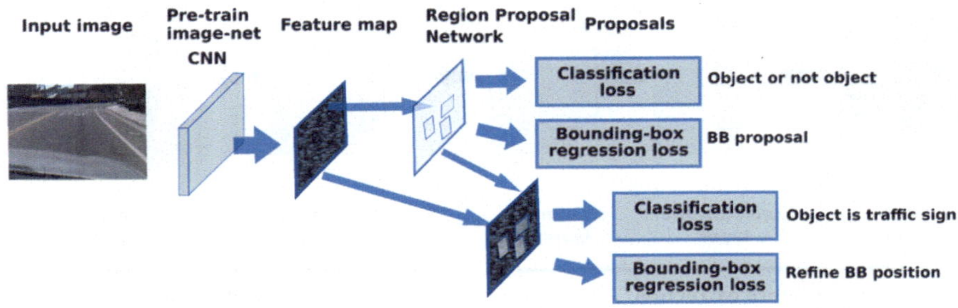


**Fig. 2** Faster R-CNN model architecture

with 'attention' mechanisms, the Region Proposal Networks (RPN) module tells the Fast R-CNN module where to look. This object detection system consists of two modules. The first module is a fully convolutional deep network that proposes regions, and the second module is the Fast R-CNN detector that uses the proposed regions. This method eliminates computational loads from selective searches at the time of the tests. The effective execution time for proposals was only 10 ms [33]. The approach of the method is not only for practical use but also an effective way to improve the accuracy of object detection. This method allows deep learning to record frame rates almost in real-time. The RPN learned also enhances the quality of the region proposal and, therefore, the overall accuracy of object detection [2].

### 2.1.3 SSD

The Single Shot Multibox Detector (SSD) [46] see Fig. 3. It is an approach based on a feed-forward convolutional network that produces a fixed-size collection of bounding boxes and scores for the presence of object class instances in those boxes, followed by a non-maximum suppression step to produce the final detections. The early network layers are based on a standard architecture used for high-quality image classification. A key feature of the model is the use of multi-scale convolutional bounding box outputs attached to multiple features maps as the top of the network.

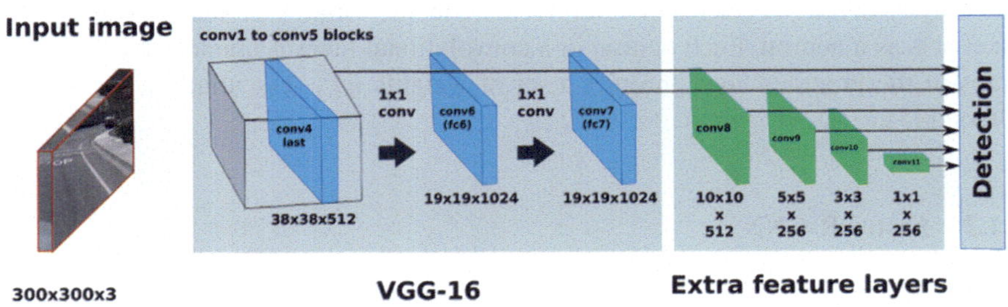


**Fig. 3** SSD model architecture

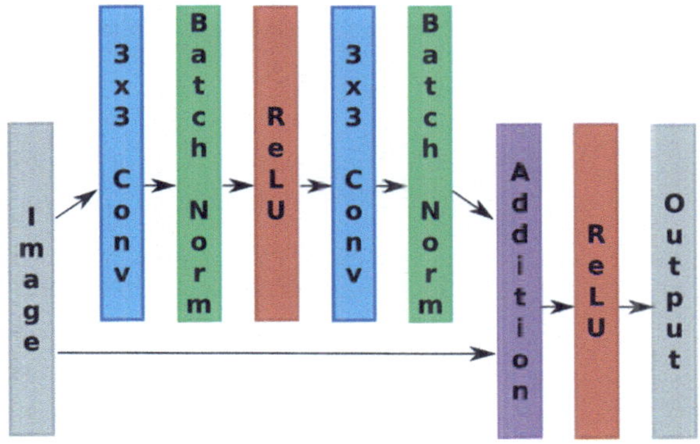


**Fig. 4** Basic ResNet block

This representation allows us to model the space of possible box shapes efficiently. The key difference between training SSD and training a typical detector that uses region proposals is that ground truth information needs to be assigned to specific outputs in the fixed set of detector outputs.

### 2.1.4 ResNet V1 101

The ResNet model [32] is based on a deep residual learning framework, which consists of multiple subsequent residual modules, see Fig. 4. The residual blocks allow the network to retain what it has previously learned. By using an identity mapping function where the output equals the input to retain what it learned previously, in case there is nothing new to learn, it will add what the network has already learned. When comparing residual networks against other flat or non-residual networks, the residual network is better. Because although they used the same parameters, depth, width, and computational costs, better results are obtained. Furthermore, this residual network increases its performance by adding more residual blocks.

### 2.1.5 MobileNet V1

The MobileNet model [47] is based on depthwise separable convolutions, which are a form of factorized convolutions that factorize a standard convolution into a depthwise convolution, and a $1 \times 1$ convolution called a pointwise convolution. The result of the depthwise convolutions significantly reduces the number of parameters compared to the network with normal convolutions with the same depth, and it thereby reduces the total number of floating-point multiplication operations, which is favorable in mobile and embedded vision applications with less computer power. This factorization has the effect of drastically reducing computation and model size. MobileNet architecture is defined in Table 1.

**Table 1** MobileNet body architecture

| Type/Stride | Filter shape | Input size |
|---|---|---|
| Conv/s2 | 3 × 3 × 3 × 32 | 224 × 224 × 3 |
| Conv/dw/s1 | 3 × 3 × 32 dw | 112 × 112 × 32 |
| Conv/s1 | 1 × 32 × 64 | 112 × 112 × 32 |
| Conv dw/s2 | 3 × 3 × 64 dw | 112 × 112 × 64 |
| Conv/s1 | 1 × 1 × 64 × 128 | 56 × 56 × 64 |
| Conv dw/s1 | 3 × 3 × 128 dw | 56 × 56 × 128 |
| Conv/s1 | 1 × 1 × 128 × 128 | 56 × 56 × 128 |
| Conv dw/s2 | 3 × 3 × 128 × 128 | 56 × 56 × 128 |
| Conv/s1 | 1 × 1 × 128 × 256 | 28 × 28 × 128 |
| Conv dw/s1 | 3 × 3 × 256 dw | 28 × 28 × 256 |
| Conv/s1 | 1 × 1 × 256 × 256 | 28 × 28 × 256 |
| Conv dw/s2 | 3 × 3 × 256 dw | 28 × 28 × 256 |
| Conv/s1 | 1 × 1 × 256 × 512 | 14 × 14 × 256 |
| Conv dw/s1 | 3 × 3 × 512 dw | 14 × 14 × 512 |
| 5 x Conv/s1 | 1 × 1 × 512 × 512 | 14 × 14 × 512 |
| Conv dw/s2 | 3 × 3 × 512 dw | 14 × 14 × 512 |
| Conv/s1 | 1 × 1 × 512 × 1024 | 7 × 7 × 512 |
| Conv dw/s2 | 3 × 3 × 512 dw | 14 × 14 × 512 |
| Conv/s1 | 1 × 1 × 1024 × 1024 | 7 × 7 × 1024 |
| Avg Pool/s1 | Pool 7 × 7 | 7 × 7 × 1024 |
| FC/s1 | 1024 × 1000 | 1 × 1 × 1024 |
| Softmax/s1 | Classifier | 1 × 1 × 1000 |

## 2.2 *Traffic Sign Detection Database*

The LISA database [19] consists of different scenes where there are cases of occlusion, illumination, blur, different sign distances, and multiple signs on each side of the road. These images were captured with different cameras obtaining information from low resolutions (640 × 480 pixels) to high resolutions (1024 × 522 pixels). The data set is composed of 47 traffic signs, which were divided into three superclasses: prohibition, mandatory, and warning. The subset consists of 2857 color images with 3232 traffic signs. Table 2 shows the classes and superclasses.

Depending on the continent and country, there are different types of traffic signs and the properties change. The database contains videos and images of traffic signs from the United States of America. This data set is composed of different images in grayscale and color (8-bit color) obtained by different cameras; each image is labeled. For the experiments, only color images were used. An example of the LISA Dataset can be seen in Fig. 5.

**Table 2** Description of traffic signal classes and superclasses available in the LISA dataset

| Prohibitory | Mandatory | Warning |
|---|---|---|
| Stop | schoolSpeedLimit25 | pedestrianCrossing |
| Yield | speedLimit15 | signalAhead |
| yieldAhead | speedLimit25 | addedLane |
| noRightTurn | speedLimit30 | Merge |
| doNotPass | speedLimit35 | stopAhead |
| noLeftTurn | speedLimit40 | School |
| doNoenter | speedLimit45 | Dip |
| – | speedLimit50 | Intersection |
| – | speedLimit55 | turnRight |
| – | speedLimit65 | curveLeft |
| – | speedLimitUrdbl | curveRight |
| – | keepRight | Slow |
| – | rightLaneMustTurn | laneEnds |
| – | zoneAhead25 | rampSpeedAdvisory20 |
| – | zoneAhead45 | rampSpeedAdvisory35 |
| – | truckSpeedLimit55 | rampSpeedAdvisory40 |
| – | – | rampSpeedAdvisory45 |
| – | – | rampSpeedAdvisory50 |
| – | – | rampSpeedAdvisoryUrdbl |
| – | – | Roundabout |
| – | – | thruMergeLeft |
| – | – | thruMergeRight |
| – | – | thruTrafficMergeLeft |
| – | – | turnLeft |

## 2.3 *Embedded System*

The use of embedded systems today has different trends in the application of industry sectors such as security, cloud connectivity, reduced energy consumption, real-time data visualization, and deep learning applications. The deep learning applications require hardware with particular features, such as memory bandwidth, several cores of graphic processing units (GPU), and Video RAM size [42].

The embedded NVIDIA systems provide GPUs to facilitate the deployment of deep learning models. One of these systems is the NVIDIA Jetson Nano see Fig. 6 which is an embedded-on-module (SoM) system and an NVIDIA Jetson family developer kit, which has an NVIDIA Maxwell GPU with a maximum throughput of 472 GFLOPS of computer performance FP16 with 5–10 watts of power consumption [4], see the hardware specifications in Table 3.

**Fig. 5** Some samples of the used traffic signs from the LISA dataset

**Fig. 6** Image of NVIDIA Jetson Nano

**Table 3** NVIDIA Jetson Nano hardware specifications

| | |
|---|---|
| GPU | NVIDIA maxwell architecture with 128 NVIDIA CUDA cores |
| CPU | Quad-core ARM Cortex-A57 MPCore processor |
| Memory | 4 GB 64-bit LPDDR4, 1600 MHz 25.6GB/s |
| Storage | 16 GB eMMC 5.1 |

## 3 Proposed Methodology

The ADAS is composed of different systems that are responsible for automating certain specific tasks to take the vehicle to different levels of autonomy. These systems involve different previous stages where sensors are required for data acquisition and an embedded system to perform the processing of these systems and to be able to deliver the appropriate output to the actuators of the vehicle [14]. The TSR systems require sensors such as cameras to obtain environmental data represented as images and thus apply the processing, see Fig. 7. Some relevant features for these sensors are camera calibration, frames per second, and resolution [7, 15].

The problem of TSD has been solved with different deep learning models focusing on obtaining better performance in the precision and recall of the models [23]. At the time of implementation, other factors must be considered for these systems to be compatible and viable in embedded systems. Some of these components are the inference time and the memory consumption of the models; therefore, in this work, a methodology design is proposed for the evaluation of TSD systems on embedded systems, see Fig. 8.

The steps of the proposed methodology for evaluating the systems are described below.

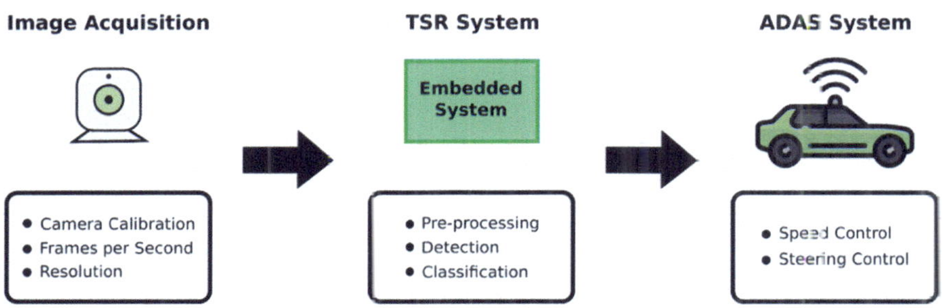


**Fig. 7** General scheme of implementation TSR

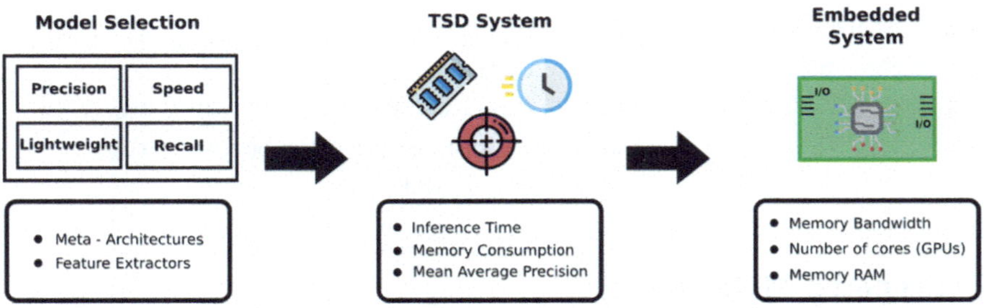


**Fig. 8** Design methodology for the evaluation of TSD systems in embedded systems

## 3.1 *Model Selection*

The selection of a model will be dependent on its properties [25, 40]. It is necessary to specify in the approach the selected model to choose the appropriate features that will make up the model. Depending on the application, a balance between the characteristics is required to obtain a desired performance and efficiency. The properties considered in this design proposal are precision, recall, lightweight, and speed. In this work, the method used for the construction of the traffic sign detection system is based on the use of meta-architectures and feature extractors [11]. Meta-architectures must be carefully selected as they influence the speed and performance of object detection. Modern meta-architectures use CNNs for object detection, having an improved infrastructure that achieves good performance. Feature extractors focus on obtaining the essence of the entire architecture. These extractors combine with the meta-architectures to extract the information of the high level of the images. Feature extractors do not directly affect the speed and memory performance of object detectors; however, capacity type selection is relevant to extractors. Therefore, the selection of a model for both meta-architectures and feature extractors will be dependent on the type of application, and this will define the properties' priority of the models.

## 3.2 *TSD System*

TSD systems require different properties to obtain good performance. The properties considered in this design proposal are the inference time of the model, the memory consumption, and the mean average precision. The importance of the inference time of the model lies in how quickly the system delivers the result, having an appropriate response in real-time. Memory consumption is that so many hardware resources require the system for the embedded device to support it. The mean average precision is centralized to select a system with good performance concerning its precision and sensitivity. Taking into consideration the above properties, a system that can be implemented in embedded systems for the use of real applications is obtained [14].

### 3.3 Embedded System

Embedded systems require specific hardware features for the implementation of these systems [39]. Some features to consider in this design proposal are RAM, the number of graphic processing units (GPUs), and memory bandwidth. The RAM capacity of the device must be specific to select the indicated models. Deep Learning-based systems use GPUs to perform massive data processing in their training and validation stages; therefore, embedded devices need to have enough or minimal GPUs for its implementation. The use of bandwidth allows higher handling of data by the amounts of cores, allowing the GPUs to carry out massive processing and, in this way, exploit the potential of the multiple processing units of the deep learning algorithms. Therefore, it is essential to reduce the hardware requirements for embedded applications since they have limited resources.

### 3.4 Evaluation Metrics

The Average Precision measure (AP) from [37] is used to evaluate the performance of the detectors. This measure calculates the area under the precision/recall curve obtaining a precision for each recall, determining the accuracy of the models over 0–1, indicating that values close to 1, better results are obtained. See Eqs. (3) and (4).

$$p(r) = \max_{r':r_i \geq r} p(r') \quad (3)$$

$$AP = \frac{1}{n} \sum_{k=1}^{N} p(k) \Delta r(k) \quad (4)$$

The inference time measures the response time of the models [21], specifically considering the time required by the model to deliver the output response upon receipt of the input image, not counting the output of the class label and its location in the bounding box of the sign. Equation (5) describes the calculation of the inference time, where time is measured in milliseconds. The inference time is calculated by the time the model needs to evaluate one sample. Indicating that at time values smaller or close to 0, a better model speed result is obtained.

$$Inference\ time = \frac{number\ of\ samples}{start\ inference - end\ inference} \quad (5)$$

The memory consumption measure of the models is an important metric to consider for its real-time applications [21]; currently, the hardware has been evolving, but that does not mean that embedded systems have unlimited resources, so it is necessary to consider the minimum memory for these devices. The sum of used memory and swap memory gives memory consumption; see Eq. (6). This metric indicates that the lower the memory consumption of the system, it will have more possibilities of being implemented on embedded devices with few resources.

$$memory = normal\ memory + swap\ memory \tag{6}$$

### 3.5 Training and Evaluation

The experimental phase was divided into two sections: training and validation in a tower and the implementation in an embedded device. In the training section, the LISA database [19] was used, where 70% of the data set was chosen for the training, and the remaining 30% for validation. The hardware used in training was a server with an Nvidia Ge Force GTX TITAN × GPU card that has 3072 CUDA cores and 12 GB of memory, and the hardware used for the implementation tests was the Nvidia Jetson Nano which has 128 CUDA Cores and 4 GB of memory. The software used was the Tensorflow Object Detection API, in which pre-trained models with the COCO database were chosen [48, 49].

## 4 Experiments and Results

The objective of the experiments was to validate the evaluation tests carried out on the tower obtained the same results on the embedded devices. The configuration of the systems is composed as follows: two meta-architectures (Faster R-CNN and SSD) and two convolutional feature extractors (MobileNet V1 and ResNet V1). System one (**S1**) is composed of SSD + MobileNet V1, and system two (**S2**) is composed of Faster R-CNN + ResNet V1.

In this work, the term superclasses were used to encompass the different kinds of traffic signs based on their shape and color, obtaining the following categories: prohibitive, warning, and mandatory. The database used is LISA [19], which is composed of 47 kinds of traffic signs, 7855 annotations in 6620 frames. It also contains different sign sizes from 6 × 6 to 167 × 168 pixels, and its frame capture is obtained from different cameras varying their sizes from 640 × 48 to 1024 × 522 pixels. These images contain one or multiple traffic signs with different conditions such as illumination, occlusion, or blur. To indicate the detection of superclasses of traffic signs, a bounding box, and a label with the percentage of belonging are used. The detection results obtained in the average precision for both systems are shown in Figs. 15 and 16, it is observed that the system **S2** obtains a better performance in the detection results for the three superclasses; however, in terms of memory consumption shown in Table 4, the other system **S1** gets the better performance, with lower memory consumption.

**Table 4** Deep Learning methods inference time and memory consumption

| Methods | Inference time | Memory consumption |
| --- | --- | --- |
| Faster R-CNN + ResNet V1 101 | **13.9 ms** | 5136 MB |
| SSD + MobileNet V1 | 14.7 ms | **4949 MB** |

## *4.1 Detection in the LISA Dataset*

The selected test samples have the following cases: blur (Figs. 9 and 10), occlusion (Figs. 11 and 12), and illumination (Figs. 13 and 14). Therefore, the results show that the models deliver good performance (98–99%), approaching the performance of a person (**98.81%**) under these conditions. It is observed that the case of a distant object from the scene (Figs. 11 and 12) give results less than **99%** which was mentioned in [2], this case is an opportunity for improvement in the detection of small objects.

**Fig. 9** Results of SSD + MobileNet V1 prohibitory class

**Fig. 10** Results of faster R-CNN + ResNet V1 prohibitory class

**Fig. 11** Results of SSD + MobileNet V1 warning class

**Fig. 12** Results of faster R-CNN + ResNet V1 warning class

**Fig. 13** Results of SSD + mobilenet V1 mandatory class

**Fig. 14** Results of faster R-CNN + ResNet V1 mandatory class

## 5 Analysis

There are some challenges in the design of the ideal object detector. These challenges focus on two properties which are: high accuracy (where location and recognition are involved) and high efficiency (which involves time, memory, and storage). In the case of detection systems based on deep learning, they use different types of CNNs for specific purposes, where depending on the design of the base architecture, it will allow obtaining high accuracy or high efficiency. In this work, a meta-architecture with fast response time (Faster R-CNN), and a feature extractor efficient in terms of memory consumption (MobileNet V1) were selected. The aim is evaluating inference time and memory consumption in **S1** and **S2** systems, with these two different combinations. It can be observed that the methods used have a good performance in their average precision for the different superclasses; however, the memory consumption and inference time of the models are different for the detection systems. The test performed validates the dependency on detection performance when deploying systems on specific hardware. The results of these metrics are indicated in Table 4 and the precision/recall curves for both nets are shown in Figs. 15 and 16.

The results shown in Table 4, prove the need to optimize the models in embedded systems with few resources to reduce their memory consumption; however, this problem can be solved using other devices with higher capacity. Due to this, the inference time could be a relevant approach to improve inference speed; hence the

**Fig. 15** SSD + MobileNet v1 curve

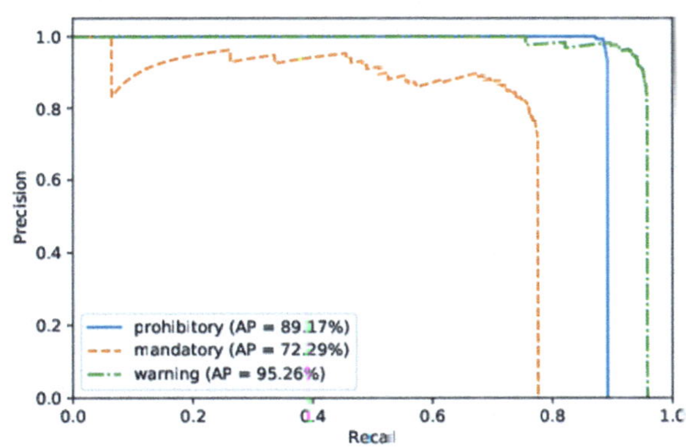


**Fig. 16** Faster R-CNN + ResNet v1 101 curve

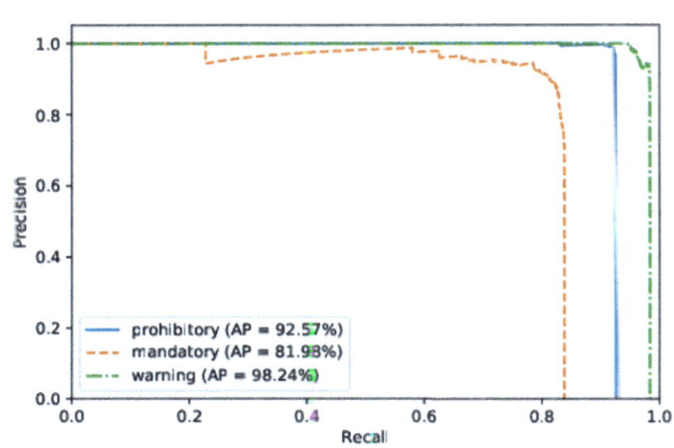

following techniques can be considered: weight pruning, quantization of weights, and weight sharing. These techniques focus on facilitating the compression of dense models, reducing the representation of weights, and speeding up multiplication and activation operations. Although this will affect the precision of the models.

The precision-sensitivity curve of both systems (**S1 and S2**) has good performance; see Figs. 15 and 16, the precision response for the prohibitive and warning class is higher than the mandatory class, this is due to the reduced number of samples for this superclass specifically, reducing their ability to discriminate against other classes. On the other hand, the implementation of the device Nvidia Jetson Nano had complications from the memory consumption of the models. The memory required for the models is higher than that available on the embedded device (4 GB). Swap memory was therefore used to supplement the missing memory so that the models could be deployed. As mentioned at the beginning of this chapter, the evolution in the problem of traffic sign detection has involved the field of artificial intelligence, where deep learning algorithms have been the methods currently used for their excellent performance for the development of these systems detection.

## 6 Conclusions and Future Work

The results show that for the implementation of detection systems in deep learning in embedded devices, it is necessary to perform an optimization of the models to reduce memory consumption. Both systems show good performance, as can be seen in their precision-sensitivity curves (see Figs. 15 and 16). However in memory consumption (see Table 4) it is greater than the total memory capacity of the embedded device, opting for using the exchange memory to solve the problem; however, this falls into two solutions: the use of embedded systems with higher capacity or the optimization of the models for embedded devices with few resources. The inference time of the Faster R-CNN + ResNet V1 101 method showed a better performance than SSD + MobileNet V1 method. However, it requires more memory, confirming that the implementation in embedded systems with few memory resources such as Nvidia Jetson Nano involves using other models or optimizing them to reduce memory consumption without wasting inference time and the performance in the detection of traffic signs.

For future works, it is required to used techniques such as data augmentation and metrics to evaluate and improve the training step of both methods. It is also required to generate or use an extensive database with different examples to have more data. Likewise, to find a better combination of meta-architectures and feature extractors to achieve better performance and use another embedded system for the experimental platform with more memory such as Jetson TX1, TX2, and others to do tests without using swap memory and avoid problems with the overflow of the memory consumption.

**Acknowledgments** This work was supported by Instituto Politécnico Nacional (institutional project SAPI 2020053) and by the Mexican National Council of Science and Technology (CONACYT, Mexico). We thank the Coordinación Institucional de Investigación of CETYS Universidad for their collaboration.

## References

1. G.E. Hinton, What kind of graphical model is the brain? in *Proceedings of the Nineteenth International Joint Conference on Artificial Intelligence* (Morgan Kaufmann Publishers Inc 2005), pp 1765–1775
2. R. Girshick, Fast R-CNN. in *2015 IEEE International Conference on Computer Vision (ICCV), 7–13 Dec 2015* (2015), pp. 1440–1448. https://doi.org/10.1109/iccv.2015.169
3. B. Oyster, *ABI Research Forecasts 8 Million Vehicles to Ship with SAE Level 3, 4 and 5 Autonomous Technology in 2025* (2018). ABI research. http://ABI Research: for Visionaries. www.abiresearch.com/press/abi-research-forecasts-8-million-vehicles-ship-sae-level-3-4-and-5-autonomous-technology-2025
4. Nvidia, Jetson Nano. https://developer.nvidia.com/embedded/jetson-nano (2019)
5. J. Shuttleworth, SAE J3016 automated-driving graphic. https://www.sae.org/news/2019/01/sae-updates-j3016-automated-driving-graphic (2019)
6. M. Lynberg, Automated Vehicles for Safety. https://www.nhtsa.gov/technology-innovation/automated-vehicles-safety (2019)
7. D.G. Costa, Visual sensors hardware platforms: a review. IEEE Sens. J. **20**(8), 4025–4033 (2020). https://doi.org/10.1109/JSEN.2019.2952447
8. K. Panetta, *5 Trends Appear on the Gartner Hype Cycle for Emerging Technologies, 2019*. https://www.gartner.com/smarterwithgartner/5-trends-appear-on-the-gartner-hype-cycle-for-emerging-technologies-2019/ (2020)
9. P. Sermanet, Y. LeCun, Traffic sign recognition with multi-scale Convolutional Networks, in *The 2011 International Joint Conference on Neural Networks, 31 July-5 Aug 2011* (2011). Pp. 2809–2813. https://doi.org/10.1109/ijcnn.2011.6033589
10. A. Malinowski, H. Yu, Comparison of embedded system design for industrial applications. IEEE Trans. Indus. Inf. **7**(2), 244–254 (2011). https://doi.org/10.1109/TII.2011.2124466
11. N. Yadav, U. Binay, *Comparative Study of Object Detection Algorithms* (2017)
12. M. Swathi, K.V. Suresh, Automatic traffic sign detection and recognition: a review, in *2017 International Conference on Algorithms, Methodology, Models and Applications in Emerging Technologies (ICAMMAET), 16–18 Feb 2017* (2017). pp. 1–6. https://doi.org/10.1109/icammaet.2017.8186650
13. Y. Saadna, A. Behloul, An overview of traffic sign detection and classification methods. Int. J. Mult. Inf. Ret. **6**(3), 193–210 (2017). https://doi.org/10.1007/s13735-017-0129-8
14. W.G. Hatcher, W. Yu, A survey of deep learning: platforms, applications and emerging research trends. IEEE Access **6**, 24411–24432 (2018). https://doi.org/10.1109/ACCESS.2018.2830661
15. X. Bangquan, W.X. Xiong, Real-time embedded traffic sign recognition using efficient convolutional neural network. IEEE Access **7**, 53330–53346 (2019). https://doi.org/10.1109/ACCESS.2019.2912311
16. J. Salvi, X. Armangué, J. Batlle, A comparative review of camera calibrating methods with accuracy evaluation. Pattern Recogn. **35**(7), 1617–1635 (2002). https://doi.org/10.1016/S0031-3203(01)00126-1
17. G.E. Hinton, S. Osindero, Y.-W. Teh, A fast learning algorithm for deep belief nets. Neural Comput. **18**(7), 1527–1554 (2006). https://doi.org/10.1162/neco.2006.18.7.1527
18. R. Raina, A. Madhavan, Ng AY large-scale deep unsupervised learning using graphics processors, in *26th Annual International Conference, 2009*. (ACM Press, Montreal, Quebec, Canada, 2009), pp 1–8. https://doi.org/10.1145/1553374.1553486

19. A. Mogelmose, M.M. Trivedi, T.B. Moeslund, Vision-based traffic sign detection and analysis for intelligent driver assistance systems: perspectives and survey. IEEE Trans. Intell. Trans. Syst. **13**(4), 1484–1497 (2012). https://doi.org/10.1109/TITS.2012.2209421
20. Y. LeCun, Y. Bengio, G. Hinton, Deep learning. Nature **521**(7553), 436–444 (2015). https://doi.org/10.1038/nature14539
21. A. Canziani, A. Paszke, E. Clurciello, An analysis of deep neural network models for practical applications. https://arxiv.org/abs/1605.07678 (2016)
22. A. Krizhevsky, I. Sutskever, G.E. Hinton, ImageNet classification with deep convolutional neural networks. Commun. ACM **60**(6), 84–90 (2017). https://doi.org/10.1145/3065386
23. Á. Arcos-García, J.A. Álvarez-García, L.M. Soria-Morillo, Evaluation of deep neural networks for traffic sign detection systems. Neurocomputing **316**, 332–344 (2018). https://doi.org/10.1016/j.neucom.2018.08.009
24. Avramović A, Tabernik D, Skočaj D Real-time Large Scale Traffic Sign Detection. In: 2018 14th Symposium on Neural Networks and Applications (NEUREL), 20–21 Nov. 2018 2018. pp 1-4. https://doi.org/10.1109/neurel.2018.8587013
25. M. Loni, M. Daneshtalab, M. Sjödin, ADONN: adaptive design of optimized deep neural networks for embedded systems, in *2018 21st Euromicro Conference on Digital System Design (DSD), 29–31 Aug. 2018* (2018), pp. 397–404. https://doi.org/10.1109/dsd.2018.00074
26. B. Yoshua, L. Pascal, P. Dan, L. Hugo, Greedy layer-wise training of deep networks, in: *Advances in Neural Information Processing Systems 19, Proceedings of the Twentieth Annual Conference on Neural Information Processing Systems* (Vancouver, British Columbia, Canada, 4–7 Dec 2006. MIT Press, 2006), pp. 153–160
27. M.A. Ranzato, C. Poultney, S. Chopra, Y. LeCun, Efficient learning of sparse representations with an energy-based model. in *Advances in Neural Information Processing Systems 19*. (MIT Press, 2007), pp 1137–1144
28. C. Rödel, S. Stadler, A. Meschtscherjakov, M. Tscheligi, Towards autonomous cars: the effect of autonomy levels on acceptance and user experience, in *6th International Conference*, 2014. (ACM Press, Seattle, WA, USA, 2014), pp 1–8. https://doi.org/10.1145/2667317.2667330
29. J.M. Lillo-Castellano, I. Mora-Jiménez, C. Figuera-Pozuelo, J.L. Rojo-Álvarez, Traffic sign segmentation and classification using statistical learning methods. Neurocomputing **153**, 286–299 (2015). https://doi.org/10.1016/j.neucom.2014.11.026
30. M. Yang, B. Li, H. Fan, Y. Jiang, Randomized spatial pooling in deep convolutional networks for scene recognition, in *2015 IEEE International Conference on Image Processing (ICIP)*, 27–30 Sep 2015. pp 402–406. https://doi.org/10.1109/icip.2015.7350829
31. Y. Yang, H. Luo, H. Xu, F. Wu, Towards real-time traffic sign detection and classification. IEEE Trans. Intell. Transp. Syst. **17**(7), 2022–2031 (2016). https://doi.org/10.1109/TITS.2015.2482461
32. K. He, X. Zhang, S. Ren, S. Jian, Deep residual learning for image recognition, in *2016 IEEE Conference on Computer Vision and Pattern Recognition (CVPR)*, 27–30 June 2016. pp 770–778. https://doi.org/10.1109/cvpr.2016.90
33. S. Ren, K. He, R. Girshick, J. Sun, Faster R-CNN: towards real-time object detection with region proposal networks. IEEE Trans. Pattern Anal. Mach. Intell. **39**(6), 1137–1149 (2017). https://doi.org/10.1109/TPAMI.2016.2577031
34. K. Lim, Y. Hong, Y. Choi, H. Byun, Real-time traffic sign recognition based on a general purpose GPU and deep-learning. PLoS ONE **12**(3), e0173317 (2017). https://doi.org/10.1371/journal.pone.0173317
35. V.K. Kukkala, J. Tunnell, S. Pasricha, T. Bradley, Advanced driver-assistance systems: a path toward autonomous vehicles. IEEE Consum. Electron. Maga. **7**(5), 18–25 (2018). https://doi.org/10.1109/MCE.2018.2828440
36. L.-M. Miguel, R. Yoshio, S.-A. Moisés, O.-R. Ulises, Evaluation of algorithms for traffic sign detection, in *Proceedings of SPIE* (2019). https://doi.org/10.1117/12.2529709
37. M. Everingham, L. Van Gool, C.K.I. Williams, J. Winn, A. Zisserman, The pascal visual object classes (VOC) challenge. Int. J. Comput. Vision **88**(2), 303–338 (2010). https://doi.org/10.1007/s11263-009-0275-4

38. R. Qian, Z. Bailing, Y. Yong, W. Zhao, F. Coenen, Robust chinese traffic sign detection and recognition with deep convolutional neural network, in *2015 11th International Conference on Natural Computation (ICNC)*, 15–17 Aug 2015, pp. 791–796. https://doi.org/10.1109/icnc.2015.7378092
39. J. Xu, B. Wang, J. Li, C. Hu, J. Pan, Deep learning application based on embedded GPU, in *2017 First International Conference on Electronics Instrumentation & Information Systems (EIIS)*, 3–5 June 2017, pp 1–4. https://doi.org/10.1109/eiis.2017.8298723
40. B. Taylor, V.S. Marco, W. Wolff, Y. Elkhatib, Z. Wang, Adaptive deep learning model selection on embedded systems. SIGPLAN Not **53**(6), 31–43 (2018). https://doi.org/10.1145/3299710.3211336
41. T. Yang, X. Long, A.K. Sangaiah, Z. Zheng, C. Tong, Deep detection network for real-life traffic sign in vehicular networks. Comput. Netw. **136**, 95–104 (2018). https://doi.org/10.1016/j.comnet.2018.02.026
42. X. Feng, Y. Jiang, X. Yang, M. Du, X. Li, Computer vision algorithms and hardware implementations: a survey. Integration **69**, 309–320 (2019). https://doi.org/10.1016/j.vlsi.2019.07.005
43. Y. Zhu, C. Zhang, D. Zhou, X. Wang, X. Bai, W. Liu, Traffic sign detection and recognition using fully convolutional network guided proposals. Neurocomputing **214**, 758–766 (2016). https://doi.org/10.1016/j.neucom.2016.07.009
44. F. Shao, X. Wang, F. Meng, T. Rui, D. Wang, J. Tang, Real-time traffic sign detection and recognition method based on simplified gabor wavelets and CNNs. Sensors **18**(10), 3192 (2018). https://doi.org/10.3390/s18103192
45. L. Liu, W. Ouyang, X. Wang, P. Fieguth, J. Chen, X. Liu, M. Pietikäinen, Deep learning for generic object detection: a survey. Int. J. Comput. Vision **128**(2), 261–318 (2020). https://doi.org/10.1007/s11263-019-01247-4
46. W. Liu, D. Anguelov, D. Erhan, C. Szegedy, S. Reed, C.-Y. Fu, Berg AC SSD: single shot MultiBox detector, in *Computer Vision—ECCV 2016*, eds by B. Leibe, J. Matas, N. Sebe, M. Welling. (Springer International Publishing, 2016), pp. 21–37
47. A.G. Howard, M. Zhu, B. Chen, D. Kalenichenko, W. Wang, T. Weyand, M. Andreetto, H. Adam, *MobileNets: Efficient Convolutional Neural Networks for Mobile Vision Applications* (2017). arXiv:170404861 [cs]
48. T.-Y. Lin, M. Maire, S. Belongie, J. Hays, P. Perona, D. Ramanan, P. Dollár, C.L. Zitnick, Microsoft COCO: common objects in context, in *Computer Vision—ECCV 2014*. eds by D. Fleet, T. Pajdla, B. Schiele, T. Tuytelaars (Springer, Cham, 2014), pp. 740–755
49. J. Huang, V. Rathod, C. Sun, M. Zhu, A. Korattikara, A. Fathi, I. Fischer, Z. Wojna, Y. Song, S. Guadarrama, K. Murphy, Speed/accuracy trade-offs for modern convolutional object detectors, in *2017 IEEE Conference on Computer Vision and Pattern Recognition (CVPR), July 21–26 2017*. (2017) pp. 3296–3297. https://doi.org/10.1109/cvpr.2017.351
50. J. Gu, Z. Wang, J. Kuen, L. Ma, A. Shahroudy, B. Shuai, T. Liu, X. Wang, G. Wang, J. Cai, T. Chen, Recent advances in convolutional neural networks. Pattern Recogn. **77**, 354–377 (2018). https://doi.org/10.1016/j.patcog.2017.10.013

# Performance Comparison of Parallel PSO-GA Algorithm with Dynamic Parameter Adjustment Using Type-1 and Interval Type-2 Fuzzy Systems

**Yunkio Kawano and Fevrier Valdez**

**Abstract** This paper reveals a comparison between different optimization methods with dynamical adjustment of parameters using Type-1 and interval Type-2 Fuzzy logic systems. In addition, the use of migration blocks in the PSO and Genetic Algorithm (GA) optimization methods is used to help each other to reach the best possible solution, which is also investigated in detail. The hybrid PSO-GA approach offers good simulation results outperforming individually the PSO and GA methods.

**Keywords** Swarm intelligence · PSO · GA · Fuzzy systems

## 1 Introduction

This paper provides a comparison of the combination of the PSO algorithm and the Genetic Algorithm which within them will have a migration block with information sharing between the algorithms. In addition, a dynamic adaptation of parameter with fuzzy logic is made with both Type-1 and Intervals Type-2 fuzzy systems

In this case, our contribution is a method for finding the global minimum defined by each of the benchmark functions that we evaluate with the use of different types of fuzzy systems. Also the use of migration between the PSO and GA algorithms is tested. In addition, the configuration of one input and six outputs of the fuzzy systems causes a small reduction in time at the time they are evaluated, since it is read less times when executing the algorithm.

The paper is organized as follows: Between Sects. 2 and 4, each of the optimization algorithms that were used and the combination of both PSO and GA are briefly described. In Sect. 5, the structure referring to fuzzy logic and with the rules used is also described.

Y. Kawano · F. Valdez (✉)
Tijuana Institute of Technology, Tijuana, BC, Mexico
e-mail: fevrier@tectijuana.mx


P. Melin et al. (eds.), *Recent Advances of Hybrid Intelligent Systems Based on Soft Computing*, Studies in Computational Intelligence 915,
https://doi.org/10.1007/978-3-030-58728-4_6

Then in Sect. 6 come the performed experiments as well as the obtained results are presented. Finally, in Sect. 7, a brief conclusion is offered after analyzing the obtained results.

## 2 Particle Swarm Optimization

The PSO algorithm is a method that was proposed to be used in the field of computation in form of an algorithm and is used for the optimization of different types of problems. It is a technique focused on search and optimization in the field of machine learning and is inspired by the behavior of insects or bird flocks.

PSO is considered as a swarm intelligence technique and has received a lot of attention recently [1–5].

The pseudocode of PSO algorithm is provided in the following Table 1.

The following mathematical equation refers to the step where the position and speed of the PSO algorithm is updated.

$$v_i^{k+1} = \omega \cdot v_i^k + \emptyset_1 \cdot rand_1 \cdot pBest_i + x_i^{k^{\Sigma}} + \emptyset_2 \cdot rand_2 \cdot g_i - x_i^{x^{\Sigma}}$$

A more detailed description of each factor is presented below:

$v_i^k$ = Is the velocity of the particle $i$ in the iteration $k$.

$\omega$ = Is the inertial factor.

$\emptyset_1, \emptyset_2$ = The learning ratios (weights) that control the cognitive and social components.

$rand_1, rand_2$ = The random numbers between 0 and 1.

$x_i^k$ = Current position of particle $i$ in iteration $k$.

$pBest_i$ = Best position (solution) found by the particle $i$ so far $g_i$ represents the position of the particle with the $pBest\_fitness$ of the environment of $p_i(localbest)$.

**Table 1** Pseudocode of particle swarm optimization algorithm

| |
|---|
| 1. Start the necessary parameters |
| 2. Evaluate the initial population of particles |
| 3. Update the positions and speeds of the particles |
| 4. Evaluate the population |
| 5. If the stop condition is met then the best solution found is printed, if not it found, return to step 3 |

**Table 2** Pseudocode of genetic algorithm

| | |
|---|---|
| 1. | Start the necessary parameters |
| 2. | Generate random initial populat on |
| 3. | Calculate the aptitude and assign the aptitude value to the entire population |
| 4. | Selection the best individual for crossing |
| 5. | Mutation applies to individuals |
| 6. | Evaluate the best solution after step 4 |
| 7. | Reinsert the best individuals resulting from step 5 |
| 8. | If the stop condition is met then the best solution found is printed, if not it found, step 2 is run again |

## 3 Genetic Algorithms

A genetic algorithm (GA) is a set of steps inspired by biological evolution. These algorithms simulate the evolution of a population of individuals by applying random genetic operations similar to those that occur in biological evolution, as shown in the following pseudocode in Table 2 [6–10].

The crossing process within the genetic algorithm is one of the most important within the algorithm since it is where individuals reproduce and inherit a certain part of the chromosome of each of them. And it is where the new generation of individuals is created.

As the name implies, the mutation process makes a certain random percentage of change in the chromosome genes, which allows no chromosome to be the same and thus to achieve more diversity within the new population of individuals after each generation.

## 4 Parallel PSO-GA Algorithm

The Parallel PSO-GA algorithm is a combination of the two above described algorithms, which in this case are being executed in parallel to allow them to share information on the best individuals or particles. In this case, a migration block is used each twenty-five iterations to exchange good information.

The used migration block makes a search of the best individual within each population of the algorithms, then a comparison of these individuals is made, and in the case that the winning individual is by the PSO algorithm then is integrated into the population of the Genetic Algorithm, the amount of the each population is one hundred individuals [4, 11–15].

Figure 1 illustrates the PSO-GA algorithm flowchart, which summarizes the flow of information in this hybrid method.

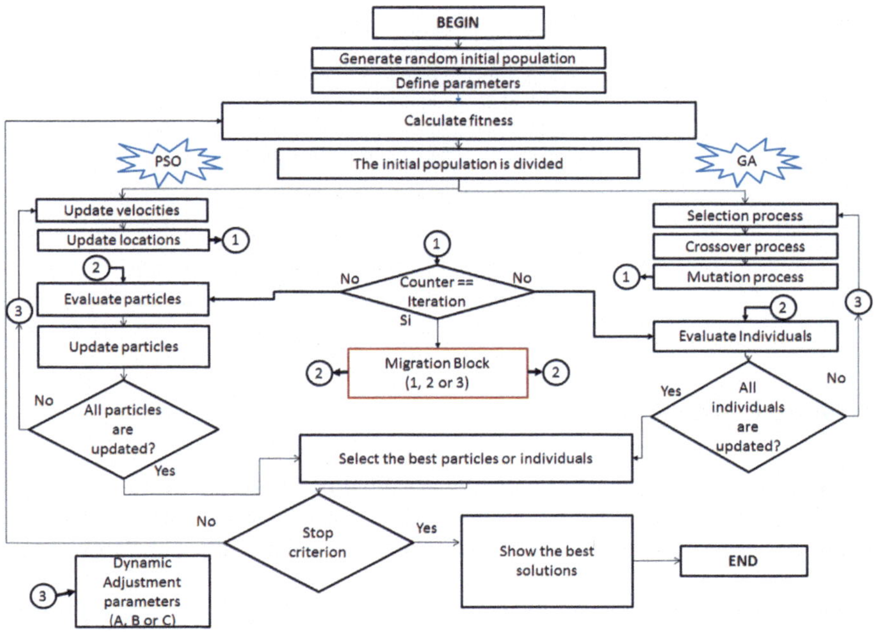


**Fig. 1** Flowchart of parallel PSO-GA algorithm

On each of the algorithms the use of fuzzy logic is made to provide a dynamic parameter setting and thus have more opportunities to obtain the best solution aimed for each of the used benchmark functions.

The use of the migration block is done every certain number of iterations and within it a comparison of the best global result of the two used algorithms is made. Once this is obtained, an exchange of a single particle of the winning algorithm towards the loser is made.

## 5 Fuzzy Logic Systems

It is based on the relative truth of the observed phenomena, this type of logic not only takes two truth values, but many values contextualized and referred to each other. For example, the height of a person, where a person is two meters tall, is clearly tall, and in the case that the person is one meter tall, he would be a short person. The two values are oriented to people and refers to a linear metric measure.

Fuzzy logic allows our expressions to be adapted to the problem we are working on, for example, the expressions could be "Low, Medium or High" and be a reference within a range of defined numbers like "Low is near to zero, High is near to one and the Medium is between Low and High" [16, 17].

Type-1 and interval type-2 fuzzy logic systems are used within the PSO-GA algorithm, which consists of two cases, one that has the triangular membership functions and the other one has Gaussians, in which, each one has as input the number of iterations and as output has the parameters that were selected to be dynamically adjusted in each of the algorithms [5, 14, 18–20].

Table 3 shows the output parameters with their acronyms.

The representation of variables in Table 3 is the one used in Table 4 for describing the fuzzy rules that are used.

Figure 2 shows the composition of one input and six outputs of the type-1 and fuzzy logic system. This shows that they are triangular membership functions although Gaussian membership functions were also used.

**Table 3** The fuzzy system output parameter are described

| Variable description | Variable representation |
|---|---|
| Iterations | Itera |
| Cognitive parameter | CoPar |
| Social parameter | SoPar |
| Inertial weight | InWei |
| Selection percentage | Select |
| Crossover percentage | Cross |
| Mutation percentage | Mut |

**Table 4** Rules used in fuzzy systems

| Rules | Input | Outputs | | | | | |
|---|---|---|---|---|---|---|---|
| | Itera | CoPar | SoPar | InWei | Select | Cross | Mut |
| 1 | Low | High | Low | High | Low | High | High |
| 2 | Mean | Mean | Mean | Mean | Mean | Mean | Mean |
| 3 | High | Low | High | Low | High | Low | Low |

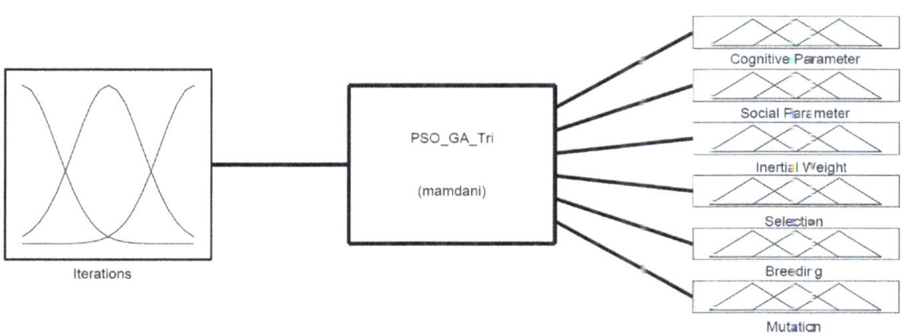


**Fig. 2** Type-1 fuzzy logic system

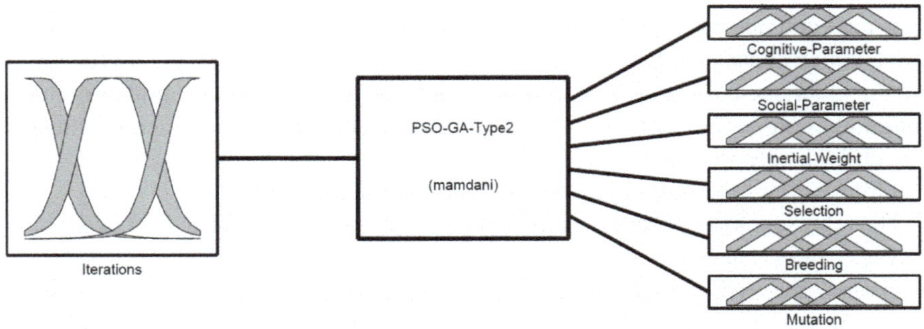


**Fig. 3** Interval type-2 fuzzy logic system

In Fig. 3, it is observed that the composition is similar to Fig. 2, but in this case the interval type-2 fuzzy system is used, which adds an uncertainty value to the membership functions.

Table 4 that includes the rules that are used within the fuzzy systems.

## 6 Experiments and Results

Each of the tables and figures that will be shown in this section are focused on each of the benchmark functions that were used in the experimentation.

They consist of the first column with the number of dimensions with which we are working, in the second column we have the result of the PSO-GA algorithm without the use of fuzzy systems, and in the next four columns we have the fuzzy logic system either type-1 and interval type-2 with Triangular and Gaussian membership functions (four cases). It is also clear that each of the results are averages of 30 runs of the PSO-GA algorithm with the parameters described above.

**Table 5** Parameters used in the PSO-GA algorithm

| PSO-GA parameters | Without dynamic parameter adjustment | T1 and interval T2 fuzzy systems |
|---|---|---|
| Population | 100 | 100 |
| Dimensions | 5, 10, 20, 40, 80, 160 | 5, 10, 20, 40, 80, 160 |
| Max iterations | Dimensions * 100 | Dimensions * 100 |
| Cognitive parameter | 0.6 | Dynamic |
| Social parameter | 0.3 | Dynamic |
| Inertial weight | 0.5 | Dynamic |

Table 5, describes the values that each of the necessary parameter values to be able to execute the algorithm with each of the methods.

Table 6 shows the acronyms that will be used to shorten the name of each of the methods used in the figures of the performed experiments.

In Fig. 4, it is shown that the results obtained by each of the methods used are increasing according to the number of dimensions used, the method where it shows the highest values are for the case where fuzzy dynamic parameters were not used. To corroborate these results, Table 7 was included with the results.

**Table 6** Acronyms shown in each results figures

| Methods used | Acronyms of methods used |
|---|---|
| PSO-GA without fuzzy logic | No fuzzy |
| PSO-GA with type-1 fuzzy logic with triangular membership functions | T1-TRI |
| PSO-GA with type-1 fuzzy logic with gaussian membership functions | T1-GAUS |
| PSO-GA with interval type-2 fuzzy logic with triangular membership functions | T2-TRI |
| PSO-GA with interval type-2 fuzzy logic with gaussian membership functions | T2-GAUS |

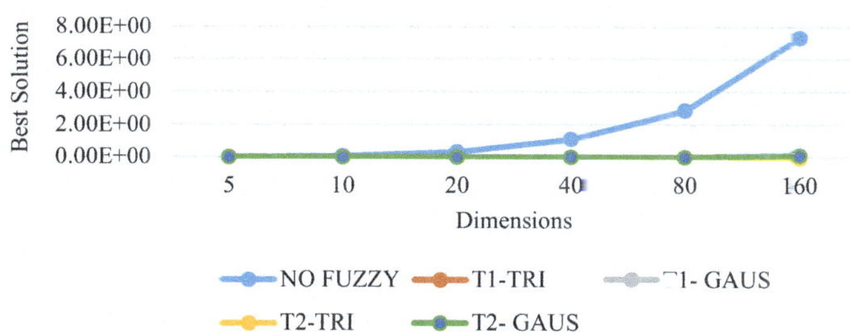


**Fig. 4** Comparison of average runtimes of genetic algorithm for 2 dimensions

**Table 7** Average results of PSO-GA with different dynamic adjustment parameters by benchmark function 1

| Dims | No fuzzy | T1-TRI | T1- GAUS | T2-TRI | T2- GAUS |
|---|---|---|---|---|---|
| 5 | 1.19E-03 | 0 | 0 | 0 | 0 |
| 10 | 6.42E-02 | 0 | 0 | 0 | 0 |
| 20 | 3.24E-01 | 0 | 0 | 0 | 0 |
| 40 | 1.111 | 0 | 0 | 0 | 0 |
| 80 | 2.890 | 0 | 0 | 0 | 1.28E-04 |
| 160 | 7.329 | 0 | 0 | 8.20E-04 | 1.11E-01 |

Figure 5, illustrates the comparison of the results of the benchmark functions. It is shown in the same way as in the previous case that the values obtained without using fuzzy systems are too high compared to the other methods as is shown in Fig. 6.

In Table 8, the results are shown in numerical form to be able to see exactly how much each of the methods produced. And as we can see there, the results without a fuzzy system are 572.845 working with 160 dimensions and with the other methods when much closer to 5 using T2-GAUS.

In Fig. 7, the improvement between the methods is not so significant because they all seem to be zeros.

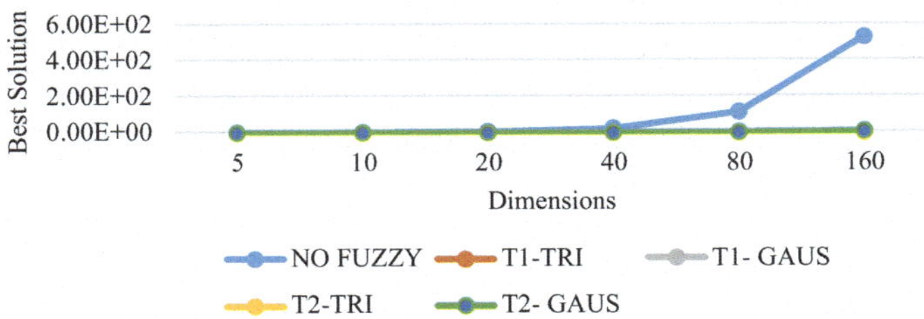


**Fig. 5** Comparison of each methods used with the variation of dimensions for the benchmark function 2

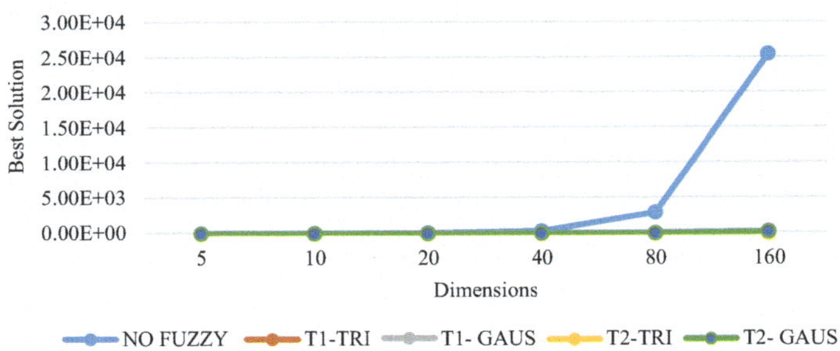


**Fig. 6** Comparison of each methods used with the variation of dimensions for the benchmark function 3

**Table 8** Average results of PSO-GA with different dynamic adjustment parameters by benchmark function 2

| Dims | NO FUZZY | T1-TRI | T1- GAUS | T2-TRI | T2- GAUS |
|---|---|---|---|---|---|
| 5 | 5.34E-03 | 0 | 0 | 0 | 0 |
| 10 | 2.13E-01 | 0 | 0 | 0 | 0 |
| 20 | 2.804 | 0 | 0 | 0 | 0 |
| 40 | 20.269 | 0 | 0 | 0 | 0 |
| 80 | 111.287 | 0 | 0 | 0 | 1.16E-02 |
| 160 | 527.845 | 2.61E-01 | 8.49E-01 | 1.093 | 4.184 |

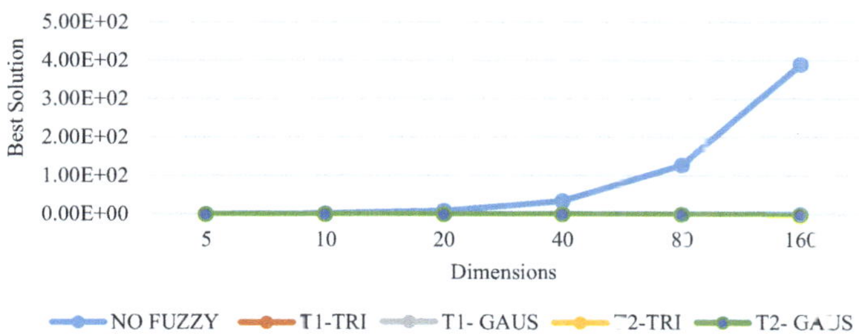


**Fig. 7** Comparison of each methods used with the variation of dimensions for the benchmark function 4

**Table 9** Average results of PSO-GA with different dynamic adjustment parameters by benchmark function 3

| Dims | No fuzzy | T1-TRI | T1-GAUS | T2-TRI | T2-GAUS |
|---|---|---|---|---|---|
| 5 | 7.23E-03 | 0 | 0 | 0 | 0 |
| 10 | 8.56E-01 | 0 | 0 | 0 | 0 |
| 20 | 16.133 | 0 | 0 | 0 | 0 |
| 40 | 268.722 | 0 | 0 | 0 | 4.29E-03 |
| 80 | 2884.409 | 4.33E-01 | 0 | 6.87E-03 | 5.046 |
| 160 | 25,511.330 | 20.522 | 24.741 | 49.140 | 201.118 |

But in Table 9 it is observed that the improvement obtained between not using and using dynamic parameter adjustment in between 100 and 1000 times better, when we are working with 160 dimensions.

In Fig. 7, for benchmark function number 4 all methods where fuzzy systems are used, an absolute zero value was obtained.

As shown in Table 10, only the column corresponding to that without fuzzy system gives numbers smaller than 400 working 160 dimensions.

In Fig. 8, there are small imperceptible variations when working with 20 or less dimensions, otherwise changes are shown between the methods in the graph.

**Table 10** Average results of PSO-GA with different dynamic parameters adjustment by benchmark function 4

| Dims | No fuzzy | T1-TRI | T1-GAUS | T2-TRI | T2-GAUS |
|---|---|---|---|---|---|
| 5 | 9.83E-03 | 0 | 0 | 0 | 0 |
| 10 | 1.434 | 0 | 0 | 0 | 0 |
| 20 | 8.592 | 0 | 0 | 0 | 0 |
| 40 | 33.748 | 0 | 0 | 0 | 0 |
| 80 | 127.746 | 0 | 0 | 0 | 0 |
| 160 | 389.120 | 0 | 0 | 0 | 0 |

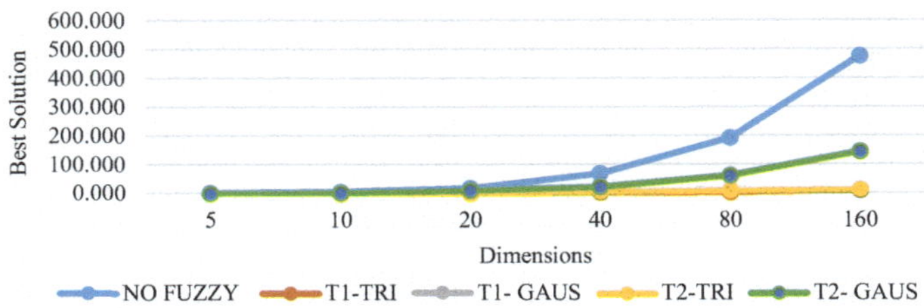


**Fig. 8** Comparison of each methods used with the variation of dimensions for the benchmark function 5

**Table 11** Average results of PSO-GA with different dynamic parameters adjustment by benchmark function 5

| Dims | No fuzzy | T1-TRI | T1-GAUS | T2-TRI | T2-GAUS |
|---|---|---|---|---|---|
| 5 | 1.358 | 3.32E-02 | 1.94E-02 | 0 | 4.90E-01 |
| 10 | 5.893 | 3.73E-02 | 4.75E-01 | 4.06E-01 | 2.691 |
| 20 | 17.832 | 1.996 | 3.416 | 1.374 | 10.982 |
| 40 | 69.152 | 2.848 | 3.867 | 6.232 | 23.431 |
| 80 | 194.129 | 4.674 | 11.296 | 10.217 | 62.588 |
| 160 | 478.119 | 9.199 | 7.614 | 13.412 | 146.611 |

On the other hand in Table 11, it can be seen that in some cases it could be said that they are close to zero if we compare them in the first column, but if we compare them to all the best ones, they are the columns that refer to the method using Type-1 Fuzzy logic systems.

In Fig. 9, it is shown that the results of all columns are increasing for increasing dimensions, not as in the other figures where only the non-fuzzy column had increasing values.

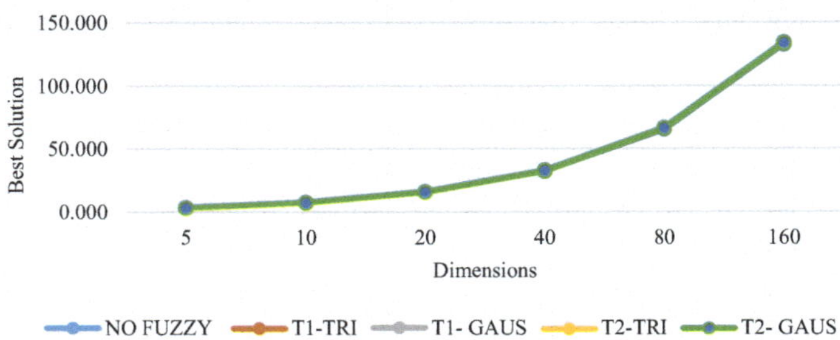


**Fig. 9** Comparison of each methods used with the variation of dimensions for the benchmark function 6

Although in the case of Table 12 it is observed that the values are very similar except for the non-fuzzy case where from 20 dimensions onwards it shows by very little that it is better than the others.

In Fig. 10, when we are working with benchmark function 7 and it is shown that when using dynamic parameter adjustment it is beneficial to reach the target or as near possible to it.

Referring to Fig. 10 it is Table 13, which shows the numerical values of the figure, where it is seem that the values obtained where non-fuzzy are very close to zero, but similarly are greater than in the other cases.

In Fig. 11, a strange behavior is observed in the line that corresponds to non-fuzzy, since when we are working with 80 dimensions or less the values decrease, although it is average of 30 runs per method and only happens in benchmark function 8.

Although in Fig. 11 a great difference is noted, here in Table 14 we can note that the difference is very small.

Figure 12 shows the comparison of the time it took to run each of the methods used for benchmark function 1 with different numbers of dimensions.

These times mentioned above were obtained with the following features of the machine that was used in the experiments, it was an Intel core i7-6700 with 4-cores and 16 gigabytes of RAM.

**Table 12** Average results of PSO-GA with different dynamic parameters adjustment by benchmark function 6

| Dims | No fuzzy | T1-TRI | T1-GAUS | T2-TRI | T2-GAUS |
|---|---|---|---|---|---|
| 5 | 4.207 | 4.207 | 4.207 | 4.207 | 4.207 |
| 10 | 8.415 | 8.415 | 8.415 | 8.415 | 8.415 |
| 20 | 16.767 | 16.829 | 16.829 | 16.829 | 16.829 |
| 40 | 32.987 | 33.659 | 33.659 | 33.659 | 33.659 |
| 80 | 65.895 | 67.318 | 67.318 | 67.318 | 67.318 |
| 160 | 133.177 | 134.635 | 134.635 | 134.635 | 134.635 |

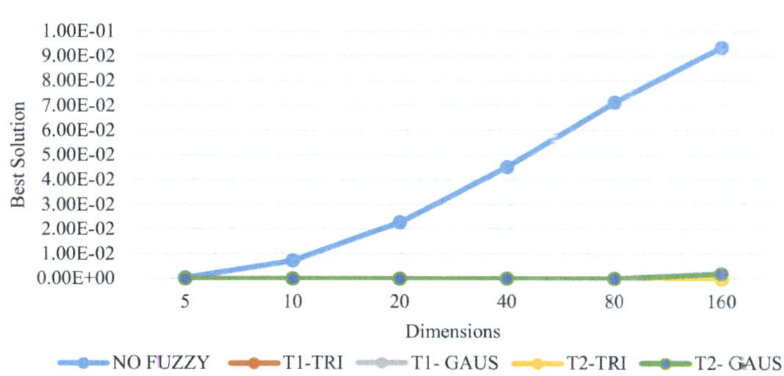


**Fig. 10** Comparison of each methods used with the variation of dimensions for the benchmark function 7

**Table 13** Average results of PSO-GA with different dynamic parameters adjustment by benchmark function 7

| Dims | No fuzzy | T1-TRI | T1-GAUS | T2-TRI | T2-GAUS |
|---|---|---|---|---|---|
| 5 | 4.71E-04 | 0 | 0 | 0 | 0 |
| 10 | 7.29E-03 | 0 | 0 | 0 | 0 |
| 20 | 2.28E-02 | 0 | 0 | 0 | 0 |
| 40 | 4.51E-02 | 0 | 0 | 0 | 0 |
| 80 | 7.13E-02 | 0 | 0 | 0 | 1.11E-05 |
| 160 | 9.35E-02 | 0 | 1.01E-06 | 2.98E-05 | 1.97E-03 |

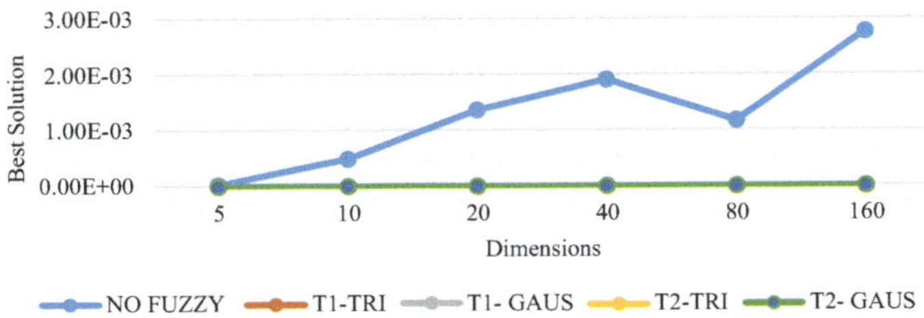


**Fig. 11** Comparison of each methods used with the variation of dimensions for the benchmark function 8

**Table 14** Average results of PSO-GA with different dynamic parameters adjustment by benchmark function 8

| Dims | No fuzzy | T1-TRI | T1-GAUS | T2-TRI | T2-GAUS |
|---|---|---|---|---|---|
| 5 | 2.61E-05 | 0 | 0 | 0 | 0 |
| 10 | 4.95E-04 | 0 | 0 | 0 | 0 |
| 20 | 1.37E-03 | 0 | 0 | 0 | 0 |
| 40 | 1.92E-03 | 0 | 0 | 0 | 9.53E-33 |
| 80 | 1.18E-03 | 0 | 1.37E-35 | 2.95E-16 | 1.28E-20 |
| 160 | 2.77E-03 | 3.07E-22 | 4.33E-17 | 9.73E-11 | 6.83E-16 |

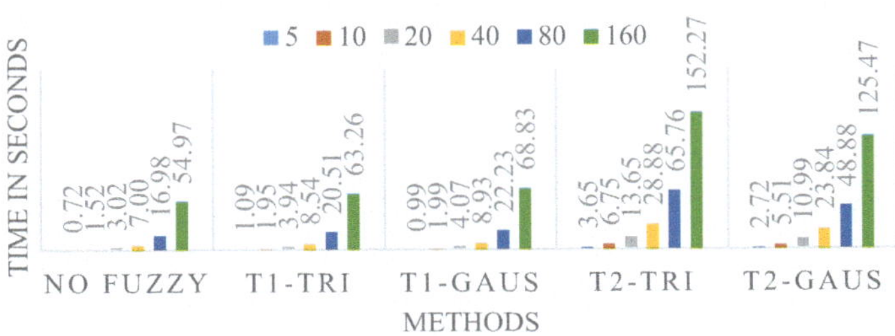


**Fig. 12** Comparison of the average run times for the benchmark function 1

Each of the times shown depends on the function being evaluated, as well as the number of dimensions and the number of iterations (which is the number of dimensions multiplied by one hundred). We had already worked with a time comparison of these two algorithms, but it was separate, and the number of iterations that had been used before was only of 500 regardless of the number of dimensions [21, 22].

## 7 Conclusions

After the experiments were carried out, we can conclude that the use of dynamic parameter adjustment allows us to reach the global minimum required by each of the benchmark functions that we use, well except for only one function. The other interesting fact is that the use of fuzzy logic systems seemed to be a better option for the difficulty of these functions as well as the amount of dimensions we use.

The use of the fuzzy system of one input and six outputs can be confusing, but within it includes all the necessary parameters. On the other hand, it can be said that it saves time when the file is read since we only have to read a single file in comparison of having to read six files for each of the output parameters shown.

The run times shown are only of one function although among the other functions they vary a little, what matters in this demonstration of time is what we can see that although it takes a slightly longer time, better results can be obtained.

As a future work, we intend to work with more complex benchmark functions, as well as a higher number of dimensions to actually find which of the used methods is more robust at the time of dynamic parameter adjustment to reach the expected global minimum values.

As well as testing other optimization algorithms and we can work on improving run times with the use of new technologies in parallel. such as the use of Nvidia CUDA Architecture. In addition, we could consider other types of applications, like in [23–30].

**Acknowledgments** The authors would like to thank CONACYT and Tecnológico Nacional de Mexico/Tijuana Institute of Technology for the support during this research work.

## References

1. Eberhart, Y. Shi, Particle swarm optimization: developments, applications and resources, in *Proceedings of the 2001 Congress on Evolutionary Computation (IEEE Cat. No.01TH8546)*, vol. 1 (Seoul, South Korea, 2001), pp. 81–86. https://doi.org/10.1109/cec.2001.934374
2. J. Kennedy, Particle swarm optimization, in *Encyclopedia of Machine Learning*. ed. by C. Sammut, G.I. Webb (Springer US, Boston, MA, 2010), pp. 760–766
3. F. Olivas, F. Valdez, and O. Castillo, Particle swarm optimization with dynamic parameter adaptation using interval type-2 fuzzy logic for benchmark mathematical functions, in *2013*

*World Congress on Nature and Biologically Inspired Computing* (Fargo, ND, USA, 2013), pp. 36–40. https://doi.org/10.1109/nabic.2013.6617875
4. Y. Zhou, Y. Tan, GPU-based parallel particle swarm optimization, in *2009 IEEE Congress on Evolutionary Computation* (Trondheim, Norway, 2009), pp. 1493–1500. https://doi.org/10.1109/cec.2009.4983119
5. J.C. Vázquez, F. Valdez, Fuzzy logic for dynamic adaptation in PSO with multiple topologies, in *Joint IFSA World Congress and NAFIPS Annual Meeting, IFSA/NAFIPS 2013* (Edmonton, Alberta, Canada, 24–28 June 2013), pp. 1197–1202. https://doi.org/10.1109/ifsa-nafips.2013.6608571
6. J. Carnahan, R. Sinha, Nature's algorithms [genetic algorithms]. IEEE Potentials **20**(2), 21–24 (2001). https://doi.org/10.1109/45.954644
7. J.G. Digalakis, K.G. Margaritis, An experimental study of benchmarking functions for genetic algorithms. Int. J. Comput. Math. **79**(4), 403–416 (2002). https://doi.org/10.1080/00207160210939
8. G.R. Harik, F.G. Lobo, D.E. Goldberg, The compact genetic algorithm. IEEE Trans.Evolut. Comput. **3**(4), 11 (1999)
9. H. Pohlheim, GEATbx: genetic and evolutionary algorithm toolbox for use with MATLAB version 1.92. User Guide, July **1**, 997 (1998)
10. J. Xuan, H. Jiang, Z. Ren, Pseudo Code of Genetic Algorithm and Multi-Start Strategy Based Simulated Annealing Algorithm for Large Scale Next Release Problem, p. 3
11. A.S. Radhamani, E. Baburaj, Performance evaluation of parallel genetic and particle swarm optimization algorithms within the multicore architecture. Int. J. Computat. Intell. Appl. **13**(4) (2014). https://doi.org/10.1142/s1469026814500242
12. F. Valdez, P. Melin, O. Castillo, Parallel particle swarm optimization with parameters adaptation using fuzzy logic, in *Advances in Computational Intelligence*, pp. 374–385 (2013)
13. Z. Dingxue, G. Zhihong, L. Xinzhi, On multi-population parallel particle swarm optimization algorithm, in *2007 Chinese Control Conference* (2007), pp. 763–765. https://doi.org/10.1109/chicc.2006.4347299
14. F. Valdez, P. Melin, O. Castillo, Fuzzy control of parameters to dynamically adapt the PSO and GA Algorithms, in *International Conference on Fuzzy Systems* (Barcelona, Spain, 2010), pp. 1–8. https://doi.org/10.1109/fuzzy.2010.5583934
15. X.H. Shi, Y.H. Lu, C.G. Zhou, H.P. Lee, W.Z. Lin, Y.C. Liang, Hybrid evolutionary algorithms based on PSO and GA, in *The 2003 Congress on Evolutionary Computation, 2003. CEC '03.*, 2003, vol. 4, pp. 2393–2399. https://doi.org/10.1109/cec.2003.1299387
16. O. Castillo et al., A high-speed interval type 2 fuzzy system approach for dynamic parameter adaptation in metaheuristics. Eng. Appl. Artif. Intell. **85**, 666–680 (2019). https://doi.org/10.1016/j.engappai.2019.07.020
17. J. Perez, F. Valdez, O. Castillo, O. Roeva, Bat algorithm with parameter adaptation using Interval Type-2 fuzzy logic for benchmark mathematical functions, in *2016 IEEE 8th International Conference on Intelligent Systems (IS)* (2016), pp. 120–127. https://doi.org/10.1109/is.2016.7737409
18. F. Olivas, F. Valdez, O. Castillo, P. Melin, Dynamic parameter adaptation in particle swarm optimization using interval type-2 fuzzy logic. Soft. Comput. **20**(3), 1057–1070 (2016). https://doi.org/10.1007/s00500-014-1567-3
19. F. Valdez, P. Melin, O. Castillo, Fuzzy control for dynamical parameter adaptation in a parallel evolutionary method combining particle swarm optimization and genetic algorithms, in *Soft Computing for Intelligent Control and Mobile Robotics* (2011), pp. 161–178
20. F. Olivas, L. Amador-Angulo, J. Pérez, C. Caraveo, F. Valdez, O. Castillo, Comparative study of type-2 fuzzy particle swarm, bee colony and bat algorithms in optimization of fuzzy controllers. Algorithms **10**(3), 101 (2017). https://doi.org/10.3390/a10030101
21. Y. Kawano, F. Valdez, O. Castillo, Performance evaluation of optimization algorithms based on GPU using CUDA architecture, in *2018 IEEE Latin American Conference on Computational Intelligence (LA-CCI)* (2018), pp. 1–6. https://doi.org/10.1109/la-cci.2018.8625236

22. Y. Kawano, F. Valdez, O. Castillo, Evaluation of parallel exploration and exploitation capabilities in two PSO variants with intra communication, in *Intuitionistic and Type-2 Fuzzy Logic Enhancements in Neural and Optimization Algorithms: Theory and Applications*, ed. by O. Castillo, P. Melin, J. Kacprzyk (Springer, Cham, 2020), pp. 169–184
23. M.A. Sanchez, O. Castillo, J.R. Castro, P. Melin, Fuzzy granular gravitational clustering algorithm for multivariate data. Inf. Sci. **279**, 498–511 (2014)
24. D. Sánchez, P. Melin, Optimization of modular granular neural networks using hierarchical genetic algorithms for human recognition using the ear biometric measure. Eng. Appl. Artif. Intell. **27**, 41–56 (2014)
25. M.A. Sanchez, J.R. Castro, O. Castillo, O. Mendoza, A. Rodriguez-Diaz, P. Melin, Fuzzy higher type information granules from an uncertainty measurement. Granul. Comput. **2**(2), 95–103 (2017)
26. O. Castillo, *Type-2 Fuzzy Logic in Intelligent Control Applications* (Springer, 2012)
27. P. Melin, O. Castillo, Intelligent control of complex electrochemical systems with a neuro-fuzzy-genetic approach. IEEE Trans. Indus. Electron. **48**(5), 951–955
28. L. Aguilar, P. Melin, O. Castillo, Intelligent control of a stepping motor drive using a hybrid neuro-fuzzy ANFIS approach. Appl. Soft Comput. **3**(3), 209–219 (2003)
29. P. Melin, O. Castillo, Adaptive intelligent control of aircraft systems with a hybrid approach combining neural networks, fuzzy logic and fractal theory. Appl. Soft Comput. **3**(4), 353–362 (2003)
30. L. Rodriguez, O. Castillo, J. Soria, P. Melin, F. Valdez, C. Gonzalez, G. Martinez, J. Soto, A fuzzy hierarchical operator in the grey wolf optimizer algorithm. Appl. Soft Comput. **57**, 315–328 (2017)

# A Novel Study of the Multi-verse Optimizer and Its Applications on Multiple Areas of Computer Science

**Lucio Amézquita, Oscar Castillo, José Soria, and Prometeo Cortes-Antonio**

**Abstract** The purpose of this paper is to study the applications of the metaheuristic multi-verse optimizer in a variety of problems involving optimization in the distinct areas of computer science, keeping a focus on fuzzy logic. This is a state-of-the-art study, so it is only showing different cases where the multi-verse optimizer has been implemented and showed positive and negative results (mostly positive), so we can conclude that it can be applied to fuzzy control with positive results in further investigations. In this review, we can observe different population-based optimization algorithms in various applications, and many of them come from other colleagues that apply these algorithms for different problems, so we can compare with the applications of MVO over his short time along other algorithms. The paper shows multiple applications of many known metaheuristics such as PSO, GWO, HSA; and has nine applications of the MVO in different works.

**Keywords** Multi-verse optimizer · Particle swarm optimization · Fuzzy logic · Study · Optimization

## 1 Introduction

Today is more common to see the use of computational techniques in distinct areas such as automatization of processes in the industry, or more accordingly to Computer Science. One of the major applications is the optimization or improvement of different activities like a manufacture process.

---

L. Amézquita · O. Castillo (✉) · J. Soria · P. Cortes-Antonio
Tijuana Institute of Technology, Tijuana, Mexico
e-mail: ocastillo@tectijuana.mx

L. Amézquita
e-mail: lucio.amezquita19@tectijuana.edu.mx

P. Cortes-Antonio
e-mail: prometeo.cortes@tectijuana.edu.mx



P. Melin et al. (eds.), *Recent Advances of Hybrid Intelligent Systems Based on Soft Computing*, Studies in Computational Intelligence 915,
https://doi.org/10.1007/978-3-030-58728-4_7

One of the main areas of optimization is control, like on fuzzy logic, where fuzzy controllers have diverse applications and these techniques have similarities in the main reasoning of people to solve these problems, because they use linguistic variables.

The metaheuristics are stochastic methods where the optimization is their main function, so they can find solutions for a specific problem or problems in conjunction [1]. These metaheuristics have different bio-inspirations or artificial inspirations where they can demonstrate if are capable of bringing good solutions to a specific problem.

There is a wide variety of bio-inspired algorithms that help us solve these many problems such as Particle Swarm Optimization (PSO) [1], Grey Wolf Optimizer (GWO) [2], Flower Pollination Algorithm (FPA) [3]; were the main objective is to find the best solution.

In this state-of-the-art study we are presenting some of the recent bio-inspired algorithms, and also different applications of the Multi-verse optimizer (MVO) since its introduction in 2015 and its possible implementation on fuzzy control problems. The main objective of this study is to analyze the possibilities of the algorithm in the wide variety of situations, mainly for applications of control and fuzzy systems.

This paper is organized as follows: Sect. 2 describes the Multi-verse Optimizer algorithm and his equations, Sect. 3 describes some applications of their bio-inspired algorithms, Sect. 4 describes nine cases of the MVO algorithm and Sect. 5 has the conclusions to this work.

## 2 Multi-verse Optimizer Algorithm

The metaheuristics have distinct inspirations over nature and artificial behavior, in optimization techniques we can observe two main modes [4], single-based and population-based solutions. In the single-based solution, algorithm starts with a random solution and evolves it in various iterations; one mayor flaw is to stagnate on a local optimum. On the other side, population-based start with multiple random solutions and they improve over iterations; they are majorly inspired in nature and try to imitate the particular behavior of multiple beings or objects.

In this study, we have the Multi-verse Optimizer (MVO) [5] algorithm proposed by Seyedali Mirjalili in 2015. It has inspiration on the big bang theory, and three concepts of cosmology: the white hole, the black hole and the wormhole. The idea comes from the creation of the universe and multiple instances on the same time-line but different space, so they can interact thanks to the holes they create (Fig. 1).

The idea of this algorithm is to move objects between the different universes that represent each one a possible solution to the problem, the way they interact is with the white holes and black holes, and to add a random change to the solution it comes the wormhole, so it can interact between the same universe or other universes.

The holes have some rules so they can be present in the universes: a higher inflation rate (this represents fitness of the universe), a higher probability of having

**Fig. 1** White hole, black hole and wormhole [5]

white holes and a lower probability to having black holes; universes with a higher inflation rate tend to send objects thought white holes and with lower inflation rate tend to receive more objects through black holes; and objects in all universes may face random movements towards the best universe via wormholes [6].

For the representation of the algorithm, it can be assumed that $U$ is the set of universes of solutions to the problem, where $d$ represents the number of parameters or variables and $n$ is the number of universes or solution candidates:

$$U = \begin{bmatrix} x_1^1 & x_1^2 & \cdots & x_1^d \\ x_2^1 & x_2^2 & \cdots & x_2^d \\ \vdots & \vdots & \vdots & \vdots \\ x_n^1 & x_n^2 & \cdots & x_n^d \end{bmatrix}$$

Each parameter is represented as follows:

$$x_i^j = \begin{cases} x_k^j & r1 < NI(U_i) \\ x_i^j & r1 \geq NI(U_i) \end{cases} \quad (1)$$

where $x_i^j$ indicates the $j$th parameter of the $i$th universe, $U_i$ represents the $i$th universe, $NI(U_i)$ is the normalized inflation rate of the $i$th universe, $r1$ is a random number between [0, 1], and $x_k^j$ indicates the $j$th parameter of the $k$th universe selected by a roulette wheel.

The Pseudocode for the black and white hole mechanism, where $-NI$ is used for minimization problems, if it is for maximization $NI$ is used (Fig. 2).

As we analyzed in the previous algorithm, here the solutions are explored through the black and white holes, if exploitation needs to be reached, each universe must present the wormholes to transport objects randomly. To represent the wormholes, it is done as follows:

```
SU=Sorted universes
NI=Normalize inflation rate (fitness) of the universes
for each universe indexed by i
    Black_hole_index=i;
    for each object indexed by j
        r1=random([0,1]);
        if r1<NI(Ui)
            White_hole_index= RouletteWheelSelection(-NI);
            U(Black_hole_index,j)= SU(White_hole_index,j);
        end if
    end for
end for
```

**Fig. 2** Pseudocode for white hole and black hole mechanism

$$x_i^j = \begin{cases} \begin{cases} X_j + TDR \times \left(\left(ub_j - lb_j\right) \times r4 + lb_j\right) & r3 < 0.5 \\ X_j - TDR \times \left(\left(ub_j - lb_j\right) \times r4 + lb_j\right) & r3 \geq 0.5 \end{cases} & r2 < WEP \\ x_i^j & r2 \geq WEP \end{cases} \tag{2}$$

where $X_j$ indicates the *j*th parameter of the best universe so far, *TDR* and *WEP* are coefficients named Travel Distance Rate and Wormhole Existence Probability respectively, $lb_j$ shows the lower limit of the *j*th variable, $ub_j$ is the upper limit of the *j*th variable, $x_i^j$ indicates the *j*th parameter of the *i*th universe, and $r2, r3, r4$ are random numbers between [0, 1]. Given this equation, the wormhole pseudocode is as follows (Fig. 3).

In the case of *TDR* and *WEP*, their behavior during the iteration's changes from the initial value, to promote exploration, and then, exploitation. A clear difference between these coefficients is that the TDR increases with the passage of the iterations to have better exploitation of the solution. The adaptive formula for both coefficients is as follows:

```
for each universe indexed by i
    for each object indexed by j
        r2=random([0,1]);
        if r2<Wormhole_existance_probability
            r3= random([0,1]);
            r4= random([0,1]);
            if r3<0.5
                U(i,j)=Best_universe(j) + Travelling_distance_rate
            * (( ub(j) - lb(j)) * r4 + lb(j));
            else
                U(i,j)=Best_universe(j) - Travelling_distance_rate
            * (( ub(j) - lb(j)) * r4 + lb(j));
            end if
        end if
    end for
end for
```

**Fig. 3** Pseudocode for wormhole mechanism

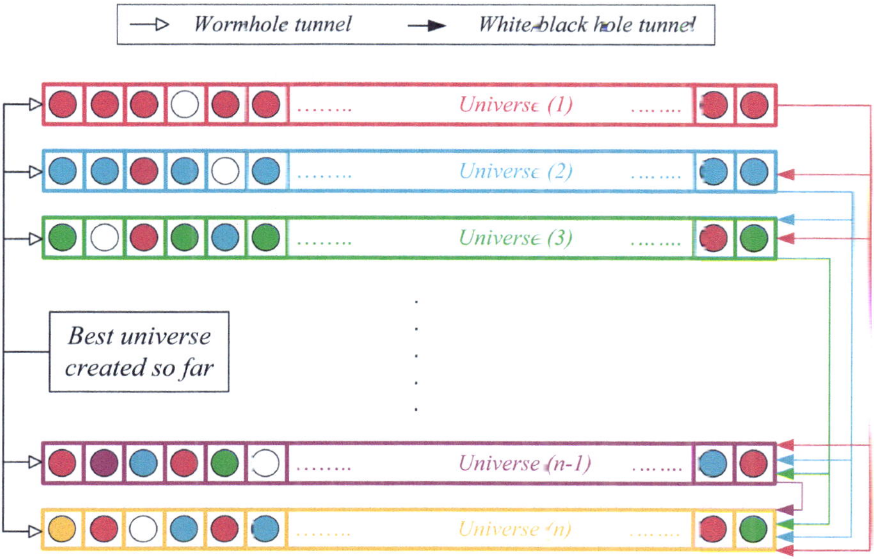


**Fig. 4** Representation of the MVO algorithm [5]

$$WEP = min + l \times \left(\frac{max - min}{L}\right) \tag{3}$$

$$TDR = 1 - \frac{l^{1/p}}{L^{1/p}} \tag{4}$$

where *min* is the minimum, *max* is the maximum, *l* indicates the current iteration, and *L* is the maximum of iterations. In the case of *TDR*, *p* is the precision of the exploitation in the iterations; the higher the *p* value, the faster and more precise the local search or exploitation.

A way to see the universes in this algorithm is shown in Fig. 4, where the white and black holes move objects between universes to promote exploration towards the universes with a higher inflation rate. The white dots represent the objects the wormholes are transporting to the best universe so far so it can promote exploitation and the search of the global optima. A view of the flowchart of MVO is shown in Fig. 5.

## 3 Other Optimization Algorithms

Some of the most applied optimization algorithms start with the ones in Swarm Intelligence, originated from the study of colonies, these studies of the social behavior of

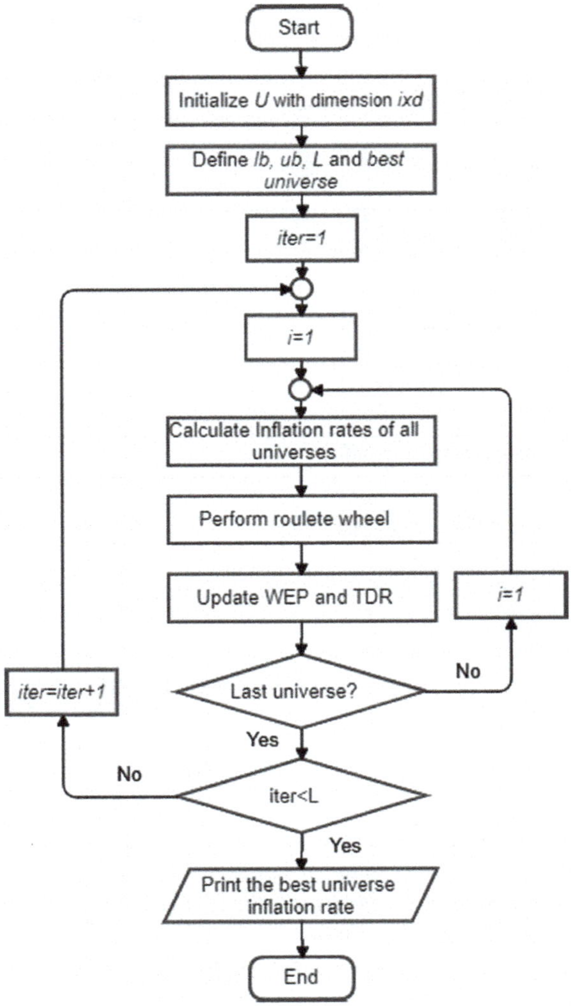


**Fig. 5** General flowchart of MVO

organisms (individuals) in swarms prompted the design of very efficient optimization and clustering algorithms such as Particle Swarm Optimization (PSO) [1].

PSO is a stochastic optimization approach, modeled on the social behavior of bird flocks that is population-based where the individuals, referred to as particles, are grouped into a swarm where each particle in the swarm represents a candidate solution, each particle "flies" through the search space, adjusting its position in search space according to its own experience and that of neighboring particles. The effect is that particles "fly" toward an optimum, while still searching a wide area around the current best solution. The performance of each particle is measured according to a predefined fitness function which is related to the problem being solved.

Thanks to the nature of this algorithm, PSO is widely used along with other optimization algorithms, because of its nature inspiration of how it searches the solutions in swarm.

In this case, the GWO has been used by several scientists, like the case of Hernández et al. [7], where the algorithm is used alongside a fuzzy controller for an autonomous mobile robot to optimize the controller, so the robot can go through a route with the minimum error in the trajectory.

Another application of metaheuristics is the Harmony Search Algorithm (HSA) that was created by Geem [8], this algorithm uses three operators for its functionality that are harmony memory accepting(HMR), the pitch adjustment and randomization. These operators are modeled in some simple equations that have a range of [0, 1].

The HSA is also widely used in different applications, were we can focus on a recent work of Peraza et al. [9], where they use HAS with a dynamic adaptation of parameters on various control problems, like the autonomous mobile robot; applying the algorithm to the fuzzy controller to adjust the membership functions parameters.

If we go to the case of population-based algorithms, we can see the Imperialist Competitive Algorithm (ICA) that was proposed by Atashpaz-Gargari and Lucas [10], which has inspiration on imperialism, where the most powerful countries aspire to make a colony from the less powerful countries, and then, make part of the whole country. It has a basis of social-political progress, unlike other metaheuristics.

One application to ICA that we can mention, is the work of Bernal et al. [11], where they use the algorithm with a type-2 fuzzy system to adjust the decades of the ICA, so it can get better results from the optimization, leading to faster convergence of the algorithm over benchmark functions thus, having better performance than the original ICA.

Leading to another metaheuristic, is the Firefly Algorithm (FA) [12], which has inspiration over the behavior of the fireflies. This algorithm has three basic rules: fireflies can be attracted to any other firefly, the attractiveness depends on the brightness and their separation, the glow of the fireflies is done by a function of adequacy.

For this bio-inspired algorithm, In [13], it is used the FA to optimize some fuzzy controllers with type-1 and type-2 fuzzy logic, using the water level control in a tank and temperature control in a shower. They compare the results with both types of fuzzy logic and with optimization with the FA.

Another nature-inspired algorithm is the Flower Pollination Algorithm (FPA) proposed by Yang [3], his motivation comes from the natural process called pollination of flowers. There are four characteristics in this method: biotic and cross-pollination, abiotic pollination and self-pollination, pollinators such as insects, and interaction or switching of local pollination and global pollination.

Metaheuristics have been used in the many application areas, for example, a FPA was implemented to optimize a fuzzy controller for the autonomous mobile robot in [14], the fuzzy inference system had two inputs and two outputs to control the behavior of the robot following a trajectory.

## 4 Some Applications of MVO

As the main application in this section, is the case of the algorithm creator Mirjalili et al. [5], where the first tests were done with 19 benchmark functions, taken from various sources; and ones from CEC 2005 special session, were the dimension from these tests were done to 50, we can observe a sample in Table 1. The way to do a comparison between the tests was to run 30 times the algorithm and apply Wilcoxon's nonparametric statistical test, we can observe a sample of results in Table 2.

For the practical application of this model, it was used in five constrained real engineering problems; they have several inequality constraints, so the algorithm can handle them in the optimization. In the cases we can observe its application are the welded beam design so it can deal with the lowest fabrication cost, were it outperforms PSO, GA, and GSA; other case is the gear train design, where the objective is the minimization of gear ratio in 4 gears and it shows similar performance in comparison with MBA [15]; continuing with cases comes the three-bar truss design, because of its popularity in meta-heuristics, were the objective is to design a truss with the minimum weight that satisfy some constraints(stress, deflection, buckling), where the results come very close to DEDS and PSO-DE; another case is the pressure vessel

**Table 1** Sample of benchmark functions for MVO tests

| Function | Dim | Range | $f_{min}$ |
|---|---|---|---|
| $F_1(x) = \sum_{i=1}^{n} x_i^2$ | 50 | [−100, 100] | 0 |
| $F_2(x) = \sum_{i=1}^{n} \lvert x_i \rvert + \prod_{i=1}^{n} \lvert x_i \rvert$ | 50 | [−10, 10] | 0 |
| $F_3(x) = \sum_{i=1}^{n} \left( \sum_{j-1}^{i} x_j \right)^2$ | 50 | [−100, 100] | 0 |
| $F_4(x) = \max_i \{\lvert x_i \rvert, 1 \le i \le n\}$ | 50 | [−100, 100] | 0 |
| $F_5(x) = \sum_{i=1}^{n-1} \left[ 100\left(x_{i+1} - x_i^2\right)^2 + (x_i - 1)^2 \right]$ | 50 | [−100, 100] | 0 |

**Table 2** Sample of test results in benchmark functions for MVO

| Function | MVO | GWO | GSA | PSO | GA |
|---|---|---|---|---|---|
| F1 | 2.08583 | 2319.19 | 2983.667 | 3.552364 | 27,187.58 |
| F2 | 15.92479 | 14.43166 | 10.96518 | 8.716272 | 68.6618 |
| F3 | 453.2002 | 7278.133 | 113,740.4 | 2380.963 | 48,530.91 |
| F4 | 3.123005 | 13.09729 | 32.2563 | 21.5169 | 62.99326 |
| F5 | 1272.13 | 3,425,462 | 7582.498 | 1132.486 | 65,361,620 |

design were the objective is the minimization of the fabrication cost of a vessel, were MVO outperforms all compared methods; the last case is the cantilever beam design were the objective is to minimize the weight of a cantilever beam with hollow square blocks, were the algorithm also outperform all compared algorithms.

With the appearance of the MVO algorithm, other scientists became to test out its promising results in a wide variety of cases. Later in the same year were MVO was published, Hu et al. [16] came with the idea of incorporating Levy flights to MVO (LFMVO), this because the original algorithm could stagnate when the wormholes re-span in the universes stochastically, so it comes to a local optimal; with the levy flights, the wormholes can be more precise in the case that it needs more exploration. The way it works is that, Levy flights update WEP and TDR of the algorithm, so it can affect the wormhole appearance on the new solutions.

The way that LFMVO was tested is using 23 benchmark functions, with 40 dimensions, and with 40 executions each one, the population size was set to 60 and 600 iterations. The comparisons were made with MVO, PSO, and MFO. Comparing with other algorithms, the objective of this application was to improve performance by taking less time to approach to a solution.

In order of appearance in the area of neural networks, Faris et al. [17] used the MVO algorithm to train feedforward neural networks; more specifically, a Multilayer Perceptron(MLP) neural network. In this case, each universe represents a NN configuration, consisting of three sections: weights between inputs and hidden layer, weights between hidden layer and output layer, and bias weights.

The main problem to solve in this case, is for classification, were it uses nine medical classification datasets (Blood, Breast cancer, Diabetes, etcetera). In means of comparison, the training of the MLP was done with GA, PSO, DE, FF and CS. The results come positive in comparison with the other training methods, coming to a robust approach comparing to the MSE of each one.

Hybridization is one of the main tests that many scientists do after first results on new algorithms, testing their behavior with other algorithms to perform better. This is the case that presents Jangir et al. [18] were it presents a new technique hybrid Particle Swarm Optimization (PSO)-Multi-verse Optimizer (MVO); this method has a fast convergence rate because it uses a roulette wheel selection method. The way it works is that the MVO replaces the Pbest value of PSO with the universe value of MVO.

The tests in PSO-MVO are done in 23 benchmark functions with 30 search agents and 500 iterations, outperforming original methods; in case of problems it works with 5 cases: Minimization of generation fuel cost, Voltage profile improvement, Voltage stability enhancement, Minimization of active power transmission losses and Minimization of reactive power transmission losses; in all cases, it has positive results except the last case, were the results are not essentially positive.

Coming in another application for engineering, the work of Karthikeyan and Dal [19] were MVO is used to stabilize voltage on an IEEE 57 bus. The voltage stability of a multi-machine system is evaluated with continuation power flow (CPF), this is done when the system undergoes unexpected overloading, affecting stability, the system is analyzed by CPF to improve this stability, here it comes the Static VAR

Compensator or SVC (a device used to improve power flow parameters) with the tuning of the MVO over the IEEE 57 bus system, so it can correct the voltage when it is needed. There is no statistical test performed to compare with other algorithms, it is only compared without MVO being used, but it has better results in terms of Loading parameter and loads.

Another applied case comes in the work that presents Fathy and Rezk [20] were they present an optimization problem involving the optimal parameters to build a proton exchange membrane fuel cell(PEMFC) under certain operating conditions. The way they do it is to optimize seven parameters to obtain polarization curves to comply with manufacturers' datasheets, doing this with less computational resources. It compares with SGA, HGA, HABC, RGA, and HADE; doing the best results with the MVO with less fitness and convergence time.

Al-Madi et al. [21] presented a binary version of the algorithm to solve problems with discrete variables called BMVO; the method uses a V-shaped transfer function to convert continuous values to binary thus updating values over the optimization. It compares with Binary Bat Algorithm, Binary Particle Swarm Optimization, Binary Dragon Algorithm, and Binary Grey Wolf Optimizer over 13 benchmark functions and solves feature selection and knapsack problems. It proves that it has higher profit and lower error compared to the mentioned algorithms.

One of the most recent studies is the work of Abasi et al. [22] where it presents a novel method called Link-based multi-verse optimizer (LBMVO) for text documents clustering. The idea comes from a text document clustering, which is a key task in text mining and unsupervised machine learning, where it divides certain documents into a collection varied in K-groups according to certain similarity or dissimilarity criteria. The enhancement comes from adding a neighbor operator to MVO and improves search capability via a novel probability factor, named neighborhood selection strategy (NSS).

The LBMVO was tested with six standard datasets on the clustering domain, and compares to other methods like Spectral, Agglomerative, DBSCAN, K-means, K-means++ clustering techniques and the optimization algorithms like HS, GA, PSO, KHA, CMAES, and COA; proving very good results over the tests.

Proving over modified versions of the MVO comes the work of Ewees and Elaziz [23] called Chaotic Multi-verse Harris Hawks Optimization (CMBHHO), which uses the chaos theory and the Harris Hawks Optimization; the objective is to determine the optimal value for the parameters of the MVO. The performance tests are done with a set of chaotic maps to determine the most suitable map, with the best map is applied to solve four well known engineering problems: Welded beam design, Tension/Compression spring design, Pressure vessel design and Rolling element bearing design. In tests, the new algorithm proves to work well under certain problems, but overall, it has a great performance compared to traditional methods.

In one last analysis we have the comparison work of Abderazek et al. [24] were it compares seven recent meta-heuristic optimization algorithms to automate the design of disk cam mechanism following four motion laws. In this comparison MVO does not come amongst the best performing algorithms, but is used in the comparison for the cam mechanism design.

## 5 Conclusions

The paper presented many recent works and applications that used MVO to find optimal solutions. Also, it includes the study of other metaheuristic algorithms that are used for different application areas and shows the used of the metaheuristics on optimization. Some of them involved the hybridization of methods, resulting in better results in most cases that they were presented.

We can see with these studies that the MVO algorithm has had very good results in several applications in these few years since its appearance on metaheuristic models, proving it can outcome a wide variety of situations, managing its different parameters over minimization, maximization and, in some cases, about the performance of the algorithm.

As per the main objective of the study, we want to motivate the use of the algorithm in other applications, such as control or robotics among others and study its behavior when hybridizing it with other metaheuristics to improve the performance of the MVO.

## References

1. A.P. Engelbrecht, *Computational Intelligence: an Introduction*. Wiley (2007)
2. S. Mirjalili, S.M. Mirjalili, A. Lewis, Grey wolf optimizer. Adv. Eng. Softw. **69**, 46–61 (2014). https://doi.org/10.1016/j.advengsoft.2013.12.007
3. X.S. Yang, *Nature-Inspired Optimization Algorithms*. Elsevier Inc. (2014). https://doi.org/10.1016/C2013-0-01368-0
4. J.S.R. Jang, C.T. Sun, E. Mizutani, *Neuro-fuzzy and Soft Computing: A Computational Approach to Learning and Machine Intelligence*. Prentice Hall (1997)
5. S. Mirjalili, S.M. Mirjalili, A. Hatamlou, Multi-verse optimizer: a nature-inspired algorithm for global optimization. Neural Comput. Appl. **27**, 495–513 (2016). https://doi.org/10.1007/s00521-015-1870-7
6. P.J. Steinhardt, N. Turok, *The Cyclic Model Simplified* (2005). https://doi.org/10.1016/j.newar.2005.01.003
7. E. Hernández, O. Castillo, J. Soria, Optimization of fuzzy controllers for autonomous mobile robots using the grey wolf optimizer, in O. Castillo, P. Melin (eds.) *Hybrid Intelligent Systems in Control, Pattern Recognition and Medicine*, pp. 289–299. Springer International Publishing, Cham (2020). https://doi.org/10.1007/978-3-030-34135-0_20
8. Z.W. Geem, J.H. Kim, G.V. Loganathan, A new heuristic optimization algorithm: harmony search. Simulation **76**, 60–68 (2001). https://doi.org/10.1177/003754970107600201
9. C. Peraza, F. Valdez, O. Castillo, Harmony search with dynamic adaptation of parameters for the optimization of a benchmark controller, in O. Castillo, P. Melin, J. Kacprzyk (eds.) *Intuitionistic and Type-2 Fuzzy Logic Enhancements in Neural and Optimization Algorithms: Theory and Applications*, pp. 157–168. Springer International Publishing, Cham (2020). https://doi.org/10.1007/978-3-030-35445-9_14
10. E. Atashpaz-Gargari, C. Lucas, Imperialist competitive algorithm: an algorithm for optimization inspired by imperialistic competition, in *2007 IEEE Congress on Evolutionary Computation, CEC*, pp. 4661–4667 (2007). https://doi.org/10.1109/CEC.2007.4425083
11. E. Bernal, O. Castillo, J. Soria, F. Valdez, Parameter adaptation in the imperialist competitive algorithm using generalized type-2 fuzzy logic, in O. Castillo, P. Melin, J. Kacprzyk (eds.)

*Intuitionistic and Type-2 Fuzzy Logic Enhancements in Neural and Optimization Algorithms: Theory and Applications*, pp. 3–10. Springer International Publishing, Cham (2020). https://doi.org/10.1007/978-3-030-35445-9_1
12. J. Pérez, F. Valdez, O. Castillo, Modification of the bat algorithm using type-2 fuzzy logic for dynamical parameter adaptation. *Studies in Computational Intelligence*, pp. 343–355. Springer Verlag (2017). https://doi.org/10.1007/978-3-319-47054-2_23
13. M.L. Lagunes, O. Castillo, F. Valdez, J. Soria, Comparison of fuzzy controller optimization with dynamic parameter adjustment based on of type-1 and type-2 fuzzy logic, in O. Castillo, P. Melin (eds.) *Hybrid Intelligent Systems in Control, Pattern Recognition and Medicine*, pp. 47–56. Springer International Publishing, Cham (2020). https://doi.org/10.1007/978-3-030-34135-0_4
14. O.R. Carvajal, O. Castillo, J. Soria, Optimization of membership function parameters for fuzzy controllers of an autonomous mobile robot using the flower pollination algorithm. J. Autom. Mob. Robot. Intell. Syst. (2018). https://doi.org/10.14313/JAMRIS_1-2018/6
15. A. Sadollah, A. Bahreininejad, H. Eskandar, M. Hamdi, Mine blast algorithm: a new population based algorithm for solving constrained engineering optimization problems. Appl. Soft Comput. **13**, 2592–2612 (2013). https://doi.org/10.1016/J.ASOC.2012.11.026
16. C. Hu, Z. Li, T. Zhou, A. Zhu, C. Xu, A multi-verse optimizer with levy flights for numerical optimization and its application in test scheduling for network-on-chip. PLoS One **11**, 1–22 (2016). https://doi.org/10.1371/journal.pone.0167341
17. H. Faris, I. Aljarah, S. Mirjalili, Training feedforward neural networks using multi-verse optimizer for binary classification problems. Appl. Intell. **45**, 322–332 (2016). https://doi.org/10.1007/s10489-016-0767-1
18. P. Jangir, S.A. Parmar, I.N. Trivedi, R.H. Bhesdadiya, A novel hybrid particle swarm optimizer with multi verse optimizer for global numerical optimization and optimal reactive power dispatch problem. Eng. Sci. Technol. Int. J. **20**, 570–586 (2017). https://doi.org/10.1016/J.JESTCH.2016.10.007
19. K. Karthikeyan, P.K. Dhal, Multi-verse optimization (MVO) technique based voltage stability analysis through continuation power flow in IEEE 57 bus. Energy Procedia **117**, 583–591 (2017). https://doi.org/10.1016/J.EGYPRO.2017.05.153
20. A. Fathy, H. Rezk, Multi-verse optimizer for identifying the optimal parameters of PEMFC model. Energy **143**, 634–644 (2018). https://doi.org/10.1016/j.energy.2017.11.014
21. N. Al-Madi, H. Faris, S. Mirjalili, Binary multi-verse optimization algorithm for global optimization and discrete problems. Int. J. Mach. Learn. Cybern. **10**, 3445–3465 (2019). https://doi.org/10.1007/s13042-019-00931-8
22. A.K. Abasi, A.T. Khader, M.A. Al-Betar, S. Naim, S.N. Makhadmeh, Z.A.A. Alyasseri, Link-based multi-verse optimizer for text documents clustering. Appl. Soft Comput. **87**, 106002 (2020). https://doi.org/10.1016/J.ASOC.2019.106002
23. A.A. Ewees, M.A. Elaziz, Performance analysis of chaotic multi-verse harris hawks optimization: a case study on solving engineering problems. Eng. Appl. Artif. Intell. **88**, 103370 (2020). https://doi.org/10.1016/J.ENGAPPAI.2019.103370
24. H. Abderazek, A.R. Yildiz, S. Mirjalili, Comparison of recent optimization algorithms for design optimization of a cam-follower mechanism. Knowl. Based Syst. **191**, 105237 (2020). https://doi.org/10.1016/J.KNOSYS.2019.105237

# Comparative Study of Type-1 and Interval Type-2 Fuzzy Systems in Parameter Adaptation of the Fuzzy Flower Pollination Algorithm

**Hector Carreon and Fevrier Valdez**

**Abstract** Combining Interval Type-2 Fuzzy Logic Systems with metaheuristics has shown in most investigations that better results are obtained than with Type-1 Fuzzy Logic Systems. In this comparative study, experiments were carried out with Type-1 and Interval Type-2 Fuzzy Logic Systems, each one in combination with the Flower Pollination Algorithm. In the modification of parameters, with this combination of hybrid methods we carried out the comparative study. Previously, experiments were carried out with the flower pollination algorithm and the Type-1 Fuzzy Logic System (T1FLS), with the results of both methods, and we have concluded that better results are obtained with the hybrid method of Interval Type-2 Fuzzy Logic System (IT2FLS) and the Flower Pollination Algorithm (FPA).

**Keywords** Type-1 fuzzy logic · Type-2 fuzzy logic · Flower pollination algorithm · Metaheuristic · Bioinspired algorithm

## 1 Introduction

Nature is a great source of inspiration to develop new optimization algorithms. Optimizing systems in real life is complex and difficult and many applications face these problems, optimization algorithms can be used to solve them, which does not guarantee that an optimal solution can be obtained [1], nature has been able through millions of years to solve these problems, many biological systems have demonstrated very effectively in problem solving [2–4].

Many difficult problems in industry, commerce, medicine, etc., have been solved with the help of bioinspired optimization algorithms [2, 4, 5].

Science has been developing nature-inspired optimization algorithms like the Bat Algorithm (BA) [1] and the Flower Pollination Algorithm (FPA) [6], to name a few. Since its origin, flower plants have been evolving and dominate the landscape [5,

H. Carreon · F. Valdez (✉)
Tijuana Institute of Technology, Tijuana, BC, Mexico
e-mail: fevrier@tectijuana.mx



P. Melin et al. (eds.), *Recent Advances of Hybrid Intelligent Systems Based on Soft Computing*, Studies in Computational Intelligence 915,
https://doi.org/10.1007/978-3-030-58728-4_8

7], there are thousands of flowering plants and of the total number of plants on the planet, flowering plants make up three quarters. The reproduction of flowering plants is done through the pollination process that can be described as the process of pollen distribution through pollinators that can be insects, birds, bats, some other animals, wind and water.

Already in other studies the adaptation of parameters has been performed dynamically using Type-1 and Type-2 Fuzzy Systems obtaining good results [8–10], so in this study the parameters of the Flower Pollination Algorithm bioinspired optimization are dynamically adapted using Type-1 and Interval Type-2 Fuzzy Systems. The main contribution is the proposed use of Type-2 fuzzy parameter adaptation in FPA, which has not lean previously proposed.

This article is organized as follows: Sect. 2 broadly describes the theory of Fuzzy Logic, Sect. 3 describes the Flower Pollination Algorithm, Sect. 4 presents the simulation of the experiments performed with the FPA Algorithm and Type-1 and Interval Type-2 Fuzzy Systems, the results are evaluated by statistical tests and finally in Sect. 5 the conclusions of this article are presented.

## 2 Fuzzy Logic

### *2.1 The Origins of Fuzzy Logic*

According to Fuzzy Logic theory, there is a transfer function derived from the characteristic function, generally called the "membership function", which extends from the universe of discourse, $U$, to the closed real unit interval, which is [0, 1]. Not so with "classic" or "crisp sets", where the range of the function is reduced to a set of only two elements, that is, $\{0, 1\}$. Therefore, Fuzzy Set theory is a generalization of classical set theory.

This concept arose from the study of several thinkers, who had a different view on the problems of traditional logic. The paradox of all the sets that are not contained, is very famous and was proposed by Bertrand Russell. Or the uncertainty principle of quantum physics by Werner Heisenberg.

The theory of "vague sets" (today, the so-called Fuzzy Sets) comes from the German quantum physicist and philosopher Max Black (1937), who also analyzed the problem of modeling "vagueness." It differs from Russell in that he proposed that traditional logic can be used by representing vagueness with an appropriate level of detail and suggests that Russell's definition of vagueness confuses the concept of vagueness in general.

In 1920 Jan Lukasiewicz, developed the first logic of vagueness, for him the sets have a possible degree of belonging with values ranging between 0 and 1, and in this interval, there are an infinite number of values.

Lofti A. Zadeh, his first work was based on the works of Max Black and in 1965 introduced the theory of Fuzzy Sets and later Fuzzy Logic. In this sense Zadeh is considered as creator of Fuzzy Sets and Fuzzy Logic and is also known as the father of possibility and Fuzzy Logic.

Fuzzy logic is a mathematical formalism that attempts to imitate the way humans make decisions. Through the concept of membership degree, Fuzzy Set theory and Fuzzy Logic, a simple mathematical expression of uncertainty is obtained [1].

Fuzzy Set theory allows an object to have partial membership in more than one set. It does this through the introduction of a function known as the membership function, which maps from the complete set of objects $X$ into a set known as the membership space. More formally, the definition of a Fuzzy Set [2] is.

If $X$ is a collection of objects denoted generically by $x$ then a Fuzzy Set $A$ in $X$ is a set of ordered pairs:

$$A = \{(x, \mu_A(x)) | x \in X\} \tag{1}$$

$\mu_A(X)$ is called the membership function or degree of membership (also degree of compatibility or degree of truth) of $x$ in $A$ which maps $X$ to the membership space $M$.

The logical connectives "and", "or", and "not" are defined as

$$\text{or} : A \cup B \rightarrow \mu_{A \cup B}(x) = \max[\mu_A(x), \mu_B(x)]$$

$$\text{or} : A \cap B \rightarrow \mu_{A \cap B}(x) = \max[\mu_A(x), \mu_B(x)]$$

$$\text{not} : A \quad \bar{B} \rightarrow \mu_B(x) = \mu_B(x)$$

## 2.2 *Type-2 Fuzzy Logic Systems*

In Type-2 Fuzzy Logic Systems, the antecedent and consequent membership functions are Type-2 Fuzzy Sets. Type-2 Fuzzy Sets are characterized by fuzzy membership functions that are three-dimensional. The concept of Type-2 Fuzzy Set was introduced by Zadeh [11] as an extension of the concept of Type-1 Fuzzy Set. In Type-2 Fuzzy Sets the membership functions are Type-1 Fuzzy Sets, and are useful for incorporating uncertainty when it is difficult to determine a crisp membership function for a Fuzzy Set. The degree of membership for each element of Type-2 Fuzzy Set is a Fuzzy Set in [0, 1]. The third dimension provides additional degrees of freedom to capture more information about the term represented. Type-2 Fuzzy Sets are useful in circumstances, where it is difficult to determine the crisp membership function for a Fuzzy Set, which is useful for incorporating uncertainties [12].

Type-2 Fuzzy Sets can model uncertainties because their membership functions are fuzzy. If there is no uncertainty, a Type-2 Fuzzy Set is reduced to a Type-1 Fuzzy Set [13].

The knowledge used to build rules in a Fuzzy Logic System (FLS) is uncertain. This uncertainty leads to rules that have uncertain antecedents and/or consequents, which in turn translates into uncertain antecedents and/or consequent membership functions [14].

Type-2 Fuzzy Sets allow us to handle linguistic uncertainties, because: "words can mean different things to different people [15]".

## 3 Flower Pollination Algorithm

The Flower Pollination Algorithm was developed by Xin She Yang in 2012, and this algorithm is basically inspired by the process of pollination of flowering plants [6].

It is estimated that there are more than 320,000 types of plants of which the vast majority of 260–290 thousand produce seeds, green plants provide oxygen in a large proportion in the world and are the basis for the vast majority of ecosystems in the world [16]. It is believed that the population of the plants on the earth was carried out through a mycorrhizal network, which is a network of fungi that live in symbiosis with the roots of the plants [17], the flowering plants in nature occupy approximately 80% of all species on the planet. The flowering plant has been evolving for more than 125 million years and the flowers have become so decisive in evolution that we cannot believe how this world would be without them [18]. The main objective of flowers is the reproduction of their species and they do so through pollination that is nothing other than pollen transfer, this transfer is done through pollinators that can be insects, birds, bats and other animals as well as wind and water [19].

Pollination is understood as the transfer of pollen from the male part of the flower to the female part of the same or another flower, this is a necessary process to maintain the continuity and genetic diversity of flowering plants, it is also improved the quality and quantity of seeds and fruits, as well as their offspring [20].

Types of pollination, there are basically two types:

Self-pollination or direct pollination, also known as autogamy occurs when the pollen of a flower pollinates the same flower or flowers of the same plant Fig. 1a, b, this type of pollination is also known as autogamy, approximately 10% of the global pollination is carried out through of self-pollination [21, 22].

Indirect or cross-pollination occurs when pollination is carried out between different individuals Fig. 1c, this type of pollination is also known as allogamy, approximately 90% of the global pollination is carried out through indirect or cross-pollination [21, 22].

Pollination is called abiotic if pollen is transported by wind or water, and it is called biotic if pollen is transported by a living being such as insects, birds and some mammals [6].

The FPA rules [6, 23, 24] are:

1. Biotic and cross-pollination is considered as global pollination [2] process with pollen-carrying pollinators performing Levy flights [5].

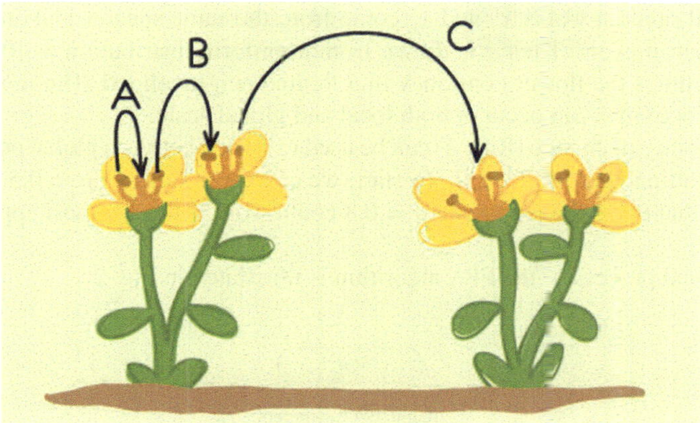


**Fig. 1** Self-pollination (**a**–**b**) and cross pollination (**c**)

2. Self-pollination occurs when the plant fertilizes itself, also known as local pollination [3].
3. Flower constancy involves the association between pollinators and flowers.
4. Probability of change $p \in [0, 1]$ controls global and local pollination, the value of $p$ is generally 0.8.

This algorithm is designed in two basic stages that are global pollination and local pollination. In global pollination, flower pollens are transported by pollinators such as insects, and pollens can travel a great distance because insects can often fly and move in a much longer range [25]. This ensures pollination and reproduction of the fittest, and therefore we represent the fittest as $g^*$. The first rule plus the constancy of the flower can be represented mathematically as:

Global Pollination [24, 26, 27] (Rule 1) is expressed as:

$$x_i^{t+1} = x_i^t + L\left(x_i^t - g^*\right) \tag{2}$$

where $x_i^t$ is the pollen $i$ or solution vector $x_i$ at iteration $t$ and $g^*$ is the current best solution found among all solutions at the current generation/iteration.

The parameter $L$ is the strength of the pollination, which essentially is a step size.

Levy Flights [26] are defined as:

$$L \sim \frac{\lambda \Gamma(\lambda) \sin(\pi \lambda/2)}{\pi} \frac{1}{s^{1+\lambda}}, \ (s \gg S_0 > 0) \tag{3}$$

Here $\Gamma(\lambda)$ is the standard gamma function and this distribution is valid for large steps $s > 0$, $\lambda$ is considered as 1.5.

Local Pollination and flower constancy [24, 27, 28] (Rule 2 and 3)

$$x_i^{t+1} = x_i^t + \varepsilon\left(x_j^t - x_k^t\right) \tag{4}$$

For local random walks $x_j^t$ and $x_k^t$ come from the same species and are pollens with different flowers. Then $\varepsilon$ is drawn from a uniform distribution as [0 1]. This basically mimics the flower constancy in a limited neighborhood. The most flower pollination activities can occur at both local and global scale.

A probability of change (Rule 4) can be used to switch between global pollination and local pollination [24, 27, 28]. To start, we can use a value of $p = 0.5$. Practice has shown that using the parameter $p = 0.8$ could work better for most applications [29].

The logical process of the FPA algorithm is illustrated in Fig. 2.

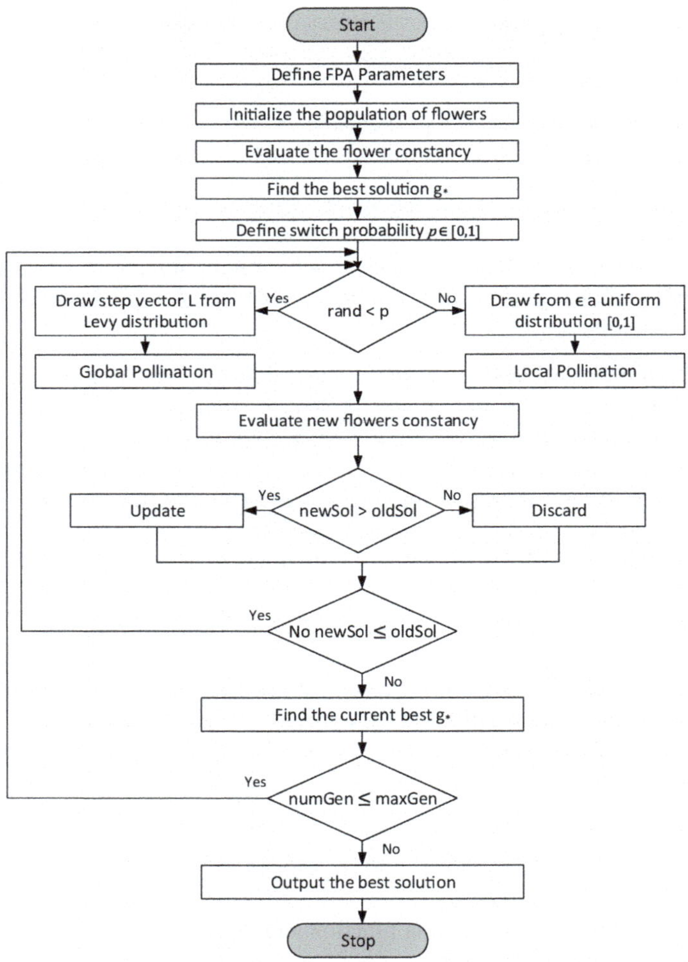


**Fig. 2** Flower pollination algorithm flowchart

# 4 Simulation

The process to obtain the results of the simulations using the bioinspired Flower Pollination Algorithm with Type-1 and Interval Type-2 Fuzzy Logic Systems is illustrated in Fig. 3, which shows the logical process of the FPA including the process with the Fuzzy Logic Systems mentioned above.

The parameters that were adjusted in the FPA algorithm were $p$ and $\varepsilon$ through iterations and the error change with the Type-1 and Interval Type-2 Fuzzy Systems and we compared their performance between both Fuzzy Systems.

The logical process of the FFPA algorithm is shown in Fig. 3.

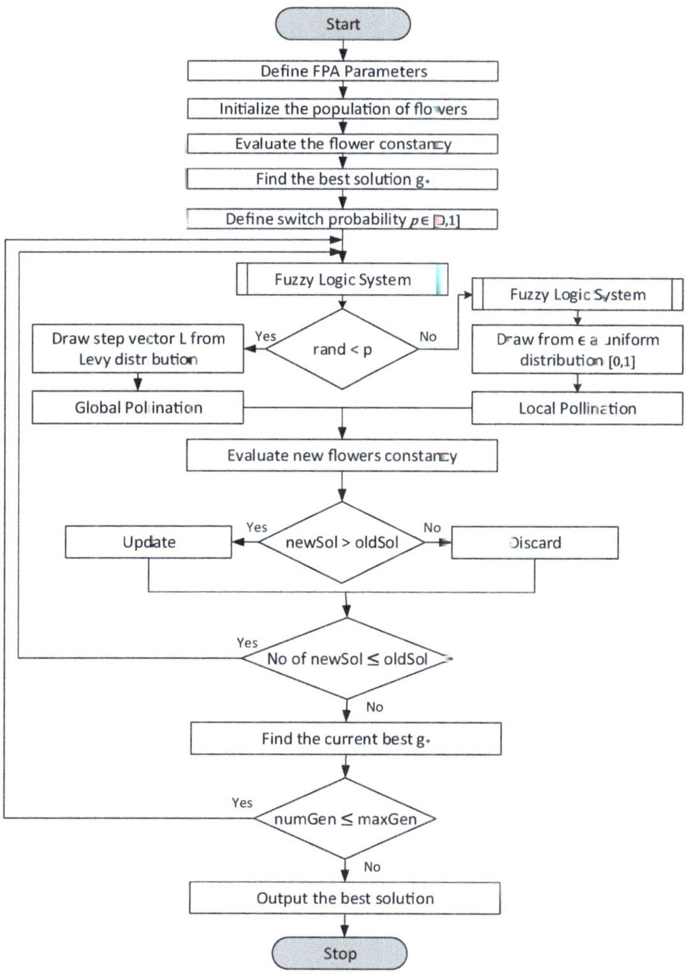


**Fig. 3** Fuzzy flower pollination algorithm flowchart

In Figs. 4, 5, 6 and 7 we show the design of architecture of the First Type-2 Fuzzy Logic System, that we use with the FPA algorithm.

In this study we use two Fuzzy Systems in which the first Fuzzy System uses an input with three Gaussian membership functions and an output with three Gaussian membership functions, and uses 3 if then rules, as shown in the following figures.

Figures 8, 9, 10, 11 and 12 show the design of architecture of the Second Type 2 Fuzzy Logic System, that we use with the FPA algorithm.

The second fuzzy system uses 2 inputs with three Gaussian membership functions each and one output with three Gaussian membership functions, and also uses 9 rules if then.

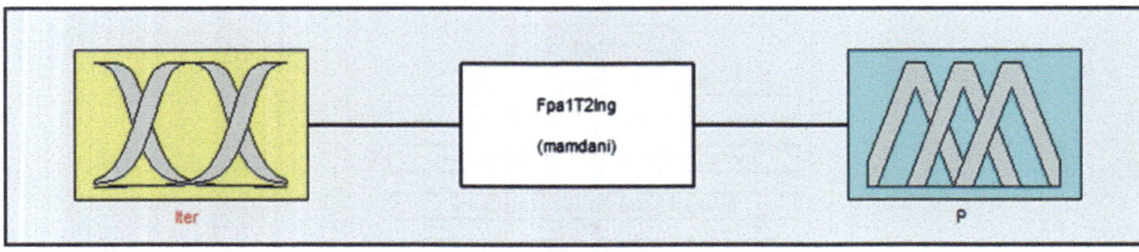


**Fig. 4** First type-2 fuzzy logic architecture

**Fig. 5** First type-2 fuzzy logic system, IF then rules

| If then rules | |
|---|---|
| 1 | If (Iter is low) then (P is low) |
| 2 | If (Iter is medium) then (P is medium) |
| 3 | If (Iter is high) then (P is high) |

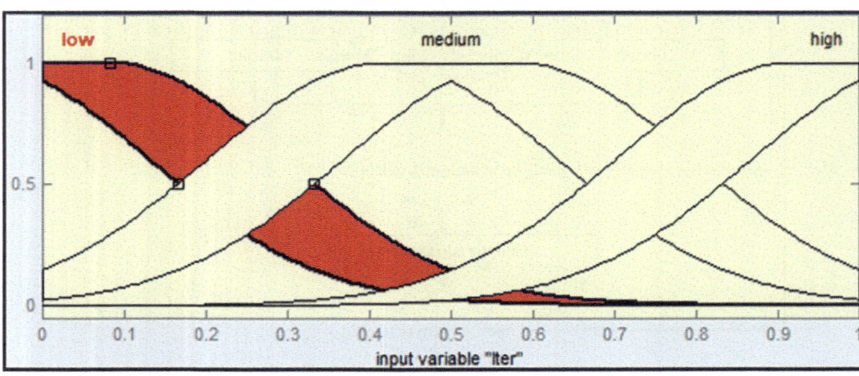


**Fig. 6** Inputs, first type-2 fuzzy logic system

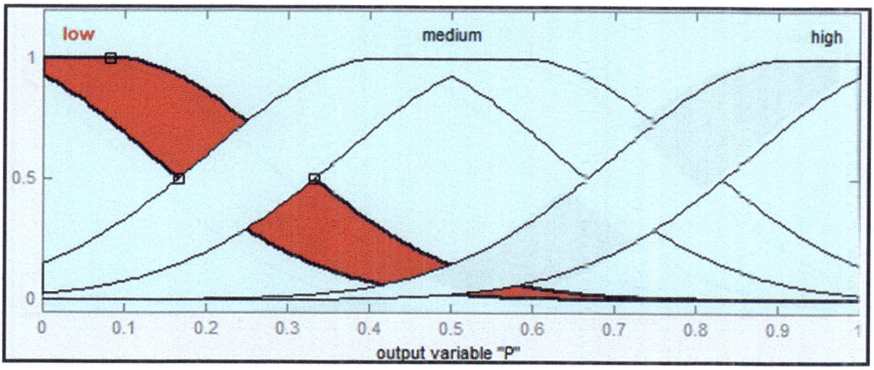


**Fig. 7** Outputs, first type-2 fuzzy logic system

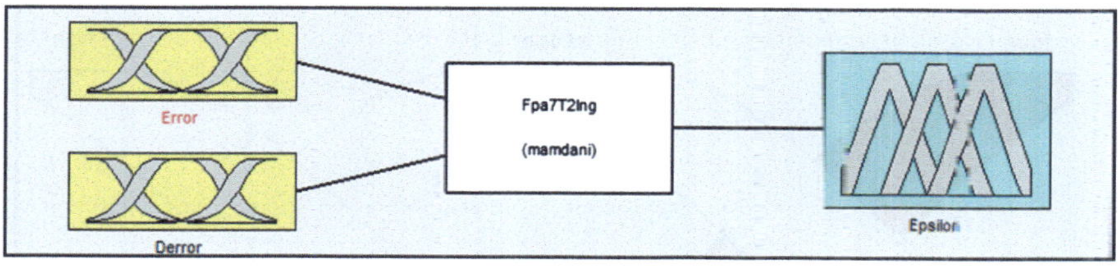


**Fig. 8** Second type-2 fuzzy logic system architecture

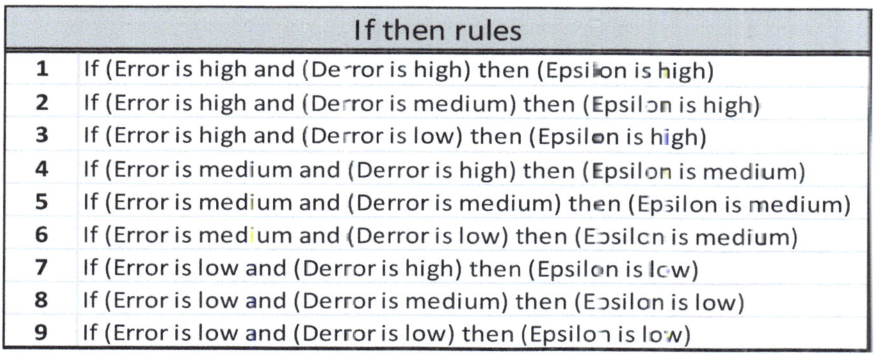

| If then rules | |
|---|---|
| 1 | If (Error is high and (Derror is high) then (Epsilon is high) |
| 2 | If (Error is high and (Derror is medium) then (Epsilon is high) |
| 3 | If (Error is high and (Derror is low) then (Epsilon is high) |
| 4 | If (Error is medium and (Derror is high) then (Epsilon is medium) |
| 5 | If (Error is medium and (Derror is medium) then (Epsilon is medium) |
| 6 | If (Error is medium and (Derror is low) then (Epsilon is medium) |
| 7 | If (Error is low and (Derror is high) then (Epsilon is low) |
| 8 | If (Error is low and (Derror is medium) then (Epsilon is low) |
| 9 | If (Error is low and (Derror is low) then (Epsilon is low) |

**Fig. 9** Second type-2 fuzzy logic system, IF then rules

## *4.1 Benchmark Functions*

In this research we use 6 Benchmark Functions that are the same as those used in our previous research and we did it for the purpose of making a comparison, the functions are:

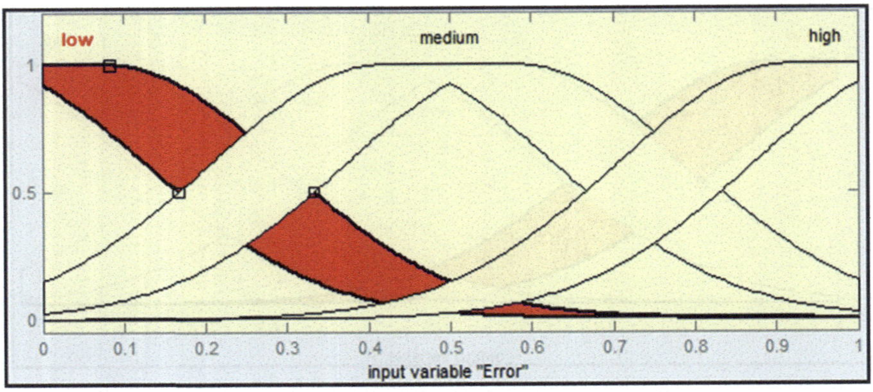


**Fig. 10** Input 1 of the second type-2 fuzzy logic system

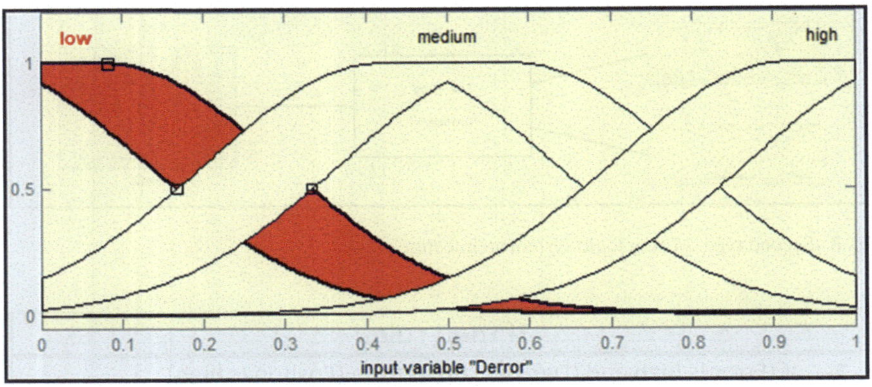


**Fig. 11** Input 2 of the second type-2 fuzzy logic system

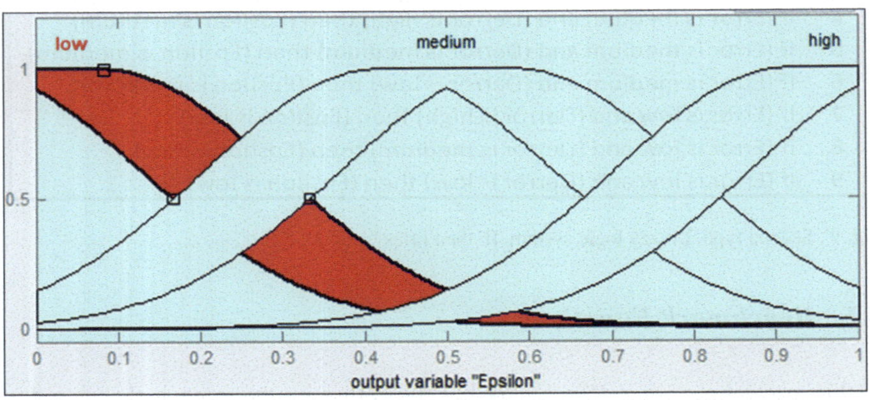


**Fig. 12** Outputs, second type-2 fuzzy logic system

| No | Function | Range | Type |
|---|---|---|---|
| **F1** | **Sphere** | **[100, 100]** | **U** |
| | 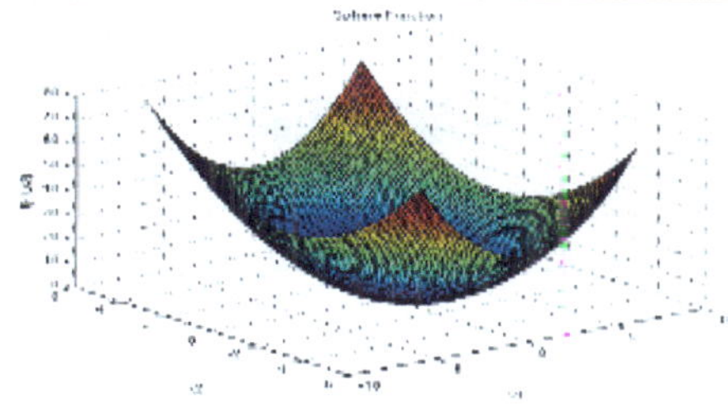 $$f(x) = \sum_{i=1}^{d} x_i^2$$ | | |
| **F2** | **Ackley** | **[-32, 32]** | **M** |
| | 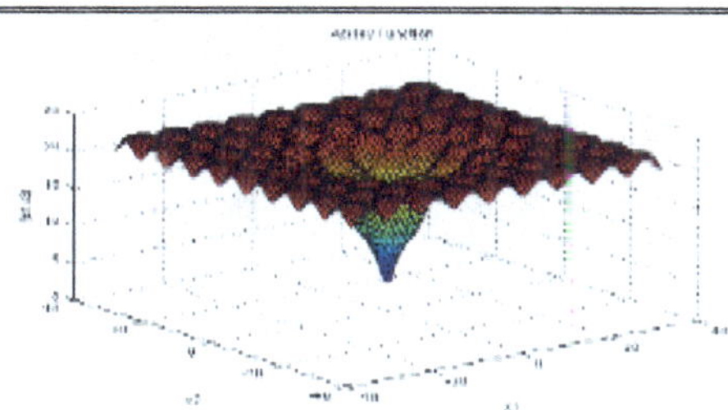 $$f(x) = -20exp\left[-0.2\sqrt{\frac{1}{d}\sum_{i=1}^{d} x_i^2}\right] - exp\left[\frac{1}{d}\sum_{i=1}^{d} \cos\,(2\pi x_i\right] + (20 + e)$$ | | |
| **F3** | **Rastrigin** | **[-5.12, 5.12]** | **M** |
| | 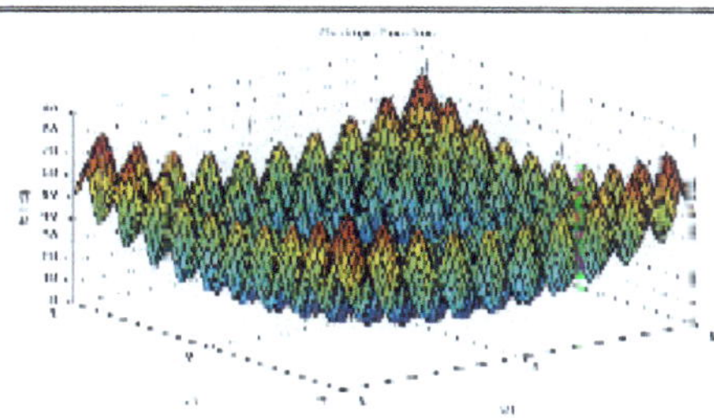 $$f(x) = 10 * (d) + \sum_{i=1}^{d} [x_i^2 - 10cos(2\pi x_i)]$$ | | |

| | | | |
|---|---|---|---|
| **F4** | **Zakharov** | **[-5, 10]** | **U** |
| | 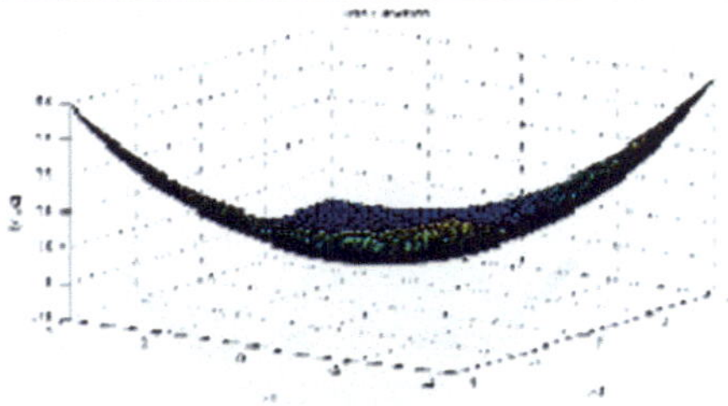 $f(x) = \sum_{i=1}^{d} x_i^2 + \left(\sum_{i=1}^{d} 0.5 i\, x_i\right)^2 + \left(\sum_{i=1}^{d} 0.5 i\, x_i\right)^4$ | | |
| **F5** | **Griewank** | **[-600, 600]** | **M** |
| | 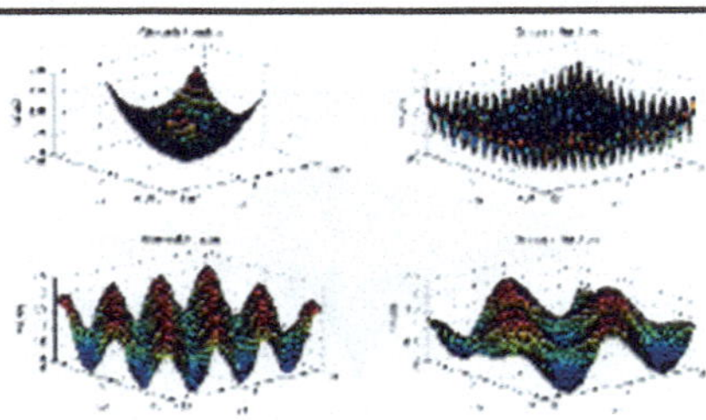 $f(x) = \frac{1}{4000}\sum_{i=1}^{d} x_i^2 - \prod_{i=1}^{d} cos\left(\frac{x^i}{\sqrt{i}}\right) + 1$ | | |
| **F6** | **Sum of Different Powers** | **[-1, 1]** | **U** |
| | 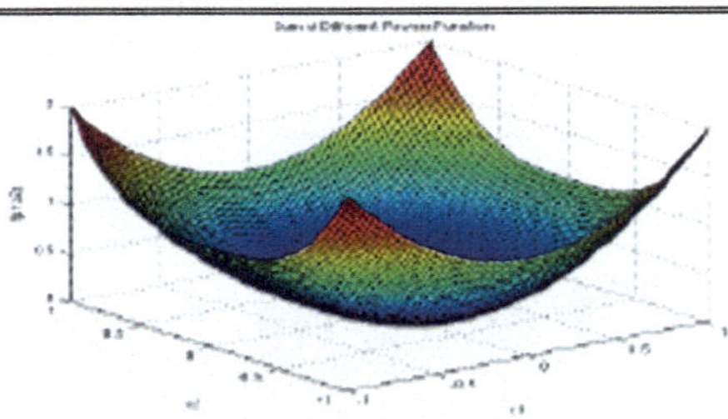 $f(x) = \sum_{i=1}^{d} \lvert x_i \rvert^{i+1}$ | | |

| | | | |
|---|---|---|---|
| **F7** | **Michalewicz** | **[0, π]** | **M** |
| | 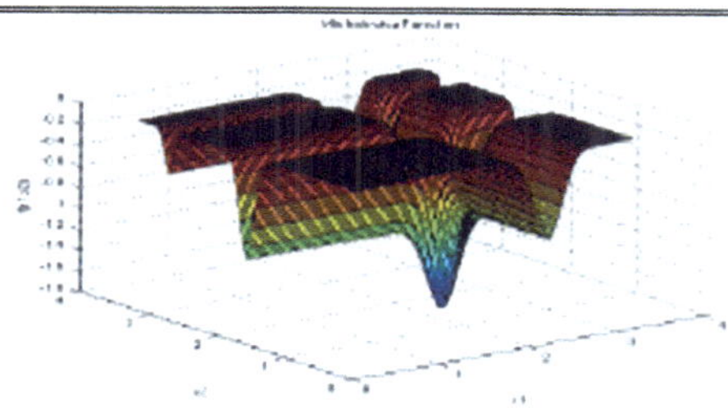 $$f(x) = -\sum_{i=1}^{d} sin(x_i)sin^{2m}\left(\frac{ix_i^2}{\pi}\right)$$ | | |
| **F8** | **Rosenbrock** | **[-5, 10]** | **U** |
| | 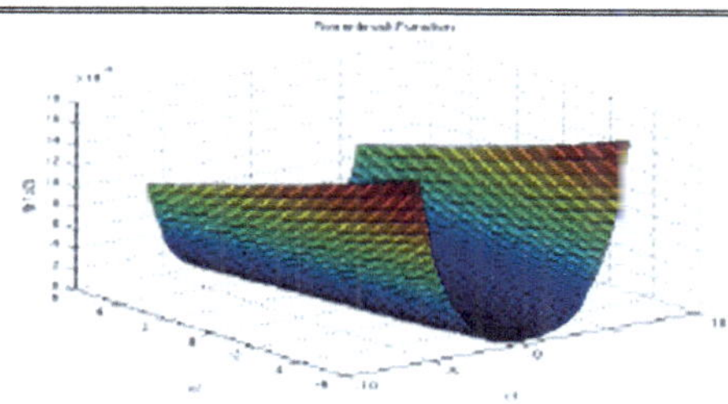 $$f(x) = \sum_{i=1}^{d-1}\left[(1-x_i)^2 + 100\left(x_{i+1}-x_i^2\right)^2\right]$$ | | |

## 4.2 Results

Tables 1 and 2 show the results of the experiments done with the FFPA algorithm, using Type-1 and Type-2 Fuzzy Systems, for 30 and 100 dimensions, respectively. The functions that were used are the following: Sphere, Ackley, Rastrigin, Zakharov, Griewank, Sum of Different Powers, Michalewicz and Rosenbrock. In addition, in Table 2 we can observe the results of the statistical test (last column) among the experiments performed with the Type-1 and Interval Type-2 Fuzzy Systems and in 4 tests the Type-2 FLS was better.

Tables 3 and 4 show the results of the experiments performed with the FFPA algorithm, using Interval Type-2 Fuzzy Systems, for 30 and 100 dimensions respectively. The functions that were used are the following: Sphere, Ackley, Rastrigin, Zakharov, Griewank, Sum of Different Powers, Michalewicz and Rosenbrock. Besides, in Table 4 we can observe the results of the statistical test (last column) among the

**Table 1** Experiments with FPA and FLS type-1

| FFPAT1–30–dimensions | | | | | |
|---|---|---|---|---|---|
| No. | Function | Best | Worst | Average | SD |
| 1. | Sphere | $1.65 \times 10^{-4}$ | $2.08 \times 10^{-2}$ | $4.35 \times 10^{-3}$ | $5.00 \times 10^{-3}$ |
| 2. | Ackley | $1.69 \times 10^{-2}$ | $2.21 \times 10^{0}$ | $7.62 \times 10^{-1}$ | $9.83 \times 10^{-1}$ |
| 3. | Rastrigin | $2.63 \times 10^{-2}$ | $1.07 \times 10^{2}$ | $9.80 \times 10^{0}$ | $2.53 \times 10^{1}$ |
| 4. | Zakharov | $2.79 \times 10^{-3}$ | $1.38 \times 10^{-1}$ | $3.89 \times 10^{-2}$ | $3.67 \times 10^{-2}$ |
| 5. | Griewank | $2.53 \times 10^{-6}$ | $6.31 \times 10^{-4}$ | $1.41 \times 10^{-4}$ | $1.32 \times 10^{-4}$ |
| 6. | Sum of D P | $5.01 \times 10^{-17}$ | $2.51 \times 10^{-11}$ | $1.12 \times 10^{-12}$ | $4.55 \times 10^{-12}$ |
| 7. | Michalewicz | $-9.52 \times 10^{0}$ | $-1.31 \times 10^{1}$ | $-1.18 \times 10^{1}$ | $8.36 \times 10^{-1}$ |
| 8. | Rosenbrock | $2.37 \times 10^{1}$ | $4.69 \times 10^{1}$ | $3.23 \times 10^{1}$ | $4.37 \times 10^{0}$ |

**Table 2** Experiments with FPA and FLS type-2

| FFPAT2–30–dimensions | | | | | | |
|---|---|---|---|---|---|---|
| No. | Function | Best | Worst | Average | SD | Z-test |
| 1. | Sphere | $3.91 \times 10^{-4}$ | $2.15 \times 10^{-2}$ | $5.10 \times 10^{-3}$ | $5.15 \times 10^{-3}$ | No |
| 2. | Ackley | $9.39 \times 10^{-2}$ | $1.22 \times 10^{-1}$ | $7.27 \times 10^{-2}$ | $3.39 \times 10^{-2}$ | Yes |
| 3. | Rastrigin | $3.17 \times 10^{1}$ | $3.84 \times 10^{1}$ | $3.50 \times 10^{1}$ | $1.88 \times 10^{0}$ | Yes |
| 4. | Zakharov | $1.01 \times 10^{-2}$ | $1.24 \times 10^{-1}$ | $5.23 \times 10^{-2}$ | $3.01 \times 10^{-2}$ | No |
| 5. | Griewank | $2.88 \times 10^{-5}$ | $6.80 \times 10^{-4}$ | $1.82 \times 10^{-4}$ | $1.52 \times 10^{-4}$ | No |
| 6. | Sum of D P | $1.21 \times 10^{-17}$ | $3.57 \times 10^{-11}$ | $1.65 \times 10^{-12}$ | $6.45 \times 10^{-12}$ | No |
| 7. | Michalewicz | $-8.51 \times 10^{0}$ | $-1.17 \times 10^{1}$ | $-9.76 \times 10^{0}$ | $7.62 \times 10^{-1}$ | Yes |
| 8. | Rosenbrock | $7.47 \times 10^{1}$ | $1.11 \times 10^{2}$ | $8.92 \times 10^{1}$ | $9.26 \times 10^{0}$ | Yes |

**Table 3** Experiments with FPA and FLS type-1

| FFPAT1–100–dimensions | | | | | |
|---|---|---|---|---|---|
| No. | Function | Best | Worst | Average | SD |
| 1. | Sphere | $3.57 \times 10^{-1}$ | $1.49 \times 10^{0}$ | $7.52 \times 10^{-1}$ | $2.83 \times 10^{-1}$ |
| 2. | Ackley | $3.96 \times 10^{-1}$ | $7.43 \times 10^{-1}$ | $5.89 \times 10^{-1}$ | $7.74 \times 10^{-2}$ |
| 3. | Rastrigin | $5.89 \times 10^{-1}$ | $1.03 \times 10^{0}$ | $7.91 \times 10^{-1}$ | $1.06 \times 10^{-1}$ |
| 4. | Zakharov | $1.11 \times 10^{0}$ | $4.10 \times 10^{0}$ | $2.08 \times 10^{0}$ | $6.09 \times 10^{-1}$ |
| 5. | Griewank | $2.66 \times 10^{-3}$ | $6.53 \times 10^{-3}$ | $4.02 \times 10^{-3}$ | $8.87 \times 10^{-4}$ |
| 6. | Sum of D P | $6.63 \times 10^{-14}$ | $1.09 \times 10^{-8}$ | $7.84 \times 10^{-10}$ | $2.17 \times 10^{-9}$ |
| 7. | Michalewicz | $-1.40 \times 10^{1}$ | $-2.35 \times 10^{1}$ | $-1.91 \times 10^{1}$ | $2.05 \times 10^{0}$ |
| 8. | Rosenbrock | $1.27 \times 10^{2}$ | $2.17 \times 10^{2}$ | $1.68 \times 10^{2}$ | $2.09 \times 10^{1}$ |

**Table 4** Experiments with FPA and FLS type-2

| FFPAT2–100–dimensions | | | | | | |
|---|---|---|---|---|---|---|
| No. | Function | Best | Worst | Average | SD | Z-test |
| 1. | Sphere | $5.71 \times 10^{-1}$ | $1.64 \times 10^{0}$ | $1.01 \times 10^{0}$ | $2.90 \times 10^{-1}$ | Yes |
| 2. | Ackley | $5.95 \times 10^{-1}$ | $4.95 \times 10^{-1}$ | $5.70 \times 10^{-1}$ | $1.09 \times 10^{-1}$ | No |
| 3. | Rastrigin | $2.26 \times 10^{2}$ | $2.54 \times 10^{2}$ | $2.36 \times 10^{2}$ | $6.47 \times 10^{0}$ | Yes |
| 4. | Zakharov | $1.30 \times 10^{0}$ | $3.45 \times 10^{0}$ | $2.19 \times 10^{0}$ | $4.74 \times 10^{-1}$ | No |
| 5. | Griewank | $5.76 \times 10^{-3}$ | $1.44 \times 10^{-2}$ | $9.69 \times 10^{-3}$ | $2.27 \times 10^{-3}$ | Yes |
| 6. | Sum of D P | $6.59 \times 10^{-15}$ | $4.51 \times 10^{-8}$ | $3.01 \times 10^{-9}$ | $8.94 \times 10^{-9}$ | No |
| 7. | Michalewicz | $-1.26 \times 10^{1}$ | $-1.64 \times 10^{1}$ | $-1.46 \times 10^{1}$ | $9.33 \times 10^{-1}$ | Yes |
| 8. | Rosenbrock | $1.08 \times 10^{3}$ | $1.32 \times 10^{3}$ | $1.21 \times 10^{3}$ | $5.39 \times 10^{1}$ | Yes |

experiments done with the Type-1 and Type-2 Fuzzy Systems and in 5 tests the Interval Type-2 FLS was better.

## 5 Conclusions

Once the experiments were carried out with the Flower Pollination Algorithm and the Type-1 and Interval Type-2 Fuzzy Systems for 30 and 100 dimensions, using the benchmark functions Sphere, Ackley, Rastrigin, Zakharov, Griewank, Sum of Different Powers, Michalewicz and Rosenbrock and performing the statistical tests between each respective pair, it has been observed that if we increase the number of dimensions, the Interval Type-2 Fuzzy System achieves better results than the Type-1. As future work it would be feasible to perform more tests by increasing the dimensions and adding more benchmark functions. Other applications can also be considered, like intelligent control, pattern recognition and time series prediction. In addition, in the future we can apply the proposed optimization method in other kinds of application areas, like in [30–38].

## References

1. Y. Xin-She, Engineering: Optimization: An Introduction with metaheuristic Application, pp. 15–16. Wiley (2010)
2. X.S. Yang, Nature-Inspired Optimization Algorithms: Elsevier: First Edition, p. 16 (2014)
3. Y. Xin-She, Nature-inspired Metaheuristic Algorithms, pp. 8–9. Luniver Press (2008)
4. K. Deb, *Multi-Objective Optimization Using Evolutionary Algorithms* (Wiley, New York, 2001), pp. 18–20
5. Y. Xin-She, K. Mehmet, H. Xingshi, Multi-objective flower algorithm for optimization. in *International Conference on Computational Science, ICCS* (2013)

6. X.S. Yang, *Flower Pollination Algorithm for Global Optimization, Unconventional Computation and Natural Computation* (Springer, Berlin Heidelberg, 2012), pp. 240–249
7. H. Carreon, F. Valdes, O. Castillo, Fuzzy Flower Pollination Algorithm to solve control problems, Hybrid Intelligent Systems in Control, Pattern Recognition and Medicine, pp 122–154. Tijuana Institute of Technology, Mexico, Springer Nature Switzerland AG (2020)
8. F. Olivas, F. Valdez, O. Castillo, P. Melin, Dynamic Parameter Adaptation for Meta-Heuristic Optimization Algorithms Through Type-2 Fuzzy Logic, Division of Graduate Studies Tijuana Institute of Technology, Springer Briefs in Computational Intelligence (2018)
9. F. Olivas, F. Valdez, O. Castillo, P. Melin, *Dynamic parameter adaptation in particle swarm optimization using interval type-2 fuzzy logic* (Division of Graduate Studies Tijuana Institute of Technology, Springer, Berlin Heidelberg, 2014)
10. V.A. Tatsis, K.E. Parsopoulos, Dynamic parameter adaptation in metaheuristics using gradient approximation and line search. Appl. Soft Comput. J. Department of Computer Science and Engineering, University of Ioannina, Greece, Elsevier (2019)
11. L.A. Zadeh, The concept of a linguistic variable and its application to approximate reasoning—1. Inf. Sci. **8**, 199–249 (1975)
12. J.M. Mendel, *Uncertain Rule-Based Fuzzy Systems Introduction and New Directions* (Prentice-Hall, Englewood Cliffs, NJ, 2001)
13. A. T. Azar, Overview of Type-2 Fuzzy Logic Systems, International Journal of Fuzzy System Applications, 2(4), 1–28, October-December 2012
14. N.N. Karnik, J.M. Mendel, Type-2 fuzzy logic systems. IEEE Trans. Fuzzy Syst. **7**(6) (December, 1999)
15. Computing with words when words can mean different things to different people, in *International ICSC Congress Computation Intelligent: Methods Application, 3rd Annual Symposium Fuzzy Logic Application*. Rochester, NY (June, 1999)
16. Wikipedia Article on Plant. https://en.wikipedia.org/wiki/Plant
17. G.A. Hoysted, J. Kowal, A. Jacob, W.R. Rimington, J.G. Duckett, S. Pressel, S. Orchard, M.H. Ryan, K.J. Field, M.I. Bidartond, A Mycorrhizal Revolution, Current Opinion y Plant Biology. Elsevier (2018)
18. M. Walker, How Flowers Conquered the World. BBC Earth News (July 10, 2009). http://news.bbc.co.uk/earth/hi/earth_news/newsid_8143000/8143095.stm
19. Wikipedia Article on Pollination. https://en.wikipedia.org/wiki/Pollination
20. M. Garc, L. Alberto, *La polinización en los sistemas de producción agrícola: revisión sistemática de la literatura Pollination in Agricultural Systems: A Systematic Literature Review* (IDESIA, Chile, 2016), pp. 53–68
21. D.P. Abrol, *Pollination Biology: Biodiversity Conservation and Agricultural Production* (Springer, Dordrecht Heidelberg, London, New York, 2012)
22. K. Balasubramani, K. Marcus, A study on flower pollination algorithm and its applications. Int. J. Appl. Innov. Eng. Manage. **3**(11) (India) (2014)
23. H. Chiroma, N.L.M. Shuib, S.A. Muaz, A.I. Abubakar, L.B. Ila, J.Z. Maitama, A Review of the applications of bio-inspired flower pollination algorithm. Procedia Comput. Sci. **62**, in *The 2015 International Conference on Soft Computing and Software Engineering*, pp. 435–441 (2015)
24. A.Y. Abdelaziz, E.S. Ali, S.M. Abd Elazim, flower pollination algorithm and loss sensitivity factors for optimal sizing and placement of capacitors in radial distribution systems. Electr. Power Energy Syst. (Elsevier) (2016)
25. P.D. Prasad Reddy, V.C. Veera Reddy, T. Gowri Manohar, Application of flower pollination algorithm for optimal placement and sizing of distributed generation in distribution systems. J. Electr. Syst. Inf. Technol. **3** (Elsevier) (2016)
26. A.Y. Abdelaziz, E.S. Ali, S.M. Abd Elazim, Flower pollination algorithm to solve combined economic and emission dispatch problems. Eng. Sci. Technol. Int. J. Elsevier (2016)
27. A, Draa, On the performances of the flower pollination algorithm—qualitative and quantitative analyses. Appl. Soft Comput. (Elsevier) (2015)

28. P. Dash, L.C. Saikia, N. Sinha, Flower pollination algorithm optimized PI-PD cascade controller in automatic generation control of a multi-area power system. Electr. Power Energy Syst. Elsevier (2016)
29. H.M. Dubey, M. Pandit, B.K. Panigrahi, Hybrid flower pollination algorithm with time-varying fuzzy selection mechanism for wind integrated multi-objective dynamic economic dispatch. Renew. Energy Elsevier (2015)
30. P. Melin, D. Sánchez, O. Castillo, Genetic optimization of modular neural networks with fuzzy response integration for human recognition. Inf. Sci. **197**, 1–19 (2012)
31. M.A. Sanchez, O. Castillo, J.R. Castro, P. Melin, Fuzzy granular gravitational clustering algorithm for multivariate data. Inf. Sci. **279**, 498–511 (2014)
32. D. Sanchez, P. Melin, Optimization of modular granular neural networks using hierarchical genetic algorithms for human recognition using the ear biometric measure. Eng. Appl. Artif. Intell. **27**, 41–56 (2014)
33. O. Castillo, *Type-2 fuzzy logic in intelligent control applications*. Springer (2012)
34. C.I. González, P. Melin, J.R. Castro, O. Mendoza, O. Castillo, An improved sobel edge detection method based on generalized type-2 fuzzy logic. Soft. Comput. **20**(2), 773–784 (2016)
35. E. Ontiveros, P. Melin, O. Castillo, High order α-planes integration: a new approach to computational cost reduction of general type-2 fuzzy systems. Eng. Appl. of AI **74**, 186–197 (2018)
36. C. Caraveo, F. Valdez, O. Castillo, Optimization of fuzzy controller design using a new bee colony algorithm with fuzzy dynamic parameter adaptation. Appl. Soft Comput. **43**, 131–142 (2016)
37. L. Aguilar, P. Melin, O. Castillo, Intelligent control of a stepping motor drive using a hybrid neuro-fuzzy ANFIS approach. Appl. Soft Comput. **3**(3), 209–219 (2003)
38. P. Melin, O. Castillo, Adaptive intelligent control of aircraft systems with a hybrid approach combining neural networks, fuzzy logic and fractal theory. Appl. Soft Comput. **3**(4), 353–362 (2003)

# A Comparative Study of the Grey Wolf Optimizer and Firefly Algorithm in Mathematical Benchmark Functions of the CEC 15 Competition

**Luis Rodríguez, Oscar Castillo, Mario García, and José Soria**

**Abstract** The main goal of this paper is to study the performance of two meta-heuristics, that are classified as belonging to Swarm Intelligence, when dealing with complex optimization problems. In this case, the Algorithms are the Grey Wolf Optimizer (GWO) algorithm and Firefly Algorithm (FA). In this case study, we are presenting in this paper the Mathematical Benchmark Functions that were given in the IEEE Congress on Evolutionary Computation 2015 (CEC' 15) competition, in order to compare these popular algorithms.

**Keywords** Firefly algorithm · Grey wolf optimizer · Constraints · Mathematical functions · Complex problems · Study · Optimization · CEC competition

## 1 Introduction

Computer science has a great impact on the optimization area and the main reason is that today it is more common that people use technology in order to make life easier. People use the new intelligence that researchers created in order to improve (optimize) common activities like driving, working, exercising, playing, etcetera.

In the optimization area, as part of Computer Science, we can find different methodologies to solve a specific problem and one of these methodologies that are based on the stochastic method are the called metaheuristics, which main features are: simple, flexible and avoid the local optimal.

Through time humans have needed to solve different types of problems, sometimes complex problems or new problems that we do not know if it has a good solution with known techniques or if it is necessary to find a new one. According to the description above, in this work, we are presenting an analysis of two popular algorithms with complex problems, in order to study the performance of each algorithm with different problems that were proposed in the original paper of the authors.

L. Rodríguez · O. Castillo (✉) · M. García · J. Soria
Tijuana Institute of Technology, Tijuana, BC, Mexico
e-mail: ocastillo@tectijuana.mx



P. Melin et al. (eds.), *Recent Advances of Hybrid Intelligent Systems Based on Soft Computing*, Studies in Computational Intelligence 915,
https://doi.org/10.1007/978-3-030-58728-4_9

Finally, in this comparative study, we are presenting a statistical test in order to formally compare the GWO and FA metaheuristics in the optimization of Benchmark Functions of CEC 15 Competition and in this way conclude which algorithm has better performance in this specific case.

This paper is organized as follows: Sect. 2 describes the Grey Wolf Optimizer algorithm, Sect. 3 shows a brief explanation of the Firefly Algorithm, in Sect. 4 we are presenting the simulation results and finally in Sect. 5 we offer the conclusions.

## 2 Grey Wolf Optimizer Algorithm

Within the optimization area we can note that each algorithm has features and techniques that can be classified according to the inspiration in evolutionary [1], based on physics [2] and Swarm Intelligence [3]. In addition the No free lunch theorem [4] for optimization proved that there is no meta-heuristic appropriately suited for solving all optimization problems.

The Grey Wolf Optimizer (GWO) [5] algorithm was proposed by Mirjalili in 2014. The main inspiration of this algorithm is based on the behavior of the Canis lupus (Grey Wolf) and the justification of this new metaheuristic is because in the literature and specifically in the Swarm Intelligence area there was not a technique that mimics the behavior of this species.

The original author of the GWO algorithm designed the mathematical model based on two main features of the grey wolf that are the following: the hunting process and the hierarchy in the members of the pack that Muro presented in his work [6] where we can find the hierarchy pyramid [7] of the pack as a leadership strategy. In other words, the best wolf is called alpha ($\alpha$) and is considered as the best solution in the algorithm, the second best solution is called beta ($\beta$) and the third best solution is delta ($\delta$) respectively. Finally, the rest of the candidate solutions are considered as the omega ($\omega$) wolves.

In order to mathematically model the two main inspirations described above, we present the following equations:

$$D = \left\| C \cdot X_p(t) - X(t) \right\| \tag{1}$$

$$X(t+1) = X_p(t) - AD \tag{2}$$

where: Eq. 1 represents the distance between the best solution with and the current individual that we are analyzing. Equation 2 represents the next position of the current individual based on the distance of Eq. 1. Finally, coefficients "A" and "C" are represented by the following equations:

$$A = 2a \cdot r_1 - a \tag{3}$$

$$C = 2 \cdot r_2 \tag{4}$$

where for Eq. 3 the “a” parameter is linearly decreasing through of the iterations, “$r_1$” and “$r_2$” are random values between 0 and 1.

$$D_\alpha = \|C_1 \cdot X_\alpha - X\|, D_\beta = \|C_2 \cdot X_\beta - X\|, D_\delta = \|C_3 \cdot X_\delta - X\| \tag{5}$$

$$X_1 = X_\alpha - A_1 \cdot (D_\alpha), X_2 = X_\beta - A_2 \cdot (D_\beta), X_3 = X_\delta - A_3 \cdot (D_\delta) \tag{6}$$

$$X(t+1) = \frac{X_1 + X_2 + X_3}{3} \tag{7}$$

In addition Eqs. 5 and 6 are the same equations as Eqs. 1 and 2 respectively, but in this case the results are based on the three best wolves (alpha, beta and delta) respectively as we mention above. Finally, in order to calculate the next position of the current individual the original author proposed Eq. 7, which we can be described as an average based on the results obtained in Eq. 6 that represents the leaders of the pack. Finally we can find the general flow of the GWO algorithm in Fig. 1.

## 3 Firefly Algorithm

The firefly algorithm [8] was created in 2008 by Xin-She Yang and was inspired by the flashing patterns and behavior of fireflies. This algorithm has basically three rules that we can find in more detail in the original paper [9]. In this algorithm we can find the two main equations, the first is the attractiveness or the light intensity of each firefly and is represented by the following equation:

$$\beta = \beta_0 \mathbf{e}^{-\gamma \mathbf{r}^2} \tag{8}$$

where $\beta_0$ and $\gamma$ are parameters that could be optimized according to the problem and **r** represents the distance among two fireflies. Finally, the movement of a firefly is calculated with the following equation:

$$\mathbf{x}_i^{t+1} = \mathbf{x}_i^t + \beta_0 \mathbf{e}^{-\gamma \mathbf{r}_{ij}^2} \left(\mathbf{x}_j^t - \mathbf{x}_i^t\right) + \alpha_t \in_i^t \tag{9}$$

where $x_i^{t+1}$ is the new position of the Firefly, $x_i^t$ is the current position of the firefly, where we are adding the attractiveness that exists between the other best solution ($x_j^t$) and the current solution ($x_i^t$) and finally adding the randomness parameter that is represented as a Gaussian distribution or uniform distribution. In addition in Fig. 2, we are presenting a general flowchart of the Firefly Algorithm

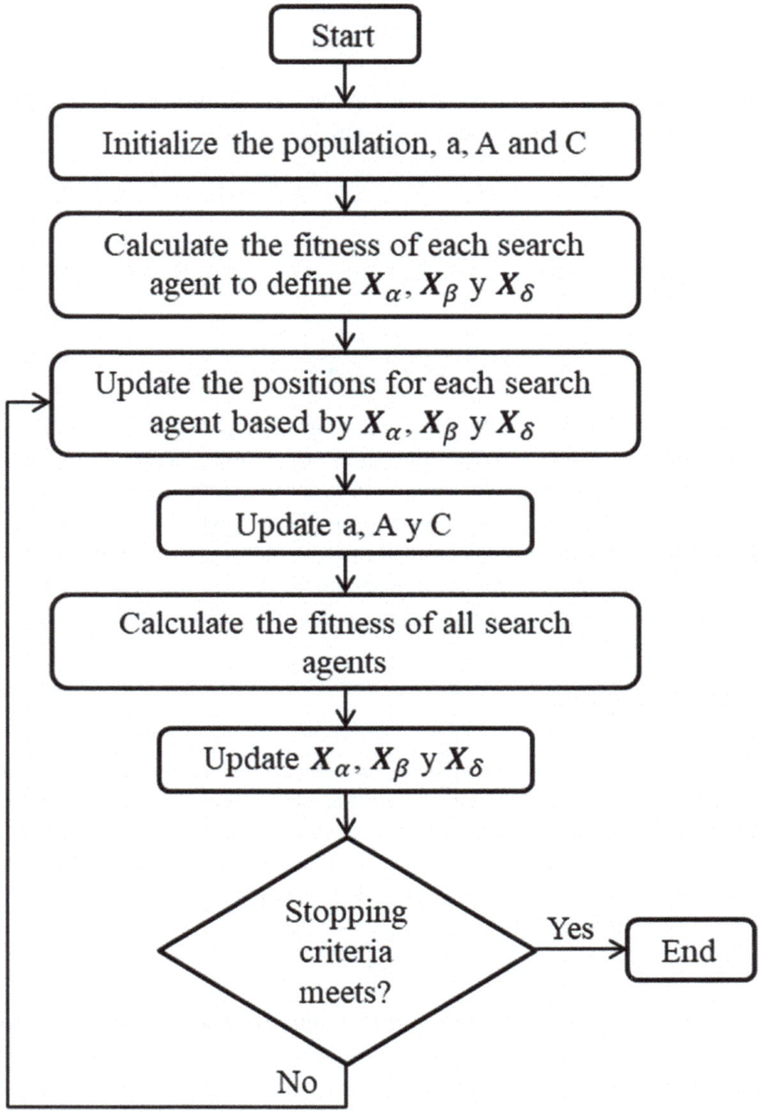


**Fig. 1** General flowchart of the GWO algorithm

## 4 Simulations Results

In this paper, we are presenting a set of problems that contain mathematical functions more difficult to solve in comparison with the conventional benchmark functions [10–12]. In this section, we are presenting the Benchmark Functions of the CEC 2015

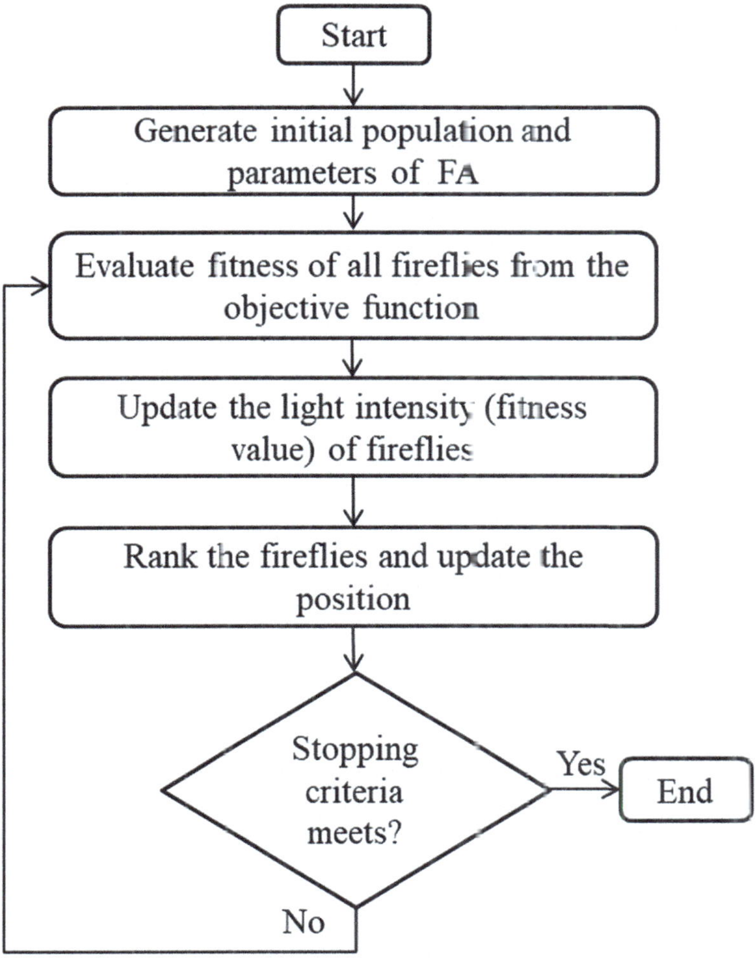


**Fig. 2** General flowchart of the Firefly algorithm

Competition [13] and we can note that all test functions are minimization problems defined as follows:

$$\text{Min } f(x), \quad x = [x_1, x_2, \ldots, x_D]^T \tag{10}$$

where the Dimensions are 10, 30, 50 and 100 respectively, but in this case we are presenting results with 30 and 50 dimensions respectively for each algorithm, the search range is $[-100, 100]^D$ and the maximum number of function evaluations are 10,000 * Dimensions, so we are free to have an appropriate population size to suit the algorithm, while not exceeding the maximum number of function evaluations. In

**Table 1** Summary of the CEC'15 learning-based benchmark suite

| Type | No. | Functions | Obj |
|---|---|---|---|
| Unimodal functions | 1 | Rotated high conditioned elliptic function | 100 |
| | 2 | Rotated cigar function | 200 |
| Simple multimodal functions | 3 | Shifted and rotated Ackley's function | 300 |
| | 4 | Shifted and rotated Rastrigin's function | 400 |
| | 5 | Shifted and rotated Schwefel's function | 500 |
| Hybrid functions | 6 | Hybrid function 1 (N = 3) | 600 |
| | 7 | Hybrid function 2 (N = 4) | 700 |
| | 8 | Hybrid function 3(N = 5) | 800 |
| Composition functions | 9 | Composition function 1 (N = 3) | 900 |
| | 10 | Composition function 2 (N = 3) | 1000 |
| | 11 | Composition function 3 (N = 5) | 1100 |
| | 12 | Composition function 4 (N = 5) | 1200 |
| | 13 | Composition function 5 (N = 5) | 1300 |
| | 14 | Composition function 6 (N = 7) | 1400 |
| | 15 | Composition function 7 (N = 10) | 1500 |
| Search range: $[-100, 100]^D$ | | | |

Table 1 we can find a brief explanation and features as name, type and the minimal optimal value of each function.

For the simulation results, we are presenting the averages and standard deviations as the result of 51 independent executions (recommended for the competition) for each problem of the CEC 15 competition with 30 individuals in both algorithms.

In the Grey Wolf Optimizer algorithm we are using the parameters that the author recommended in the original paper [14], basically the values for the parameter "a" decrease between 2 and 0 and for the "C" parameter the values is in the range [0, 2].

In addition in the FA metaheuristic we are presenting tests with values for the alpha parameter decreasing trough iterations from 0.8 to 0, the minimum value of beta was 0.2 and for the gamma value we considered a value of 1, and these parameters were used based on the paper of Lagunes et al. [15].

Is important to mention that the parameters for the algorithms that were mentioned above, were the parameters that we used in all experimentations that we are presenting in this work. Finally, we can say that these parameters were not optimized or dynamically adjusted with other technique.

In order to statistically demonstrate which metaheuristic has better performance in this set of benchmark functions (CEC 2015) we are also presenting a hypothesis test [16], specifically the z-test in order to show statistically a comparison between the performance of the GWO and FA algorithm respectively. In addition we can mention that if the z-value is less than $-1.645$ we can conclude that the FA algorithm has better performance than the GWO algorithm.

**Table 2** Comparison between FA and GWO algorithms with 30 Dimensions for CEC'15 benchmark functions

| 30 Dimensions | | | | | |
|---|---|---|---|---|---|
| Function | FA | STD | GWO | STD | Z-value |
| F1 | **2.37E+06** | **1.02E+06** | 2.98E+07 | 2.11E+07 | −9.2730 |
| F2 | **8.19E+03** | **5.09E+03** | 2.41E+09 | 1.79E+09 | −9.6150 |
| F3 | **20.0197** | **2.71E−02** | 20.9484 | 4.93E−02 | −117.8910 |
| F4 | **88.9176** | **2.22E+01** | 101.6677 | 2.83E+01 | −2.5315 |
| F5 | 3215.76 | 6.27E+02 | **2917.71** | **9.41E+02** | 1.8824 |
| F6 | **1.03E+05** | **6.68E+04** | 1.38E+06 | 1.24E+06 | −7.3439 |
| F7 | **15.5413** | **2.13E+00** | 21.2706 | 8.63E+00 | −4.6029 |
| F8 | **6.64E+04** | **3.17E+04** | 2.66E+05 | 2.41E+05 | −5.8641 |
| F9 | **102.865** | **2.75E−01** | 138.1955 | 6.59E+01 | −3.8287 |
| F10 | **1.15E+05** | **6.87E+04** | 1.50E+06 | 1.10E+06 | −8.9742 |
| F11 | **608.8026** | **2.55E+02** | 776.7721 | 1.03E+02 | −4.3617 |
| F12 | 110.8674 | 1.83E+01 | **117.6998** | **3.08E+01** | −1.3619 |
| F13 | **0.0441** | **1.07E−02** | 0.0712 | 3.16E−02 | −5.8009 |
| F14 | **3.41E+04** | **1.25E+03** | 3.58E+04 | 2.67E+03 | −4.1180 |
| F15 | **100.0136** | **1.15E−03** | 180.829 | 2.07E+02 | −2.7881 |

Bold indicates best results achieved

In the following Tables we can find the averages and standard deviations of both algorithms (GWO and FA). In addition, we can find also the result of the hypothesis test realized between the two algorithms that we presented above, we can find the Z-value in the last columns.

In Table 2 we can find the results of the GWO and FA algorithms for the 15 mathematical benchmark functions when the problems have 30 dimensions.

In addition, we are presenting the results of hypothesis test in a graphical way, so in Fig. 3 we can find the results of the hypothesis tests for the 15 problems that were analyzed in the CEC 15 competition. According to these results, we can mention that the red bars represent problems where the Firefly Algorithm has poor performance with respect to the Grey Wolf Optimizer Algorithm and the blue bars are the results where the FA method has better performance than the GWO method.

Finally, according to Table 2 and Fig. 3 we can conclude that the Firefly Algorithm has better performance in 13 of the 15 benchmark functions that were analyzed in this work, and this means that the GWO algorithm is better than FA only in a 13% of the total functions. In this case, the GWO algorithm has good results only in F5 and F12 functions respectively.

On the other hand Table 3 and Fig. 4 show the results of the statistical test, but in this case when the problem has 50 variables or dimensions. According to the results we can conclude that for 50 dimensions, the Firefly Algorithm has also better performance than Grey Wolf Optimizer in 12 of the 15 functions that were analyzed

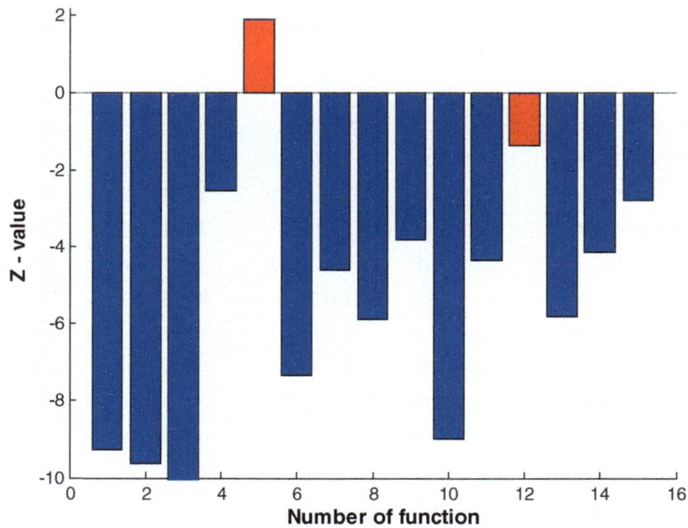


**Fig. 3** Hypothesis tests in a graphical way with 30 dimensions

**Table 3** Comparison between FA and GWO algorithms with 50 Dimensions for CEC'15 benchmark functions

| 50 Dimensions | | | | | |
|---|---|---|---|---|---|
| Function | FA | STD | GWO | STD | Z-value |
| F1 | **1.16E+07** | **4.08E+06** | 8.95E + 07 | 7.13E+07 | −7.7897 |
| F2 | **2.14E+04** | **1.61E+04** | 1.01E +10 | 4.83E+09 | −14.9334 |
| F3 | **20.0167** | **2.26E−02** | 21.1301 | 4.21E−02 | −166.4053 |
| F4 | **206.0428** | **4.31E+01** | 220.8364 | 3.47E+01 | −1.9093 |
| F5 | 5938.56 | 7.74E+02 | **5253.02** | **7.31E+02** | 4.5985 |
| F6 | **5.11E+05** | **2.79E+05** | 5.02E+06 | 3.68E+06 | −8.7252 |
| F7 | **57.4436** | **1.19E+01** | 109.2351 | 3.00E+01 | −11.4602 |
| F8 | **3.40E+05** | **1.62E+05** | 2.53E+06 | 1.37E+06 | −11.3369 |
| F9 | **104.763** | **3.65E−01** | 183.751 | 1.14E+02 | −4.9481 |
| F10 | **2.33E+05** | **1.49E+05** | 6.36E+06 | 4.86E+06 | −8.9990 |
| F11 | 1228.9662 | 1.25E+02 | **1184.1528** | **1.13E+02** | 1.8992 |
| F12 | 195.1389 | 2.09E+01 | **177.2905** | **3.86E+01** | 2.9038 |
| F13 | **0.2036** | **6.06E−02** | 0.4364 | 2.27E−01 | −7.0761 |
| F14 | **6.65E+04** | **7.48E+03** | 8.06E+04 | 4.40E+03 | −11.6032 |
| F15 | **100.0362** | **1.67E−01** | 443.3092 | 3.52E+02 | −6.9644 |

Bold indicates best results achieved

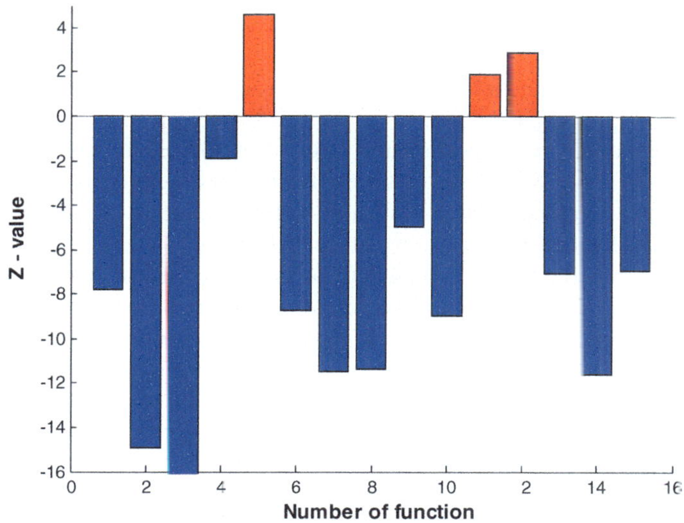


**Fig. 4** Hypothesis tests in a graphical way with 50 dimensions

in this work. Finally, in this set of problems the GWO algorithm has only good performance in F5, F11 and F12 functions respectively.

It is important to mention that these results can be improved, if the values of the parameters change, these values of the parameters that we used were obtained of the papers that we mentioned above. In other words, we can create a manual adjustment or dynamic adjustment in order to improve the results of both metaheuristics in these set of problems and we can present a new study of the results and performance that the algorithms have with the new improved. Fuzzy logic is a very important area in computational intelligence and has shown to improve the results of the algorithms with a great set of problems.

## 5 Conclusions

In this work, we have presented a comparative study between two popular algorithms that were the Grey Wolf Optimizer (GWO) and Firefly Algorithm (FA) and these algorithms were tested with a set of the 15 problems of the CEC 2015 Competition, in order to study their performance in complex problems.

In addition, we have presented a hypothesis test in order to demonstrate statistically which algorithm has better performance with the complex problems that we presented in this work (CEC 15), and for these problems we presented tests with 30 and 50 dimensions respectively and we presented these general results in tables with information that help to analyze the results that we presented above, when we can note the averages and deviation standard for each benchmark function in 51 independent executions.

In addition we can conclude that for 30 dimensions the FA algorithm has better performance in approximately 87% of the problems analyzed and for 50 dimensions the FA shows that has better performance around of 80% of the problems that we presented in this paper. So we can conclude that for these set of problems (CEC 15 Competition) the Firefly algorithm has better performance than the GWO algorithm according to the hypothesis test that we presented above.

Finally, it is important to mention that today is more common that the computers solved mathematical operations in a few seconds and this is the main reason and inspiration of computer science to create new complex problems in order to test the performance of the metaheuristics when the problems simulate the real world problems and its complexity. Finally, when the problems are difficult to solve with conventional algorithms, the researchers choose a hybridization of two or more metaheuristics or create new algorithms based on the behavior of nature.

In future work it is important improve the results in the experiments of the CEC 2015 competition on Learning-based Real-Parameter Single Objective Optimization using other metaheuristics, for example GSA [17], ICA [18], FWA [19, 20] and include fuzzy logic [21] in order to dynamically adjust the parameters in the other algorithms [22, 23]. In addition, instead of using type-1 fuzzy logic we could consider type-2 fuzzy as in [24–32], or consider other areas of application as in [33–41].

## References


1. H.R. Maier, Z. Kapelan, Evolutionary algorithms and other metaheuritics in water resources: Current status, research challenges and future directions. Environ. Model Softw. **62**, 271–299 (2014)
2. U. Can, B. Alatas, Physics based Metaheuristic algorithms for global optimization. Am. J. Inf. Sci. Comput. Eng. **1**, 94–106 (2015)
3. X. Yang, M. Karamanoglu, *Swarm Intelligence and Bio-Inspired Computation: An Overview, Swarm Intelligence and Bio-Inspired Computation*, pp. 3–23. Elsevier (2013)
4. D.H. Wolpert, W.G. Macready, No free lunch theorems for optimization. Evolut. Comput. IEEE Trans. **1**, 67–82 (1997)
5. S. Mirjalili, M. Mirjalili, A. Lewis, Grey wolf optimizer. Adv. Eng. Softw. **69**, 46–61 (2014)
6. C. Muro, R. Escobedo, L. Spector, R. Coppinger, Wolf-pack (Canis lupus) hunting strategies emerge from simple rules in computational simulations. Behav. Process. **88**, 192–197 (2011)
7. L. Rodriguez, O. Castillo, J. Soria, P. Melin, F. Valdez, C. Gonzalez, G. Martinez, J. Soto, A fuzzy hierarchical operator in the grey wolf optimizer algorithm. Appl. Soft Comput. **57**, 315–328 (2017)
8. X.S. Yang, *Firefly Algorithm, Lévy Flights and Global Optimization* (2010). arXiv:1003.1464v1
9. X.S. Yang, Firefly Algorithm: Recent Advances and Applications (2013). arXiv:1308.3898v1
10. J. Digalakis, K. Margaritis, On benchmarking functions for genetic algorithms. Int. J. Comput. Math. **77**, 481–506 (2001)
11. M. Molga, C. Smutnicki, *Test Functions for Optimization Needs* (2005)
12. X.S. Yang, *Test Problems in Optimization* (2010). arXiv:1008.0549
13. W. Guohua, R. Mallipeddi, P.N. Suganthan, *Problem Definitions and Evaluation Criteria for the CEC 2017 Competition on Constrained Real-Parameter Optimization* (2017)

14. L. Rodriguez, O. Castillo, M. Garcia, J. Soria, F. Valdez, P. Melin, *Dynamic Simultaneous Adaptation of Parameters in the Grey Wolf Optimizer Using Fuzzy Logic, FUZZ-IEEE*. Naples Italy (2017)
15. M. Lagunes, O. Castillo, J. Soria, *Optimization of Membership Functions Parameters for Fuzzy Controller of an Autonomous Mobile Robot Using the Firefly Algorithm, Fuzzy Logic Augmentation of Neural and Optimization Algorithms*, pp. 199–205 (2018)
16. R. Larson, B. Farber, *Elementary Statistics Picturing the World*, pp. 428–433. Pearson Education Inc. (2003)
17. B. Gonzalez, P. Melin, F. Valdez, G. Prado-Arechiga, *Ensemble Neural Network Optimization Using a Gravitational Search Algorithm with Interval Type-1 and Type-2 Fuzzy Parameter Adaptation in Pattern Recognition Applications, Fuzzy Logic Augmentation of Neural and Optimization Algorithms: Theoretical Aspects and Real Applications*, pp. 17–27. Springer (2018)
18. E. Bernal, O. Castillo, J. Soria, Imperialist Competitive Algorithm with Dynamic Parameter Adaptation Applied to the Optimization of Mathematical Functions, Nature-Inspired Design of Hybrid Intelligent Systems, pp. 329–341 Springer (2017)
19. J. Barraza, P. Melin, F. Valdez, C.I. Gonzalez, Fuzzy fireworks algorithm based on a sparks dispersion measure. Algorithms **10** (2017)
20. J. Barraza, P. Melin, F. Valdez, C. Gonzalez, Fuzzy FWA with Dynamic Adaptation of Parameters, pp. 4053–4060. IEEE CEC 2016, Vancouver, Canada (2016)
21. L. Rodríguez, O. Castillo, M. García, J. Soria, *A Comparative Study of Dynamic Adaptation of Parameters in the GWO Algorithm Using Type-1 and Interval Type-2 Fuzzy Logic, Fuzzy Logic Augmentation of Neural and Optimization Algorithms: Theoretical Aspects and Real Applications*, pp. 3–16. Springer (2018)
22. C. Caraveo, A. Fevrier, O. Castillo, optimization mathematical functions for multiple variables using the algorithm of self-defense of the plants. *Nature-Inspired Design of Hybrid Intelligent Systems*, pp. 631–640. Springer (2017)
23. M. Guerrero, O. Castillo, M. Garcia, Cuckoo search algorithm via Lévy flight with dynamic adaptation of parameter using fuzzy logic for benchmark mathematical functions. *Design of Intelligent Systems Based on Fuzzy Logic, Neural Networks and Nature-Inspired Optimization. Studies in Computational Intelligence*, pp. 555–571. Springer (2016)
24. C. Leal Ramírez, O. Castillo, P. Melin, A. Rodríguez Díaz, Simulation of the bird age-structured population growth based on an interval type-2 fuzzy cellular structure. Inf. Sci. **181**(3), 519–535 (2011)
25. N.R. Cázarez-Castro, L.T. Aguilar, O. Castillo, Designing type-1 and type-2 fuzzy logic controllers via fuzzy lyapunov synthesis for nonsmooth mechanical systems. Eng. Appl. of AI **25**(5), 971–979 (2012)
26. E. Rubio, O. Castillo, F. Valdez, P. Melin, C. I. González, G. Martinez: an extension of the fuzzy possibilistic clustering algorithm using type-2 fuzzy logic techniques. Adv. Fuzzy Syst. 7094046:1–7094046:23 (2017)
27. O. Castillo, P. Melin, Intelligent systems with interval type-2 fuzzy logic. Int. J. Innov. Comput. Inf. Control **4**(4), 771–783 (2008)
28. G.M. Mendez, O. Castillo, in *The 14th IEEE International Conference on Type-2 TSK fuzzy logic systems using hybrid learning algorithm, Fuzzy Systems, FUZZ'05*, pp. 230–235 (2005). https://scholar.google.com.mx/citations?view_op=view_citation&hl=en&user=1C8gb8IAAAAJ&cstart=40&citation_for_view=1C8gb8IAAAAJ:qxL8FJ1GzNcCInterval
29. P. Melin, C.I. González, J.R. Castro, O. Mendoza, O. Castillo, Edge-detection method for image processing based on generalized type-2 fuzzy logic. IEEE Trans. Fuzzy Syst. **22**(6), 1515–1525 (2014)
30. C.I. González, P. Melin, J.R. Castro, Oscar Castillo, Olivia Mendoza: Optimization of interval type-2 fuzzy systems for image edge detection. Appl. Soft Comput. **47**, 631–643 (2016)
31. C.I. González, P. Melin, J.R. Castro, Olivia Mendoza, Oscar Castillo: An improved Sobel edge detection method based on generalized type-2 fuzzy logic. Soft. Comput. **20**(2), 773–784 (2016)

32. E. Ontiveros, P. Melin, O. Castillo, High order $\alpha$-planes integration: a new approach to computational cost reduction of general type-2 fuzzy systems. Eng. Appl. of AI **74**, 186–197 (2018)
33. P. Melin, O. Castillo, Intelligent control of complex electrochemical systems with a neuro-fuzzy-genetic approach. IEEE Trans. Ind. Electron. **48**(5), 951–955
34. L. Aguilar, P. Melin, O. Castillo, Intelligent control of a stepping motor drive using a hybrid neuro-fuzzy ANFIS approach. Appl. Soft Comput. **3**(3), 209–219 (2003)
35. P. Melin, O. Castillo, Adaptive intelligent control of aircraft systems with a hybrid approach combining neural networks, fuzzy logic and fractal theory. Appl. Soft Comput. **3**(4), 353–362 (2003)
36. P. Melin, J. Amezcua, F. Valdez, O. Castillo, A new neural network model based on the LVQ algorithm for multi-class classification of arrhythmias. Inf. Sci. **279**, 483–497 (2014)
37. P. Melin, O. Castillo, *Modelling, simulation and control of non-linear dynamical systems: an intelligent approach using soft computing and fractal theory* (CRC Press, USA and Canada, 2002)
38. P. Melin, D. Sánchez, O. Castillo, Genetic optimization of modular neural networks with fuzzy response integration for human recognition. Inf. Sci. **197**, 1–19 (2012)
39. M.A. Sanchez, O. Castillo, J.R. Castro, P. Melin, Fuzzy granular gravitational clustering algorithm for multivariate data. Inf. Sci. **279**, 498–511 (2014)
40. D. Sanchez, P. Melin, Optimization of modular granular neural networks using hierarchical genetic algorithms for human recognition using the ear biometric measure. Eng. Appl. Artif. Intell. **27**, 41–56 (2014)
41. O. Castillo, *Type-2 Fuzzy Logic in Intelligent Control Applications*. Springer (2012)

# Optimization of Fuzzy Controllers for Autonomous Mobile Robots Using the Stochastic Fractal Search Method

**Marylu L. Lagunes, Oscar Castillo, Fevrier Valdez, Jose Soria, and Patricia Melin**

**Abstract** This paper aims at measuring the improvement in performance and efficiency of fuzzy controllers after being optimized. The Stochastic Fractal Search (SFS) method is used to optimize the membership functions of a fuzzy controller for an autonomous mobile robot. The experimentation is based on getting the method to reduce the error that exists between the desired trajectory and the real trajectory of the robot. First, the efficiency of the SFS method is tested with benchmark functions, which aims at finding the overall minimum for each of the functions. The SFS method is inspired by fractals, a kind of fragmented structure, which grows at every scale with self-similarity. After testing with different dimensions in the Unimodal functions and Multimodal functions, we proceed with the optimization of the fuzzy controller. Simulation results show the advantages of SFS in fuzzy control optimization.

**Keywords** Stochastic fractal search (SFS) · Optimization · Unimodal benchmark functions · Multimodal benchmark functions · Fixed-dimension multimodal benchmark functions

## 1 Introduction

Nowadays fuzzy systems are widely used, as they are a very powerful tool in making a robot manage the uncertainty between one decision and another, as we humans do when reasoning in real decision-making. For this reason, fuzzy logic originally developed by Zadeh [1–3] has come to revolutionize the world, to be used from the comfort of our home in the amount of water needed by a washing machine with certain weight of clothes, even in an operating room using a robot that performs a complex medical operation, for example in [4–7]. In addition, fuzzy logic is being merged with other methods to achieve more effective tasks depending on the area

M. L. Lagunes · O. Castillo (✉) · F. Valdez · J. Soria · P. Melin
Tijuana Institute of Technology, Tijuana, BC, Mexico
e-mail: ocastillo@tectijuana.mx



P. Melin et al. (eds.), *Recent Advances of Hybrid Intelligent Systems Based on Soft Computing*, Studies in Computational Intelligence 915,
https://doi.org/10.1007/978-3-030-58728-4_10

where it is used, for example in the area of smart computing, fuzzy controllers are being optimized by nature-inspired methods [8–12], which help to improve the performance of these systems. In this work the optimization of a fuzzy controller is performed using the SFS algorithm. To measure the effectiveness of the method it was first tested in benchmark functions. These functions are usually used to measure the effectiveness of proposed algorithms, there are different articles where they have been used [13–15] and they range from the simplest to the more complex ones [16, 17]. This paper experiments with the idea that a fuzzy controller can produce better results if it is optimized. In this case, only the parameter values of the input and output membership functions are optimized.

The paper is organized as follows. Section 2 describes the SFS method for the optimization of the parameters of the membership functions of the fuzzy system. Section 3 shows the proposed methodology for fuzzy control optimization. Section 4 shows the experimentation and results obtained from the tests and Sect. 5 describes the conclusions of this work.

## 2 Stochastic Fractal Search (SFS)

Developed by Salimi [18, 19] inspired in the nature of fractals, which are a self-similar structure growing in different scales, as illustrated Fig. 1. This method is not like some other bio-inspired ones that work together in populations. Instead, based on an existing point is generating another similar point, to form a scaled structure, with which aims at finding the best point that generates the best value of an objective function.

The SFS has the following equations.

$$P = LB + \varepsilon * (UP - \text{LB}) \tag{1}$$

The population of particles P is generated randomly based on the constraints of the problem after setting the lower limit LB and the upper limit UB.

The equation defining the ***Diffusion*** process (exploitation) is:

$$GW_1 = Gaussian(\mu_{BP}, \sigma) + \left(\varepsilon * BP - \varepsilon' * P_i\right) \tag{2}$$

$$GW_2 = Gaussian(\mu_P, \sigma) \tag{3}$$

where $\varepsilon$, $\varepsilon'$ represent random numbers in the range [0, 1], *BP* is the best position of the point, *i*th point in the group is represented by $P_i$ and *Gaussian* is a function that generates a random number from the normal distribution with a mean $\mu$ parameter and a standard deviation with the $\sigma$ parameter:

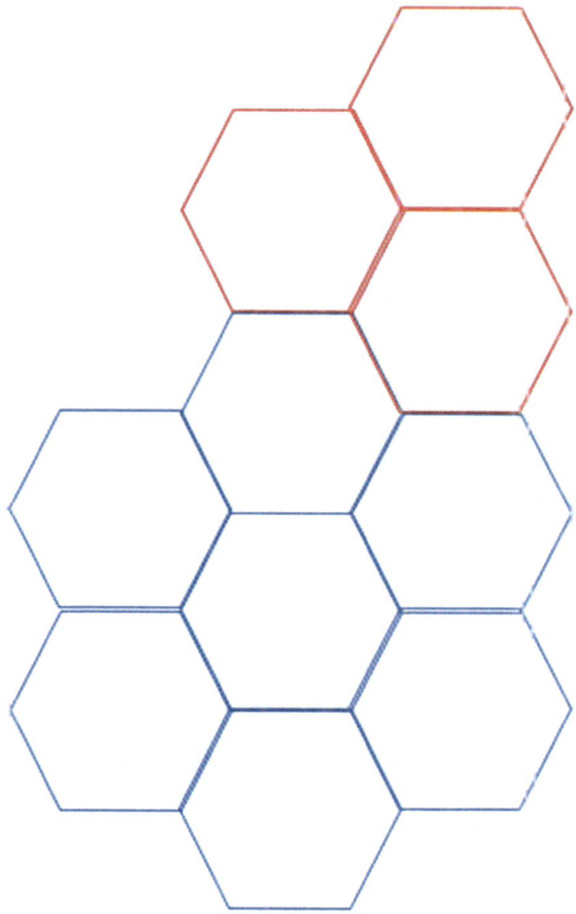

**Fig. 1** Example of a fractal object

$$\sigma = \frac{\log g}{g} * |P_i - BP| \tag{4}$$

where $\frac{\log g}{g}$ represents a ratio approaching zero as the generation number g increases.

The equation defining the ***Update*** Processes (exploration) is:

$$Pa_i = \frac{rank(P_i)}{N} \tag{5}$$

where $N$ is the number of particles in the group and $Pa_i$ is the computed probability value for a particle whose *rank* among the others is given by the *rank* function. The particles are classified with respect to their fitness value and then a probability value is assigned to each particle *i*.

$$P_i'(j) = P_x(j) - \varepsilon * \left(P_y(j) - P_i(j)\right) \tag{6}$$

where the modified component is represented by $P_i'(j)$, and $P_x$, $P_y$ are different points randomly chosen from the group. $P_i'$ replaces $P_i$ if it has a better fitness value.

## 3 Methodology

The methodology describes the flow of data that passes from the method to the fuzzy controller. The method sends each point that forms its population to create a new optimized fuzzy controller, that is if we have population of 20, then 20 times a vector will be created in this case of random values from the range $-1$ to 1 that will optimize the membership values after generating each fuzzy controller will be simulated one by one in each iteration until the best one is found according to the MSE metric. The method takes the objective function as the error of control, and based on that, it aims at finding the global minimum to solve the problem.

As described by the SFS method, it updates each of its points and ranks which is the best fitness function to discard the one that does not produce the best desired value. With this, the optimization process is carried out with a sequence of steps as shown in Fig. 2.

Figure 3 illustrates the plant, where the fuzzy controller is simulated, as noted, the fuzzy logic controller component is responsible for receiving the fuzzy system and executing it, then passing the information through the other components, which are part of the functions for the movement of the robot. There are also equations established for defining the desired trajectory, which includes the linear velocity, angular velocity, torque 1 and torque 2, and in general this is how the data flows in the plant created in Simulink.

## 4 Experimentation and Results

The development of the experimentation was based on the optimization of a fuzzy logic controller [20–22],which controls the movement of an autonomous mobile robot. The movement of the robot is governed by the angular velocity ($e_w$) and linear velocity ($e_v$) to produce the outputs of torque 1 (t1) and torque 2 (t2). The union of these inputs and outputs form the structure of the fuzzy system, which has 9 if-then rules, which trigger the values that each of the outputs take for an optimum controller performance. Figure 4 shows the fuzzy system with inputs and outputs [23–25]. The optimization is carried out by the SFS method, which takes the objective function as the error of control and executes all the established iterations, to generate in each one the best value that substitutes the values of the membership functions, until obtaining the fuzzy controller generates the best mobile robot performance, trying to reach a

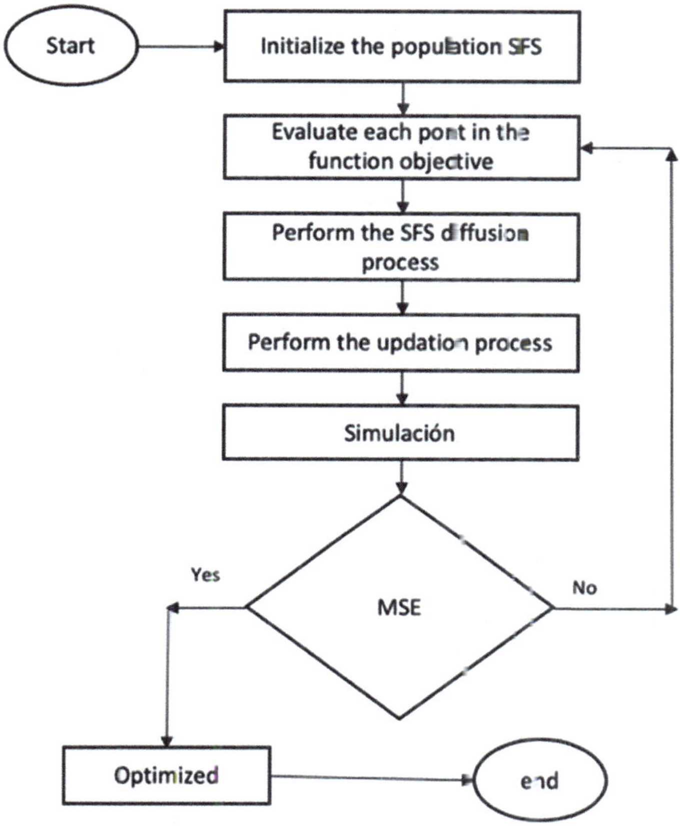


**Fig. 2** Optimization process

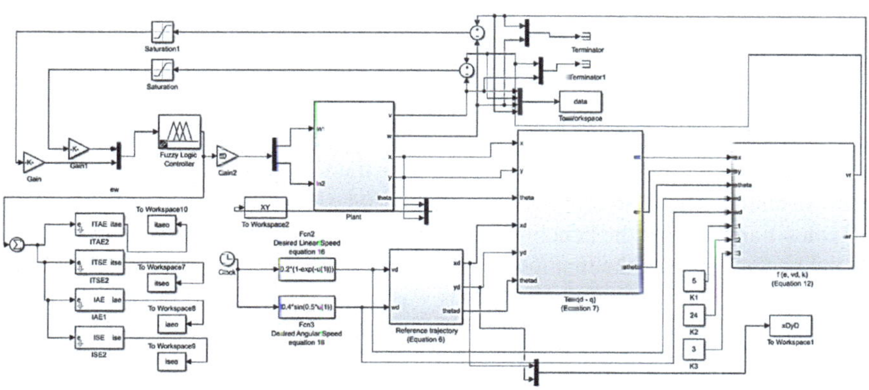

**Fig. 3** Simulation plant

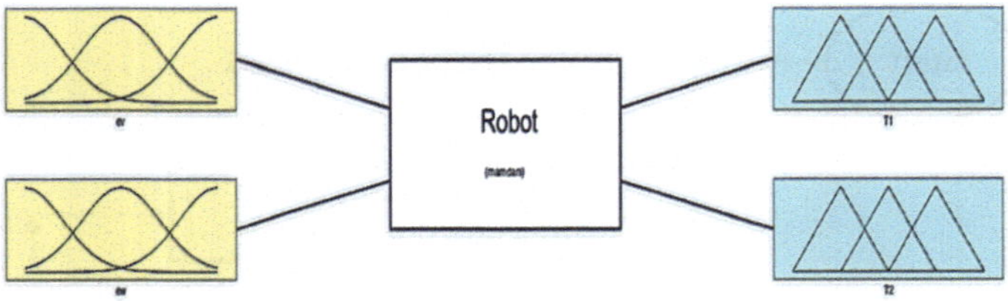


**Fig. 4** Fuzzy system structure

zero difference value between the desired trajectory and the real trajectory of the robot, using the MSE as a metric.

***The rules of the fuzzy system are***:

1. If (ev is N) and (ew is N) then (T1is N) (T2 is N)
2. If (ev is N) and (ew is Z) then (T1is N) (T2 is Z)
3. If (ev is N) and (ew is P) then (T1is N) (T2 is P)
4. If (ev is Z) and (ew is N) then (T1is Z) (T2 is N)
5. If (ev is Z) and (ew is Z) then (T1is Z) (T2 is Z)
6. If (ev is Z) and (ew is P) then (T1is Z) (T2 is P)
7. If (ev is P) and (ew is N) then (T1is P) (T2 is N)
8. If (ev is P) and (ew is Z) then (T1is P) (T2 is Z)
9. If (ev is P) and (ew is P) then (T1is P) (T2 is P).

The error is represented by the following equation of the *Mean Squared Error (MSE)*

$$MSE = \frac{1}{n}\sum_{i=1}^{n}(\hat{Y}_i - Y_i)^2 \tag{7}$$

where $\hat{Y}$ is a vector of $n$ predictions, and $Y$ is the vector of true values.

First we experimented with some benchmark functions to observe the behavior of the method, and then the fuzzy controller was optimized. For testing the fuzzy system we did consider 3 different paths, which the robot managed to match with an error of near to 0%. Figures 5, 6 and 7 illustrate each of the paths used in the experimentation, respectively.

Tables 1 and 2 show the benchmark functions used in the experiments in order to test the performance of the method.

Each of the above mentioned functions was executed 30 times for each different number of dimensions 30, 64, and 128 with the intention of examining the results and thus determining with how many dimensions the method works best. This without diminishing importance to the value of the population and generations, which are also very important in experimentation, because based on that the method may or may not have a good performance. As observed in Table 3, each column shows the

Fig. 5 Trajectory 1

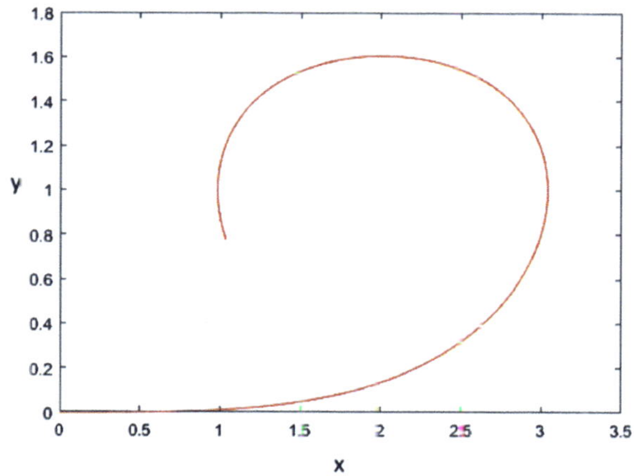


Fig. 6 Trajectory 2

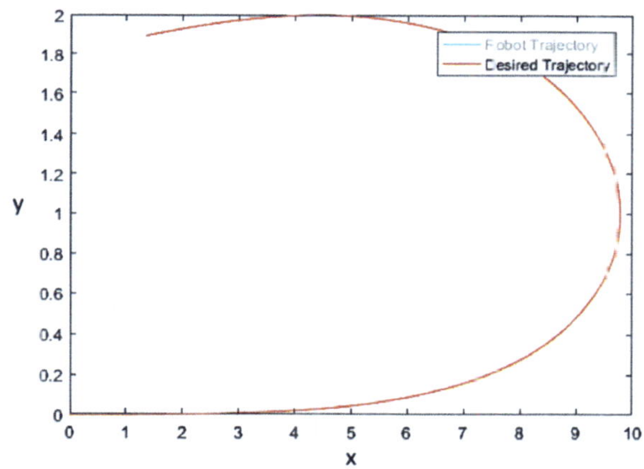


Fig. 7 Trajectory 3

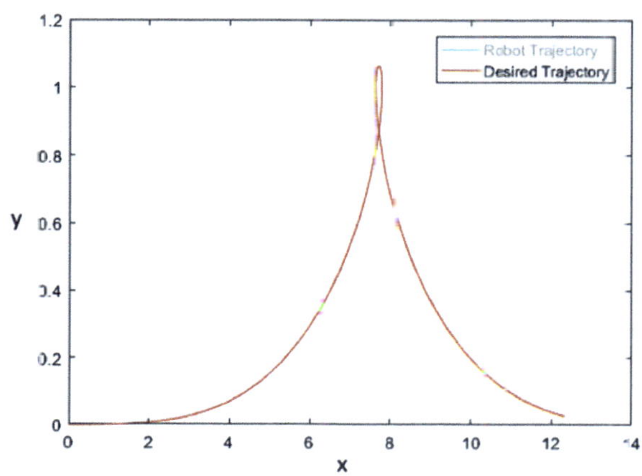

**Table 1** Benchmark functions

| Benchmark functions *f1–f7* |
|---|
| $f_1(x) = \sum_{1=1}^{n} x_i^2$ <br> *Search space* $x_j \in [-100, 100]$ *and* $f(x^*) = 0$ |
| $f_2(x) = \sum_{i=1}^{n} \lvert x_i \rvert + \prod_{i=1}^{n} \lvert x_i \rvert$ <br> *Search space* $x_j \in [-10, 10]$ *and* $f(x^*) = 0$ |
| $f_3(x) = \sum_{i=1}^{n} \left( \sum_{j-1}^{i} x_j \right)^2$ <br> *Search space* $x_j \in [-100, 100]$ *and* $f(x^*) = 0$ |
| $f_4(x) = max_i \{ \lvert x_i \rvert, 1 \leq i \leq n \}$ <br> *Search space* $x_j \in [-100, 100]$ *and* $f(x^*) = 0$ |
| $f_5(x) = \sum_{i=1}^{n-1} \left[ 100 \left( x_{i+1} - x_i^2 \right)^2 + (x_i - 1)^2 \right]$ <br> *Search space* $x_j \in [-30, 30]$ *and* $f(x^*) = 0$ |
| $f_6(x) = \sum_{i=1}^{n} ([x_i + 0.5])^2$ <br> *Search space* $x_j \in [-100, 100]$ *and* $f(x^*) = 0$ |
| $f_7(x) = \sum_{i=1}^{n} i x_i^4 + random[0, 1]$ <br> *Search space* $x_j \in [-1.28, 1.28]$ *and* $f(x^*) = 0$ |

**Table 2** Benchmark functions

| Benchmark functions *f9–f11* |
|---|
| $f_9(x) = \sum_{i=1}^{n} \left[ x_i^2 - 10\cos(2\pi x_i) + 10 \right] \mathrm{f}_9(x) = \sum_{\mathrm{i}=1}^{\mathrm{n}} \left[ \mathrm{x}_\mathrm{i}^2 - 10\cos(2\pi \mathrm{x}_\mathrm{i}) + 10 \right]$ <br> *Search space* $x_j \in [-5.12, 5.12]$ *and* $f(x^*) = 0$ <br> $f_{10}(x) = -20exp\left( -0.2\sqrt{\frac{1}{n} \sum_{i=1}^{n} x_i^2} \right) - exp\left( \frac{1}{n} \sum_{i=1}^{n} cos(2\pi x_i) \right) + 20 + e$ |
| $f_{10}(x) =$ *Search space* $x_j \in [-32, 32]$ *and* $f(x^*) = 0$ <br> $f_{11}(x) = \frac{1}{400} \sum_{i=1}^{n} x_i^2 - \prod_{i=1}^{n} cos\left( \frac{x_i}{\sqrt{i}} \right) + 1$ <br> *Search space* $x_j \in [-600, 600]$ *and* $f(x^*) = 0$ |

**Table 3** Benchmark function results

| Function | 30 dim | 64 dim | 128 dim |
|---|---|---|---|
| | Average | Average | Average |
| f1 | 0 | 0 | 0 |
| f2 | 0 | 0 | 0 |
| f3 | 0 | 0 | 0 |
| f4 | 0 | 0 | 0 |
| f5 | 5.01E−01 | 4.22E−01 | 9.86E−01 |
| f6 | 1.53E−08 | 4.53E−03 | 6.68E−01 |
| f7 | 9.56E−04 | 1.97E−03 | 1.34E−03 |
| f9 | 0 | 0 | 0 |
| f10 | 3.00E−02 | 4.21E−02 | 3.85E−02 |
| f11 | 0 | 0 | 0 |

average obtained by the SFS in the optimization of the functions, with a population of 15 points and 500 generations.

In most of the optimized functions, the SFS method showed that it has good performance, as observed in the previous results, there are 5 functions that reached an average of their global minimum 0. But we can also observe that in f5, f6, f7, f10 the results were not so satisfactory, because they were very far from their global minimum compared to the other functions. In this type of experimentation is where the quality of a method in versability to adapt from one problem to another is described. and also how much it affects the number of dimensions in which we have to work.

In Table 4, we can note a total of 10 experiments, only one produced a value near E-04 where the others were between E-02 and E-03, which pending more experimentation to increase the number of generations, to test if this improves the performance of the method (Fig. 8).

**Table 4** Results with trajectory 1

| Experiment | Generation | Point | MSE |
|---|---|---|---|
| 1 | 500 | 15 | 1.03E−02 |
| 2 | 500 | 15 | 2.22E−02 |
| 3 | 500 | 15 | 1.18E−03 |
| 4 | 500 | 15 | 1.05E−02 |
| 5 | 500 | 15 | 9.00E−03 |
| 6 | 500 | 15 | 2.04E−04 |
| 7 | 500 | 15 | 3.94E−03 |
| 8 | 500 | 15 | 3.73E−03 |
| 9 | 500 | 15 | 3.79E−02 |
| 10 | 500 | 15 | 3.15E−02 |

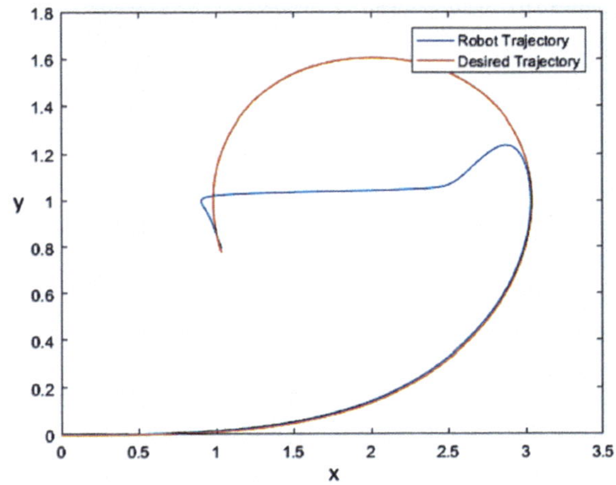


**Fig. 8** Results of optimization of trajectory 1

Table 5 shows a population of 15 points, and a total of 500 generations for each experiment, since this method is characterized by being fast and not requiring much execution time, even so, the method managed to give a good result in the optimization since it obtained values between E-03 and E-04.

In Fig. 9 it is observed how the robot follows the desired path, having an MSE of E-03 as the best result in optimization.

As illustrated in Table 6, we have a summary of 10 experiments for the optimization of the fuzzy controller, where it is observed that the results had a stagnation between E-02 and E-03, analyzing this it can be said that the method had difficulty to generate the vector of optimal data for this optimization and therefore the fuzzy system does not show great performance (Fig. 10).

**Table 5** SFS results of trajectory 2

| Experiment | Generation | Points | MSE |
|---|---|---|---|
| 1 | 500 | 15 | 5.21E−04 |
| 2 | 500 | 15 | 3.77E−03 |
| 3 | 500 | 15 | 1.51E−03 |
| 4 | 500 | 15 | 2.35E−03 |
| 5 | 500 | 15 | 2.33E−03 |
| 6 | 500 | 15 | 6.33E−03 |
| 7 | 500 | 15 | 7.66E−04 |
| 8 | 500 | 15 | 4.61E−03 |
| 9 | 500 | 15 | 3.53E−03 |
| 10 | 500 | 15 | 1.30E−03 |

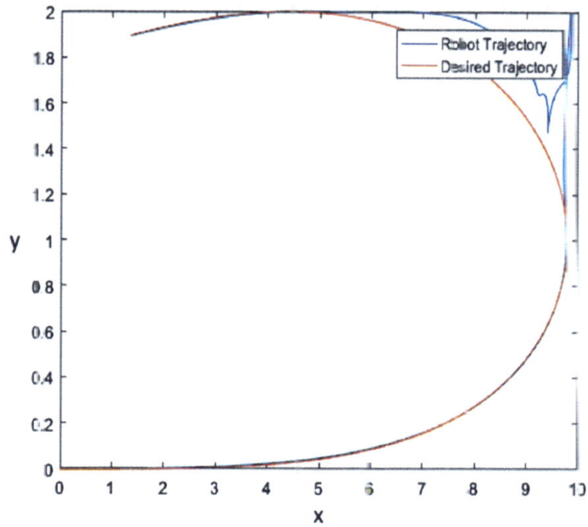


**Fig. 9** Trajectory 2, optimized with SFS

**Table 6** Results for trajectory 3 using SFS

| | Generation | Points | MSE |
|---|---|---|---|
| 1 | 500 | 15 | 1.15E−02 |
| 2 | 500 | 15 | 6.27E−02 |
| 3 | 500 | 15 | 2.01E−02 |
| 4 | 500 | 15 | 1.28E−02 |
| 5 | 500 | 15 | 5.93E−03 |
| 6 | 500 | 15 | 1.93E−03 |
| 7 | 500 | 15 | 2.76E−03 |
| 8 | 500 | 15 | 5.15E−02 |
| 9 | 500 | 15 | 3.35E−02 |
| 10 | 500 | 15 | 1.02E−03 |

## 5 Conclusions

Analyzing the results obtained in each of the optimizations, it can be concluded that the SFS method helped to improve the performance in general in the experimentation. In most of the functions it reached its global minimum, although in some more complex functions it had some difficulty.

When it comes to the optimization of fuzzy controllers, we must know that it is not as simple as the functions, since optimizing the fuzzy system entails more computational work, and requires more time. In addition to the method also requiring more to achieve the best possible result, for each generation and simulation executed. For this reason it can be concluded that, a satisfactory result was obtained, although

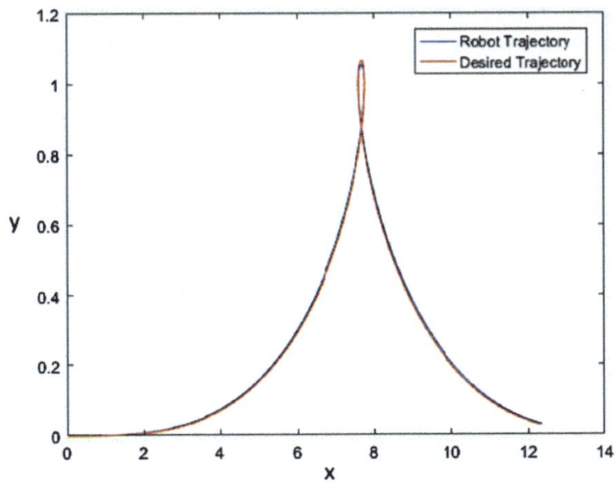


**Fig. 10** Trajectory 3 results

0% error was not reached in some trajectories, even so, the error was reduced in some cases close to E-04.

As future work, the SFS algorithm can be dynamically adjusted to improve optimization results, in addition to testing it with other optimization problems, increasing the size of the population, adjusting generations, etc., all this, so that the method better adapts of the problems that we have to solve. We can also consider other applications, as in [26–29].

**Acknowledgements** We would like to express our gratitude to CONACYT and to Tijuana Institute of Technology for the help and resources provided for the development of this research.

## References

1. L.A. Zadeh, The concept of a linguistic variable and its application to approximate reasoning-III. Inf. Sci. (Ny) **9**(1), 43–80 (1975)
2. L.A. Zadeh, Fuzzy sets. Inf. Control **8**(3), 338–353 (1965)
3. L.A. Zadeh, Fuzzy logic. Comput. (Long. Beach. Calif) **21**(4), 83–93 (April 1988)
4. C. Cecati, F. Ciancetta, P. Siano, A multilevel inverter for photovoltaic systems with fuzzy logic control. IEEE Trans. Ind. Electron. **57**(12), 4115–4125 (2010)
5. H.A. Hagras, A hierarchical type-2 fuzzy logic control architecture for autonomous mobile robots. IEEE Trans. Fuzzy Syst. **12**(4), 524–539 (2004)
6. C.T. Lin, G.C.S. Lee, Neural-network-based fuzzy logic control and decision system. IEEE Trans. Comput. **40**(12), 1320–1336 (1991)
7. P. Martin Larsen, Industrial applications of fuzzy logic control. Int. J. Man. Mach. Stud. **12**(1), 3–10 (1980)
8. A.K. Bhandari, V.K. Singh, A. Kumar, G.K. Singh, Cuckoo search algorithm and wind driven optimization based study of satellite image segmentation for multilevel thresholding using Kapur's entropy. Expert Syst. Appl. **41**(7), 3538–3560 (2014)

9. F. Olivas, F. Valdez, O. Castillo, C.I. Gonzalez, G. Martinez, P. Melin, Ant colony optimization with dynamic parameter adaptation based on interval type-2 fuzzy logic systems. Appl. Soft Comput. J. **53**, 74–87 (2017)
10. L. Rodríguez et al., A fuzzy hierarchical operator in the grey wolf optimizer algorithm. Appl. Soft Comput. J. **57**, 315–328 (2017)
11. C. Soto, F. Valdez, O. Castillo, A review of dynamic parameter adaptation methods for the Firefly algorithm, in *Nature-Inspired Desing of Hybrid Intelligent Systems*, pp. 285–295. Springer, Cham (2017)
12. J. Pérez, F. Valdez, O. Castillo, Modification of the bat algorithm using type-2 fuzzy logic for dynamical parameter adaptation, in *Nature-Inspired Desing of Hybrid Intelligent Systems*, pp. 343–355. Springer, Cham (2017)
13. C. Peraza, F. Valdez, O. Castillo, Harmony search with dynamic adaptation of parameters for the optimization of a benchmark set of functions, in *Hybrid Intelligent Systems in Control, Pattern Recognition and Medicine*, pp. 97–108 (2020)
14. P. Ochoa, O. Castillo, J. Soria, The differential evolution algorithm with a fuzzy logic approach for dynamic parameter adjustment using benchmark functions, in *Hybrid Intelligent Systems in Control, Pattern Recognition and Medicine*, pp. 169–179 (2020)
15. C. Solano-Aragón, O. Castillo, Optimization of benchmark mathematical functions using the Firefly algorithm, in *Fuzzy Logic Aumentation of Nature-Inspired Optimization Metaheuristic*, pp. 177–189 Springer, Cham (2014)
16. E. Bernal, O. Castillo, J. Soria, F. Valdez, Galactic swarm optimization with adaptation of parameters using fuzzy logic for the optimization of mathematical functions, in *Studies in Computational Intelligence*, vol. 749, pp. 131–140. Springer Verlag (2018)
17. C. Caraveo, F. Valdez, O. Castillo, A new optimization metaheuristic based on the self-defense techniques of natural plants applied to the CEC 2015 benchmark functions. Adv. Intell. Syst. Comput. **641**, 380–388 (2018)
18. H. Salimi, Stochastic fractal search: a powerful metaheuristic algorithm. Knowl. Based Syst. **75**, 1–18 (2015)
19. S. Khalilpourazari, S. Khalilpourazary, A robust stochastic fractal search approach for optimization of the surface grinding process. Swarm Evol. Comput. **38**, 173–186 (2018)
20. M.L. Lagunes, O. Castillo, F. Valdez, J. Soria, Multi-metaheuristic competitive model for optimization of fuzzy controllers. Algorithms **12**(5), 90 (2019)
21. M.L. Lagunes, O. Castillo, J. Soria, M. Garcia, F. Valdez, Optimization of granulation for fuzzy controllers of autonomous mobile robots using the Firefly Algorithm. Granul. Comput. 1–11 (July 2018)
22. A. De Luca, G. Oriolo, M. Vendittelli, *Control of Wheeled Mobile Robots: An Experimental Overview* (Ramsete, Springer, Berlin, Heidelberg, 2001), pp. 181–226
23. M.L. Lagunes, O. Castillo, F. Valdez, J. Soria, P. Melin, Parameter optimization for membership functions of type-2 fuzzy controllers for autonomous mobile robots using the Firefly algorithm, in *North American Fuzzy Information Processing Society Annual Conference*, pp. 569–579. Springer, Cham (2018)
24. M.L. Lagunes, O. Castillo, J. Soria, Optimization of membership function parameters for fuzzy controllers of an autonomous mobile robot using the Firefly algorithm, in *Fuzzy Logic Augmentation of Neural and Optimization Algorithms: Theoretical Aspects and Real Applications*, pp. 199–206. Springer, Cham (2018)
25. L. Astudillo, P. Melin, O. Castillo, Optimization of a fuzzy tracking controller for an autonomous mobile robot under perturbed torques by means of a chemical optimization paradigm, in *Recent Advances on Hybrid Intelligent Systems*, pp. 3–20. Springer, Berlin, Heidelberg (2013)
26. P. Melin, D. Sánchez, O. Castillo, Genetic optimization of modular neural networks with fuzzy response integration for human recognition. Inf. Sci. **197**, 1–19 (2012)
27. M.A. Sanchez, O. Castillo, J.R. Castro, P. Melin, Fuzzy granular gravitational clustering algorithm for multivariate data. Inf. Sci. **279**, 498–511 (2014)

28. D. Sanchez, P. Melin, Optimization of modular granular neural networks using hierarchical genetic algorithms for human recognition using the ear biometric measure. Eng. Appl. Artif. Intell. **27**, 41–56 (2014)
29. O. Castillo, *Type-2 fuzzy logic in intelligent control applications*. Springer (2012)

# Fuzzy Dynamic Parameter Adaptation for Particle Swarm Optimization of Modular Granular Neural Networks Applied to Time Series Prediction

**Daniela Sánchez, Patricia Melin, and Oscar Castillo**

**Abstract** In this paper the optimization of Modular Granular Neural Networks (MGNNs) using a Particle Swarm Optimization (PSO) approach is proposed. PSO uses dynamic parameter adaptation based on Fuzzy Logic (FL). The MGNN is applied to time series prediction. To perform comparisons with the achieved results using the fuzzy dynamic adaptation of the PSO, tests with non-optimized trainings and runs with a conventional PSO are performed. The fuzzy dynamic parameters of the PSO are $c_1$, $c_2$ and w, as these parameters are essential to have a better PSO performance. The optimization finds optimal parameters of the MGNNs such as: the number of sub modules, number of hidden layers with their number of neurons and goal error. The effectiveness of the proposed method is tested using 800 data points of the Mackey-Glass time series. Tests using 500 and 300 data points for the training phase are performed for non-optimized and optimized methods.

**Keywords** Fuzzy logic · Neural networks · Particle swarm optimization

## 1 Introduction

Nowadays, intelligence techniques have emerged and allowed to solve complex problems that have required combining two or more of these techniques. These combinations, known as hybrid intelligence systems have been proposed in many works to solve important problems, such as: human recognition [1, 2], classification [3], control problems [4, 5], time series prediction [6] among others. Some of the most

D. Sánchez (✉) · P. Melin · O. Castillo
Tijuana Institute of Technology, Tijuana, Mexico
e-mail: danielasanchez.itt@hotmail.com

P. Melin
e-mail: pmelin@tectijuana.mx

O. Castillo
e-mail: ocastillo@tectijuana.mx


P. Melin et al. (eds.), *Recent Advances of Hybrid Intelligent Systems Based on Soft Computing*, Studies in Computational Intelligence 915
https://doi.org/10.1007/978-3-030-58728-4_11

important optimization algorithms were proposed some years ago such as: genetic algorithm (GA) [7], ant colony system (ACO) [8] and particle swarm optimization (PSO) [9, 10]. Others more recently, such as: firefly algorithm (FA) [11], grey wolf optimizer (GWO) [12] or cuckoo optimization algorithm (COA) [13] have also demonstrated to achieve good results. In this work, an improvement to the particle swarm optimization is proposed to achieve better results than a conventional PSO. This improvement consists in adjustments of parameters during the execution of the algorithm. This allows adjusting parameters values depending on their current behavior.

This paper is organized as follows: In Sect. 2, the basic concepts used in this research work are described. Section 3 contains the general architecture. In Sect. 4 the experimental results are presented and in Sect. 5, the conclusions and future work are presented.

## 2 Basic Concepts

In this section the basic concepts used in this research work are presented.

### *2.1 Modular Neural Networks*

A modular neural network (MNN) is a kind of a conventional artificial neural network (ANN) [14], but MNNs are based on the principle of divide and conquer [15]. The modular neural networks are composed of modules (or sub modules), where their results are combined using an integrating unit. The main idea of this kind of neural network is to divide a general task into subtasks, where each sub module will be an expert in its assigned subtask, working independently of others. Many works have demonstrated advantages obtained using MNN due to its ability to improve significantly learning [16].

### *2.2 Fuzzy Logic*

In 1965, Zadeh introduced the term fuzzy logic and described the mathematics of fuzzy set theory [17]. Fuzzy logic is an area of soft computing that enables a computer system to reason with uncertainty. The main motivation of fuzzy logic was the need to express a partial membership of an element in a set, i.e., in classical logic an element belongs or does not belong in a set (0 or 1), meanwhile, in fuzzy logic an element has a degree of membership (a value between 0 and 1) [18].

## 2.3 Granular Computing

Granular computing (GrC) was originally proposed by Zadeh [18]. Some key issues in this area of soft computing are granules, information granulation and causation. Finer granules compose a granule, and a granule is defined as a class, object or cluster of a universe [19]. The information granulation involves decomposition of whole into parts; the organization involves integration of parts into whole; and the causation involves association of causes with effects [20, 21].

## 2.4 Particle Swarm Optimization

Eberhart and Kennedy in [9, 10] proposed the particle swarm optimization based on the social behaviors of fish schooling or birds flocking. Where a swarm contains particles and each particle represents a solution. The next position of a particle in the search space is determined by:

$$\mathrm{x}_{id}(t+1) = \mathrm{x_i(t)} + \mathrm{v}_i(t+1) \tag{1}$$

where $\mathrm{x}_{id}(t)$ denotes a current position of the particle $i$, dimension $d$, at time $t$, $\mathrm{v}_i$ $(t + 1)$ is assigned to calculate the next position. The star topology is used in this work. In this topology, the neighborhood of each particle is the entire swarm. The original algorithm was improved in [22, 23], where inertia weight ($w$) is used and its value decreases linearly. The particle velocity is determined by:

$$v_{\mathrm{id}}(t+1) = wv_{id}(t) + c_1 r_{1d}(t) * \left[\mathrm{y}_{id}(t) - \mathrm{x_{id}}(t)\right] + \mathrm{c_2 r_{2d}(t)} * \left[\hat{\mathrm{y}}_{\mathrm{d}}(t) - \mathrm{x_{id}(t)}\right] \tag{2}$$

where $r_1$ and $r_2$ are random values from 0 to 1, $w$ is the inertia weight with a decrease value. The $c_1$ and $c_2$ parameters represent cognitive and social components, the best position of a particle $i$ in dimension $d$ is represented by $y_{id}$(t) and the best position of the swarm in $d$ dimension is denoted by $\hat{y}_{\mathrm{d}}$(t).

# 3 Proposed Method

The proposed method consists of modular granular neural networks for time series prediction. The MGNNs parameters are optimized by a PSO with fuzzy dynamic adaptation. The combination of these intelligence techniques are explained in this section.

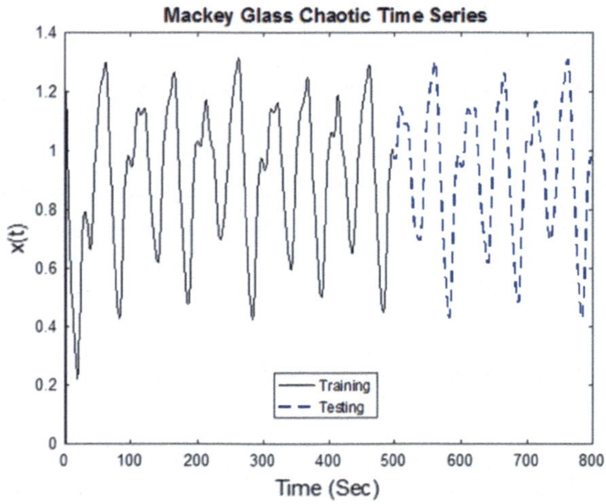


**Fig. 1** Division of data points of a time series

## 3.1 *Modular Granular Neural Networks Applied to Time Series Prediction*

The modular granular neural networks were proposed in [24, 25] applied to human recognition. In this work, this kind of neural network is applied to time series prediction. The division of the data points used in the proposed method is shown in Fig. 1, where a time series is divided into a sub set of data points used to train the MNGGs and the rest of data points are used for the testing phase. The general architecture of this application is illustrated in Fig. 2, where the main idea is to divide the data points used for the training phase into *m* sub datasets. The data points used for the testing phase are simulated in each sub module and an average is calculated using these results.

## 3.2 *Description of the Fuzzy Dynamic Parameters Adaptation for Particle Swarm Optimization*

In a previous work [26], two fuzzy inference systems were proposed to adjust PSO parameters for MGNNs optimization applied to human recognition. In this work, the best fuzzy inference system was replicated and applied to time series prediction. This fuzzy inference system has two inputs and three outputs. Each variable uses three triangular membership functions, and their linguistic labels are respectively "LOW", "MEDIUM" and "HIGH". This fuzzy inference system is shown in Fig. 3.

The fuzzy inference system is of Mamdani type and the variables are described below:

The inputs variables are:

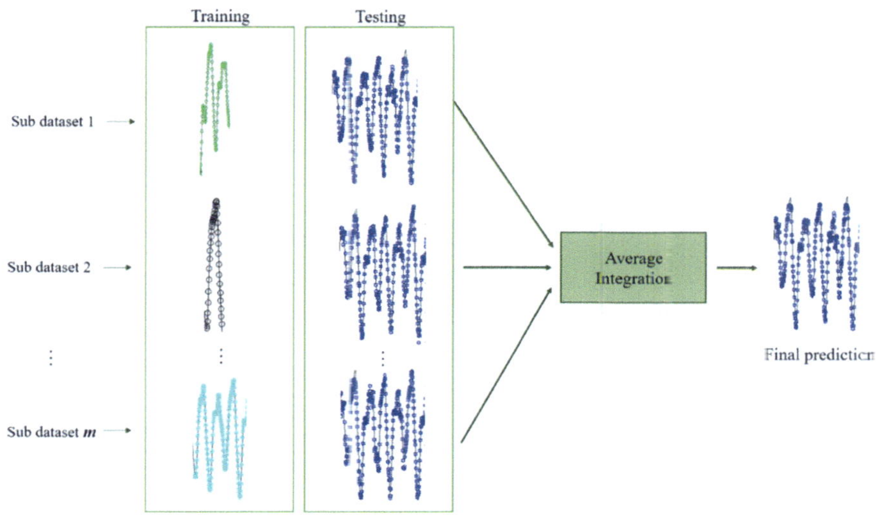


**Fig. 2** The general architecture of the proposed method

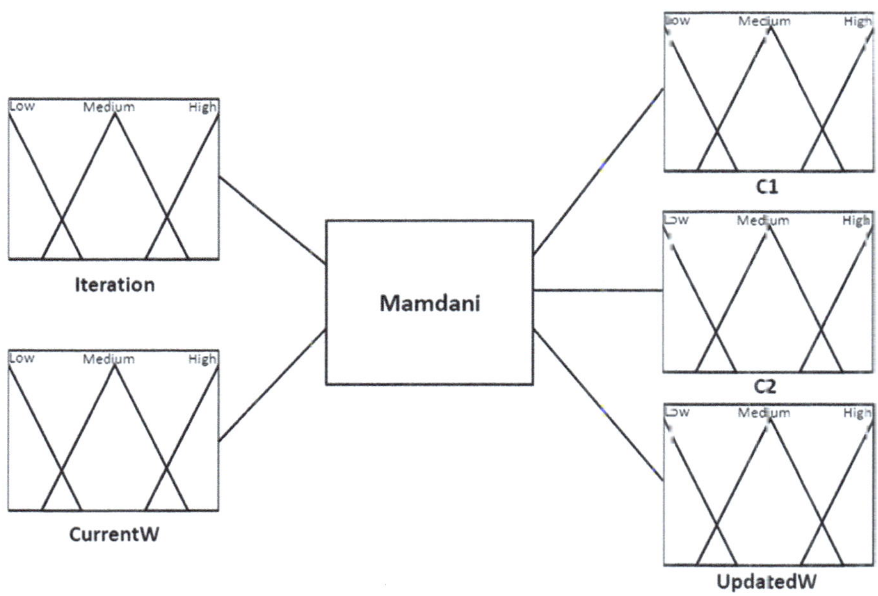


**Fig. 3** Fuzzy inference system

```
1. If Iteration is LOW and ActualW is LOW then c1 is LOW c2 is LOW UpdateW is HIGH
2. If Iteration is LOW and ActualW is MEDIUM then c1 is HIGH c2 is HIGH UpdateW is LOW
3. If Iteration is LOW and ActualW is HIGH then c1 is HIGH c2 is HIGH UpdateW is LOW
4. If Iteration is MEDIUM and ActualW is LOW then c1 is LOW c2 is LOW UpdateW is HIGH
5. If Iteration is MEDIUM and ActualW is MEDIUM then c1 is LOW c2 is HIGH UpdateW is MEDIUM
6. If Iteration is MEDIUM and ActualW is HIGH then c1 is HIGH c2 is HIGH UpdateW is LOW
7. If Iteration is HIGH and ActualW is LOW then c1 is LOW c2 is LOW UpdateW is HIGH
8. If Iteration is HIGH and ActualW is MEDIUM then c1 is HIGH c2 is HIGH UpdateW is LOW
9. If Iteration is HIGH and ActualW is HIGH then c1 is HIGH c2 is HIGH UpdateW is LOW
```

**Fig. 4** Fuzzy if-then rules

**Table 1** Ranges of variables

| Variable | Range |
|---|---|
| Iteration | 1–5 |
| ActualW | 0.1–1 |
| $C_1$ | 0.5–2 |
| $C_2$ | 0.5–2 |
| UpdatedW | 0.1–1 |

- **Iteration**: how many iterations the best particle maintains the same value of prediction error.
- **CurrentW**: Current value of *w*.

The outputs variables are:

- $\mathbf{C_1}$: this variable updates $c_1$ value to be used by the PSO.
- $\mathbf{C_2}$: this variable updates $c_2$ value to be used by the PSO.
- **UpdatedW**: this variable updates *w* value to be used by the PSO.

The fuzzy inference system has 9 fuzzy if-then rules shown in Fig. 4. These fuzzy rules were obtained by experimental knowledge.

The ranges of these variables are shown in Table 1. These values were established based on other works [26, 27], where PSO has been successfully used.

## 3.3 Application to Time Series Prediction

The objective function is to minimize the prediction error and is given:

$$\text{error} = \frac{\sum_{i=1}^{T}(|a_i - x_i|)}{T} \tag{3}$$

The average obtained by the sub modules is represented by *a* in time *i*, *x* represents real data of the testing phase, *T* is the number of data points used for testing phase. The learning algorithm uses to perform MGNNs training was the Levenberg-Marquardt (LM) backpropagation, this learning algorithm has demonstrated to have a better

**Table 2** Initial parameters of the PSO

| Parameter | Value |
|---|---|
| Particles ($n$) | 10 |
| Maximum iterations | 30 |
| $C_1$ | 2 |
| $C_2$ | 2 |
| $w$ | 0.8 |

**Table 3** Parameters for the search space

| Parameter | Value |
|---|---|
| Sub-modules ($m$) | 1–5 |
| Neurons | 5–20 |
| Hidden layers | 1–5 |
| Error goal | 0.000001–0.001 |

```
Initialize population of particles
while ( t < Max number of iterations)
    for each particle
            Calculate the fitness value of the particle (3) (evaluate the MGNN)
            If the fitness value is better than the best fitness value (pBest) in history
            set current value as the new pBest
    end for
    Choose the particle with the best fitness value of all the particles as the gBest
    for each particle
           Update c1, c2 and w using the fuzzy inference system
           Update particle velocity  (2)
           Update particle position  (1)
    end
end while
```

**Fig. 5** Pseudo code of the proposed PSO

performance with neural networks applied to time series prediction [28, 29]. The initial parameters of the PSO are shown in Table 2. For fuzzy adaptation, these parameters are only to start the algorithm, because the fuzzy inference system will update them during the execution. To compare results achieved with this approach, also results with a conventional PSO and non-optimized trainings will be shown.

The values to establish the MGNNs architectures in non-optimized trainings and optimizations (with fuzzy dynamic adaptation and conventional PSO) are shown in Table 3. In Fig. 5, the pseudo code of the proposed method is shown.

## 3.4 Description of Mackey-Glass Time Series

The Mackey-Glass time series [30] was used to prove the proposed method. These data points are corresponding to a period from 09/11/05 to 15/01/09, 800 data points are used and shown in Fig. 6.

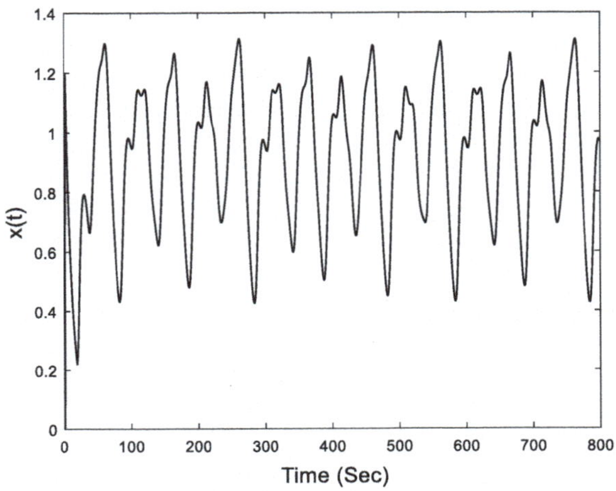


**Fig. 6** Example of the Mackey-glass time series

# 4 Experimental Results

In this section, experimental results with non-optimized training, conventional PSO and the proposed PSO with a fuzzy dynamic adaptation applied to time series prediction are shown.

## *4.1 Non-optimized Results*

The non-optimized trainings were performed establishing randomly architectures of the MGNNs using values shown in Table 3. Two tests were performed; the first test performs 30 non-optimized trainings using 500 data points for the training phase. The best result was obtained by training #9, where a prediction error of 0.0185 was achieved. The architecture of this training is shown in Table 4.

The best architecture has 3 sub modules, and the individual simulation of each one is shown in Fig. 7. The average integration is shown in Fig. 8.

The second test performs 30 non-optimized trainings using 300 data points for the training phase. The best result was obtained by training #2, where a prediction error of 0.0271 was achieved. The architecture of this training is shown in Table 5.

**Table 4** Best non-optimized architecture (500 data points)

| Module | Neurons | Data points per module | Prediction error |
|---|---|---|---|
| Module #1 | 30, 25 | 1–260 | 0.0185 |
| Module #2 | 14, 22 | 261–446 | |
| Module #3 | 29, 25 | 447–500 | |

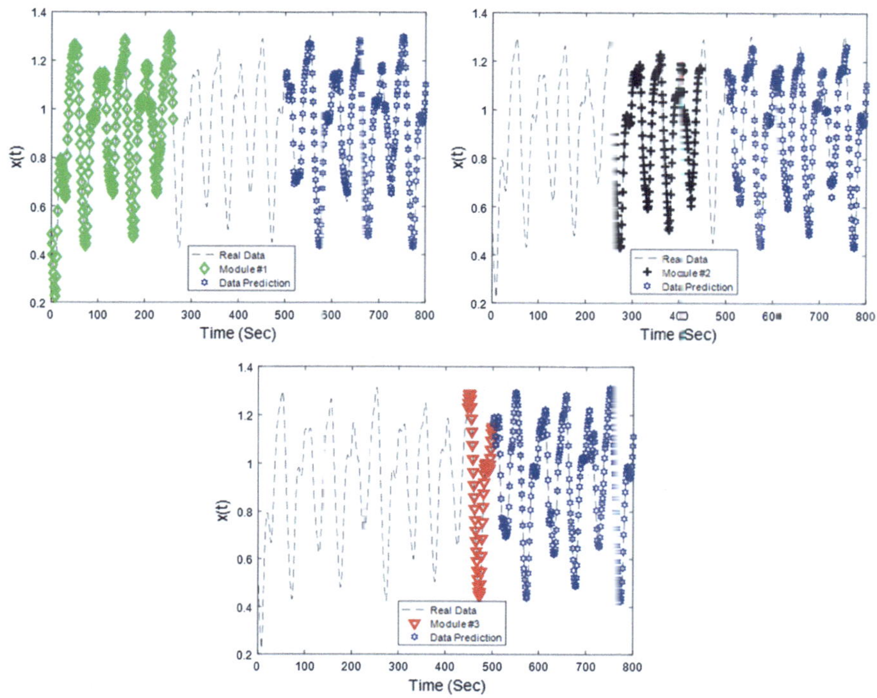


**Fig. 7** Non-optimized training #9

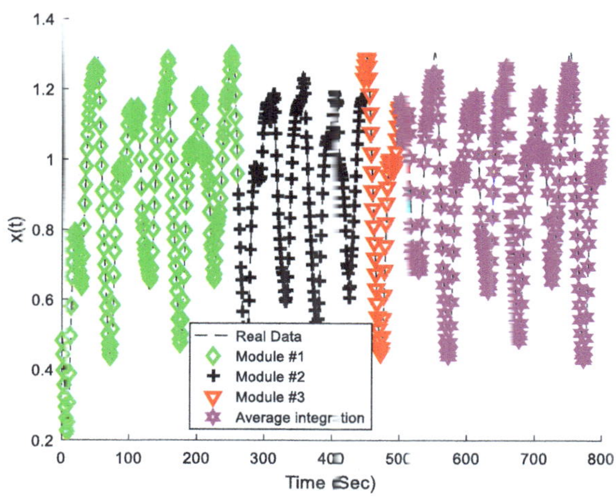


**Fig. 8** Final integration (non-optimized training #9)

**Table 5** Best non-optimized architecture (300 data points)

| Module | Neurons | Data points per module | Prediction error |
|---|---|---|---|
| Module #1 | 12, 23 | 1–213 | 0.0271 |
| Module #2 | 22, 15 | 214–300 | |

The best architecture has two sub-modules, the individual simulation of each one is shown in Fig. 9. The average integration is shown in Fig. 10.

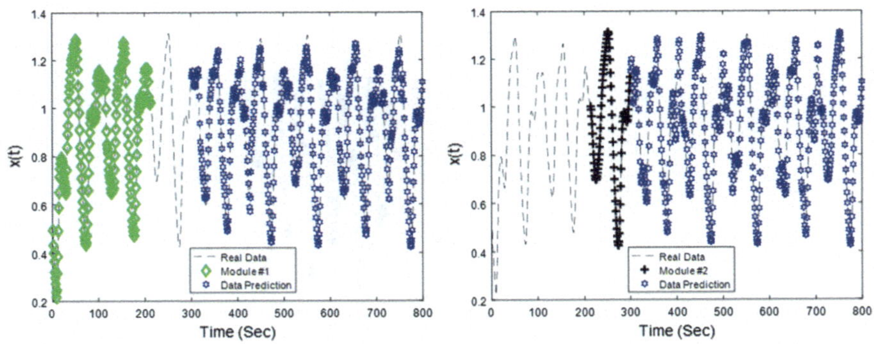


**Fig. 9** Non-optimized training #2

**Fig. 10** Final integration (non-optimized training #2)

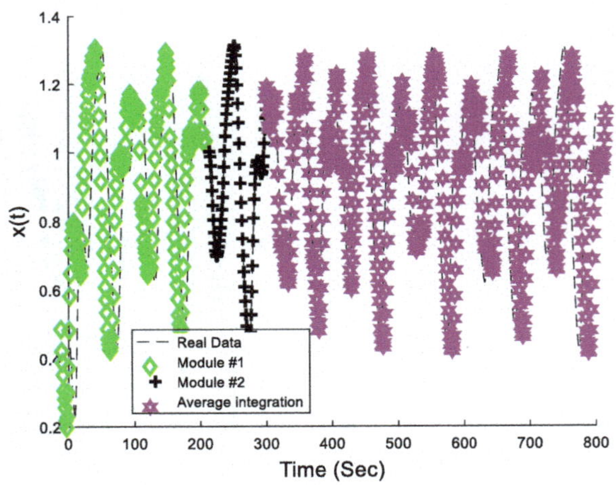

## 4.2 Optimized Results

The optimized results were performed using a conventional PSO and the proposed PSO with a fuzzy dynamic adaptation. 30 runs were performed for both optimization using 500 data points for the training phase. The best result of each optimization is shown in Table 6. The best result was obtained by the proposed PSO (run #4). Both optimizations had as best architecture a MGNN with 2 sub modules.

The best architecture has two sub modules, and the individual simulation of each one is shown in Fig. 11. The average integration is shown in Fig. 12. The convergence of the run #4 is shown in Fig. 13.

The optimized results using 300 data points for the training phase were also obtained. The best result of each optimization is shown in Table 7. The best result was also obtained by the proposed PSO (run #8). Both optimizations had as best architecture a MGNN with 2 sub modules.

The best architecture has 2 sub modules, and the individual simulation of each one is shown in Fig. 14. The average integration is shown in Fig. 15. The convergence of the run #8 is shown in Fig. 16.

**Table 6** Best optimization results (500 data points)

| Method | Module | Neurons | Data points per module | Prediction error |
|---|---|---|---|---|
| PSO (run #22) | Module #1 | 18, 8, 14 | 1–263 | 0.0130 |
| | Module #2 | 14 | 264–500 | |
| PSO + FL (run #4) | Module #1 | 13, 18 | 1–292 | 0.0130 |
| | Module #2 | 15, 10 | 293–500 | |

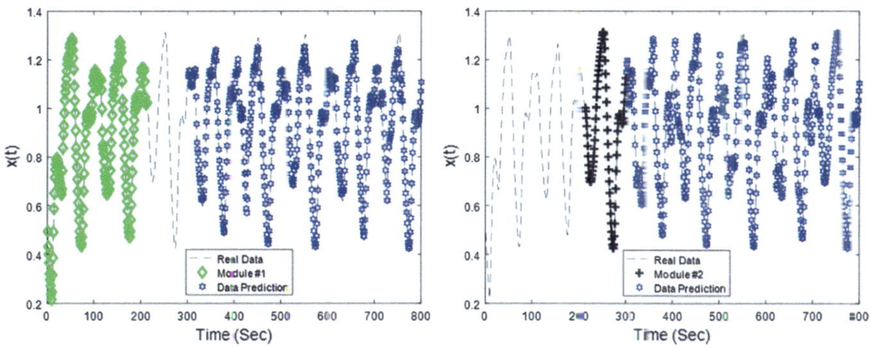


**Fig. 11** Run #4 (PSO with fuzzy dynamic adaptation)

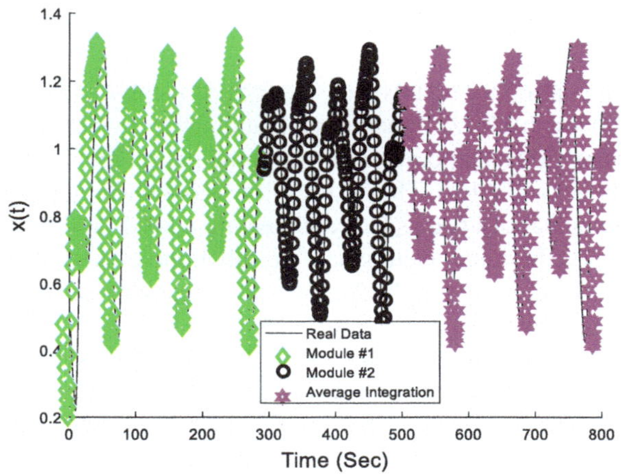


**Fig. 12** Final integration (run #4)

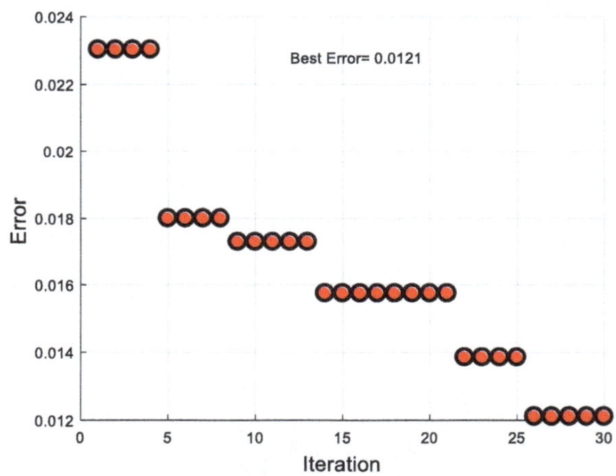


**Fig. 13** Convergence of run #4

**Table 7** Best optimization results (300 data points)

| Method | Module | Neurons | Data points per module | Prediction error |
|---|---|---|---|---|
| PSO (run #22) | Module #1 | 11 | 1–93 | 0.0175 |
| | Module #2 | 14, 8 | 94–300 | |
| PSO (run #22) | Module #1 | 15 | 1–89 | 0.0175 |
| | Module #2 | 10, 17 | 90–300 | |

## *4.3 Comparison Results*

A summary of non-optimized and optimized results is shown in Table 8. As results shown, when 500 and 300 data points for training phase are used, better averages

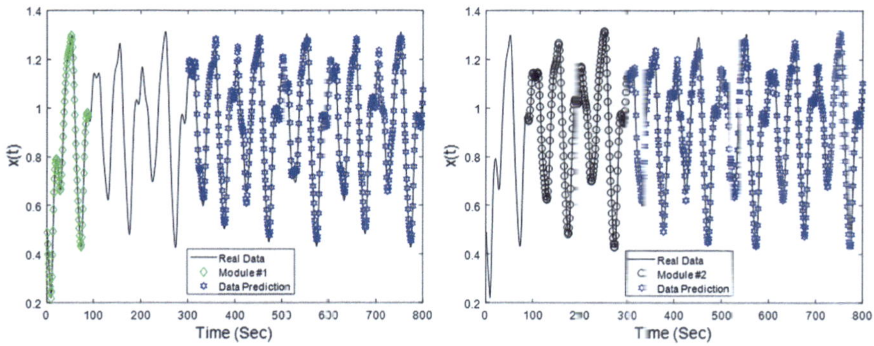


**Fig. 14** Run #8 (PSO with fuzzy dynamic adaptation)

**Fig. 15** Final integration (run #8)

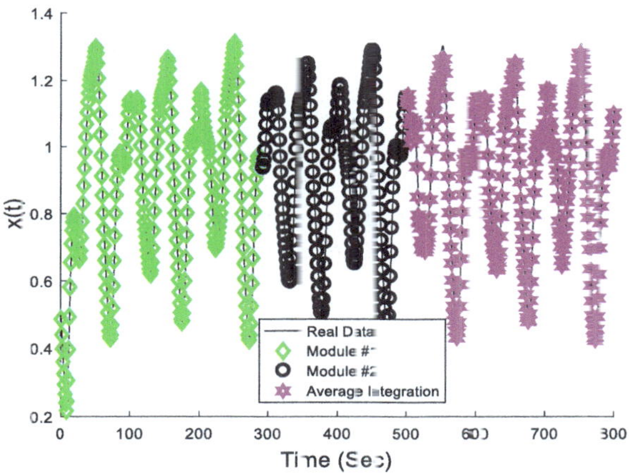


**Fig. 16** Convergence of run #8

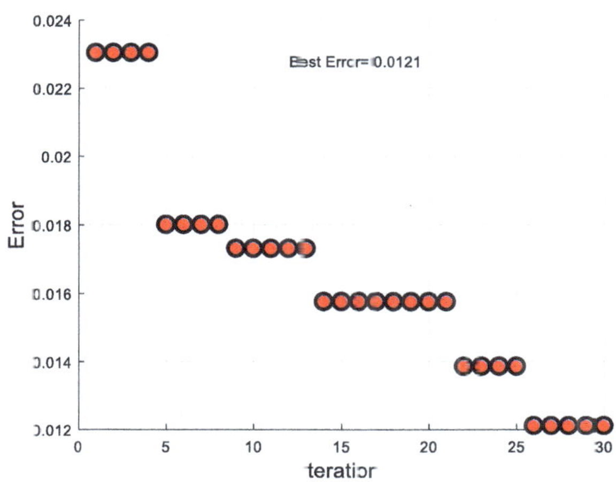

**Table 8** Comparison results

| Method | Data points for training | Best | Average | Worst |
|---|---|---|---|---|
| Non-optimized trainings | 500 | 0.0185 | 0.0608 | 0.1975 |
| | 300 | 0.0271 | 0.1320 | 0.4165 |
| PSO | 500 | 0.0130 | 0.0151 | 0.0175 |
| | 300 | 0.0175 | 0.0192 | 0.0208 |
| PSO + FIS | 500 | 0.0121 | 0.0144 | 0.0180 |
| | 300 | 0.0166 | 0.0187 | 0.0210 |

**Table 9** Execution times

| Method | Data points for training | Best | Average | Worst |
|---|---|---|---|---|
| PSO | 500 | 00:07:03 | 00:15:59 | 00:49:06 |
| | 300 | 00:02:44 | 00:07:12 | 00:14:41 |
| PSO + FIS | 500 | 00:03:36 | 00:15:33 | 00:35:16 |
| | 300 | 00:01:35 | 00:05:36 | 00:09:21 |

are obtained using the fuzzy dynamic parameters adaptation for the PSO. In Table 9, comparisons of execution times between both PSO versions are shown (30 runs) where, it can be observed that better times are obtained by the proposed PSO.

## 5 Conclusions

In this paper, a fuzzy dynamic parameter adaptation of the particle swarm optimization for modular granular network design applied to time series prediction is proposed. Different tests using non-optimized trainings, conventional PSO and the proposed fuzzy PSO were performed. These results have demonstrated than better results are obtained by the proposed method. The effectiveness of the proposed method was proved using 800 data points of the Mackey-Glass time series. The tests were performed using 500 and 300 data points for the training phase. We conclude that the performance of intelligence techniques, such as: modular granular neural networks and particle swarm optimization can be improved when the parameters of PSO are dynamic adjusted by a fuzzy inference system. As future work, other parameters can be dynamic adjusted and another optimization technique can be applied to compare its performance against our fuzzy dynamic adaptation in PSO. In addition, we can also consider other applications, like in [31, 32], and other hybrid combinations [33–36].

## References


1. D. Sánchez, P. Melin, O. Castillo, *A Grey Wolf Optimizer for Modular Granular Neural Networks for Human Recognition*, vol. 2017, pp. 4180510:1–4180510:26 (2017)
2. L. Zhang, K. Mistry, S. ChinNeoh, C.P. Lim, Intelligent facial emotion recognition using moth-firefly optimization. Knowl. Based Syst. **111**(1), 248–267 (2016)
3. J. Rajeshkumar, K. Kousalya, Diabetes data classification using whale optimization algorithm and backpropagation neural network. Int. Res. J. Pharm. **8**(11), 219–222 (2017)
4. Z. Cai et al., Quadrotor trajectory tracking and obstacle avoidance by chaotic grey wolf optimization-based active disturbance rejection control. Mech. Syst. Signal Process. **128**(1), 636–654 (2019)
5. M. Wang, H. Dong, X. Li, Y. Zhang, J. Yu, Control and optimization of a bionic robotic fish through a combination of CPG model and PSO. Neurocomputing **337**, 144–152 (2019)
6. M. Pulido, O. Castillo, P. Melin, Genetic optimization of ensemble neural networks for complex time series prediction of the Mexican exchange. Int. J. Innov. Comput. Inf. Control **9**(10), 4151–4166 (2013)
7. D.E. Goldberg, Genetic Algorithms in Search Optimization and Machine Learning: Addison-Wesley (1989)
8. M. Dorigo, Optimization, Learning and Natural Algorithms. Politecnico di Milano, Italy, PhD Thesis (1992)
9. R.C. Eberhart, J. Kennedy, A new optimizer using particle swarm, in *Sixth International Symposium on Micro Machine and Human Science*, pp. 39–43 (1995)
10. J. Kennedy, R.C. Eberhart, Particle swarm optimization, in *Proceedings of the IEEE international Joint Conference on Neural Networks*, pp. 1942–1948 (1995)
11. X.S. Yang, Firefly algorithms for multimodal optimization, in Proceedings of 5th Symposium on Stochastic Algorithms, Foundations and Applications, vol. 5792, pp. 169–178 (2009)
12. S. Mirjalili, S.M. Mirjalili, A. Lewis, Grey Wolf Optimizer. Adv. Eng. Softw. **69**, 46–61 (2014)
13. R. Rajabioun, Cuckoo Optimization Algorithm. Appl. Soft Comput. J. **11**, 5508–5518 (2011)
14. A. Sharkey, *Combining Artificial Neural Nets: Ensemble and Modular Multi-net Systems*. Springer (1999)
15. F. Azamm, *Biologically Inspired Modular Neural Networks*. Virginia Polytechnic Institute and State University, Blacksburg, Virginia, PhD thesis (2000)
16. P. Melin, O. Castillo, *Hybrid Intelligent Systems for Pattern Recognition Using Soft Computing: An Evolutionary Approach for Neural Networks and Fuzzy Systems*, 1st edn. Springer (2005)
17. L.A. Zadeh, Fuzzy sets. Inf. Control **8**(3), 338–353 (1965)
18. L.A. Zadeh, Toward a theory of fuzzy information granulation and its centrality in human reasoning and fuzzy logic. Fuzzy Sets Syst. **90**(2), 111–127 (1997)
19. Y.Y. Yao, On modeling data mining with granular computing, in *25th International Computer Software and Applications Conference (COMPSAC)*, pp. 638–649 (2001)
20. A. Bargiela, W. Pedrycz, The roots of granular computing, in *IEEE International Conference on granular computing (GrC)*, pp. 806–809 (2006)
21. Y. Qian, H. Zhang, F. Li, Q. Hu, J. Liang, Set-based granular computing: a lattice model. Int. J. Approx. Reason. **55**, 834–852 (2014)
22. R.C. Eberhart, Y. Shi, Comparing inertia weights and constriction factors in particle swarm optimization, in *Proceedings of the IEEE Congress on Evolutionary Computation*, vol. 1, pp. 84–88 (2000)
23. Y. Shi, R. Eberhart, Parameter selection in particle swarm optimization, in *International Conference on Evolutionary Programming*, pp. 591–600 (1998)
24. D. Sánchez, P. Melin, Optimization of modular granular neural networks using hierarchical genetic algorithms for human recognition using the ear biometric measure. Eng. Appl. Artif. Intell. **27**, 41–56 (2014)
25. D. Sánchez, P. Melin, O. Castillo, Optimization of modular granular neural networks using a firefly algorithm for human recognition. Eng. Appl. Artif. Intell. **64**, 172–186 (2017)

26. D. Sánchez, P. Melin, O. Castillo, Fuzzy adaptation for particle swarm optimization for modular neural networks applied to iris recognition, in *North American Fuzzy Information Processing Society Annual Conference*, pp. 104–114 (2017)
27. F. Valdez, J.C. Vázquez, F. Gaxiola, Fuzzy dynamic parameter adaptation in ACO and PSO for designing fuzzy controllers: the cases of water level and temperature control. Adv. Fuzzy Syst. **2018**, 1274969:1–1274969:19 (2018)
28. P. Melin, M. Pulido, Optimization of ensemble neural networks with type-2 fuzzy integration of responses for the Dow Jones time series prediction. Intell. Autom. Soft Comput. **20**(3), 403–418 (2014)
29. M. Pulido, P. Melin, O. Castillo, Particle swarm optimization of ensemble neural networks with fuzzy aggregation for time series prediction of the Mexican Stock Exchange. Inf. Sci. **280**, 188–204 (2014)
30. M.C. Mackey, L. Glass, Oscillation and chaos in physiological control systems. Science **197**, 287–289 (1997)
31. P. Melin, D. Sánchez, O. Castillo, Genetic optimization of modular neural networks with fuzzy response integration for human recognition. Inf. Sci. **197**, 1–19 (2012)
32. M.A. Sanchez, O. Castillo, J.R. Castro, P. Melin, Fuzzy granular gravitational clustering algorithm for multivariate data. Inf. Sci. **279**, 498–511 (2014)
33. P. Melin, O. Castillo, Intelligent control of complex electrochemical systems with a neuro-fuzzy-genetic approach. IEEE Trans. Ind. Electron. **48**(5), 951–955
34. L. Aguilar, P. Melin, O. Castillo, Intelligent control of a stepping motor drive using a hybrid neuro-fuzzy ANFIS approach. Appl. Soft Comput. **3**(3), 209–219 (2003)
35. P. Melin, O. Castillo, Adaptive intelligent control of aircraft systems with a hybrid approach combining neural networks, fuzzy logic and fractal theory. Appl. Soft Comput. **3**(4), 353–362 (2003)
36. P. Melin, J. Amezcua, F. Valdez, O. Castillo, A new neural network model based on the LVQ algorithm for multi-class classification of arrhythmias. Inf. Sci. **279**, 483–497 (2014)

# Review of Fuzzy Control for Path Tracking in the Robotino System

**Marcos Fuentes, Oscar Castillo, and Prometeo Cortés-Antonio**

**Abstract** The main goal of this paper is to study works performed with Robotino, mainly by using fuzzy logic and other metaheuristics for achieving path tracking. After this, a short comparison will be made between the previously performed works. In order to help in the orientation of envisioning future works, and to analyze how efficient it is to apply fuzzy logic (including type-2 fuzzy logic), in this specific area of robotics, regardless of whether it works in conjunction with other techniques, tools, software, hardware, methodologies. Or which of the sensors of the robot are being used at the time of the works, being able to use only the infrared sensors or all the sensors that it has. The works that are considered under this study are mainly focused on the use of the Robotino, or for the path tracking by implementing omnidirectional robots.

**Keywords** Fuzzy logic · Path tracking · Robotino · Study · Fuzzy controller · Omni-directional robot · Type-2 fuzzy logic · Matlab · Simulink · Review

## 1 Introduction

Computer science is a set of disciplines that are responsible for studying the bases of information and computation that are represented in the form of data. These disciplines are mainly programming languages, mathematical, and logical foundations. This science has many areas, such as optimization, and development, among others.

M. Fuentes · O. Castillo (✉) · P. Cortés-Antonio
Tijuana Institute of Technology, Tijuana, BC, Mexico
e-mail: ocastillo@tectijuana.mx

M. Fuentes
e-mail: marcos.fuentes@tectijuana.edu.mx

P. Cortés-Antonio
e-mail: prometeo.cortes@tectijuana.edu.mx


P. Melin et al. (eds.), *Recent Advances of Hybrid Intelligent Systems Based on Soft Computing*, Studies in Computational Intelligence 915,
https://doi.org/10.1007/978-3-030-58728-4_12

Computer science since its creation has been of vital importance in humanity, because it has been a key factor for technological advancement creating an impact. It is implemented in things that make our lives easier. Such implementations cover different areas including medicine, the automobile industry, robotics, telecommunications, and aeronautics, just to mention a few examples.

Fuzzy logic is a methodology that provides a result or conclusion based on information obtained, and that is imprecise, based on human reasoning for decision making. It is currently used in a large number of devices, mainly controller-based ones.

Robotino is a robot developed by Festo, it is an omnidirectional mobile robot system. It is mainly composed of a CPU, 3 motors, 9 infrared sensors to measure distances, 2 optical sensors, 1 inductive sensor, 1 anti-collision sensor, 1 web camera, 1 wireless modem, and 3 omnidirectional rollers.

This paper is organized as follows: Sect. 2 describes the kinematic of Robotino, Sect. 3 will explain some of the previous works, Sect. 4 shows the comparisons from their results, and finally in Sect. 5 we offer the conclusions.

## 2 Kinematic of Robotino

Robots are used in academics, in the industry, in science, among others, however, they have certain limitations. Such as the lack of freedom when it comes to movement since they cannot be moved laterally, unlike them, omnidirectional robots can move in any direction, one of the most used is the Robotino. To better understand its operation, it is necessary to know its kinematics.

The mathematical development of the kinematic model is carried out through the relationship between wheel speed, and the robot itself.

The Robotino location as well as its orientation can be represented by a vector $[x, y, \emptyset]^{\mathrm{T}}$ (Fig. 1).

The global speeds of the robot is represented by the vector $\left[\dot{x}, \dot{y}, \dot{\phi}\right]^{\mathrm{T}}$ and the angular speeds of each wheel are represented by vector $\left[\dot{\theta}_1, \dot{\theta}_2, \dot{\theta}_3\right]^{\mathrm{T}}$.

As reference for the orientation and location of the Robotino, (his kinematic) was taken from [1], as it can be seen in the Eq. (1), it can be explained in Table 1.

$$\begin{bmatrix} \dot{\theta}_1 \\ \dot{\theta}_2 \\ \dot{\theta}_3 \end{bmatrix} = \frac{1}{r} \begin{bmatrix} V_1 \\ V_2 \\ V_3 \end{bmatrix} = \frac{1}{r} \underbrace{\begin{bmatrix} -\sin(\phi + \varphi_1) & cos(\phi + \varphi_1) & R \\ -\sin(\phi + \varphi_2) & cos(\phi + \varphi_2) & R \\ -\sin(\phi + \varphi_3) & cos(\phi + \varphi_3) & R \end{bmatrix}}_{P(\phi)} \begin{bmatrix} \dot{x} \\ \dot{y} \\ \dot{\phi} \end{bmatrix} \tag{1}$$

Robot speeds could be determined from speed wheels inverting the matrix, as in:

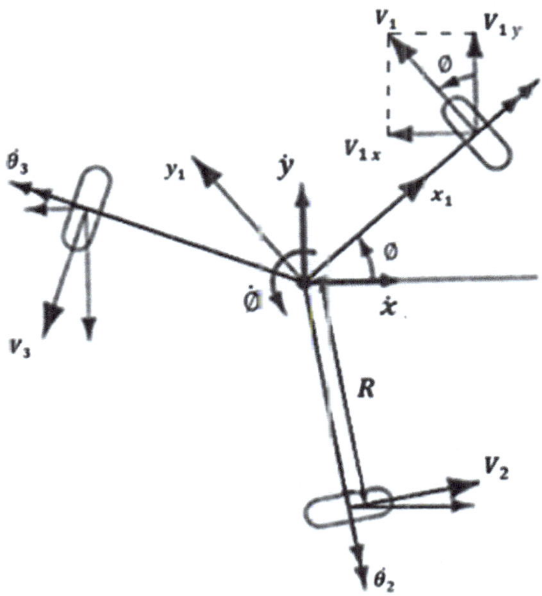


**Fig. 1** Kinematic structure of the Robotino

**Table 1** Meaning of kinematic variables about Robotino

| Nomenclature | Meaning |
|---|---|
| $\dot{\theta}_1, \dot{\theta}_2, \dot{\theta}_3$ | Angular speed of wheels 1, 2, 3 |
| $V_1, V_2, V_3$ | Linear speed of wheels 1, 2, 3 |
| $r$ | Radius of wheel |
| $R$ | Distance between each wheel from the center of the Robotino |
| $\dot{\phi}$ | Angular speed of Robotino |
| $\varphi_1, \varphi_2, \varphi_3$ | Angular locations of wheels 1, 2, 3 |
| $P(\phi)$ | Transformation matrix between the angular speeds of wheels and the global velocity vector |

$$\begin{bmatrix} \dot{x} \\ \dot{y} \\ \dot{\phi} \end{bmatrix} = rp^{-1}(\phi) \begin{bmatrix} \dot{\theta}_1 \\ \dot{\theta}_2 \\ \dot{\theta}_3 \end{bmatrix} \tag{2}$$

Using the kinematic control of the robot you can estimate the speeds required to navigate the omnidirectional mobile robot $\left[\dot{x}, \dot{y}, \dot{\phi}\right]^{\mathrm{T}}$, current position coordinates, and the target position coordinates $[x_d, y_d, \phi_d]^{\mathrm{T}}$.

The current and final position is given by:

$$\begin{bmatrix} x_e \\ y_e \\ \phi_e \end{bmatrix} = \begin{bmatrix} x \\ y \\ \phi \end{bmatrix} - \begin{bmatrix} x_d \\ y_d \\ \phi_d \end{bmatrix} \tag{3}$$

Making the derivative of (3), we obtain:

$$\begin{bmatrix} \dot{x}_e \\ \dot{y}_e \\ \dot{\phi}_e \end{bmatrix} = \begin{bmatrix} \dot{x} \\ \dot{y} \\ \dot{\phi} \end{bmatrix} = p^{-1}(\phi) \begin{bmatrix} V_1 \\ V_2 \\ V_3 \end{bmatrix} \tag{4}$$

To control the position of the robot, a stabilizer is normally used, which is given by the following control law:

$$\begin{bmatrix} V_1 \\ V_2 \\ V_3 \end{bmatrix} = p(\phi) \left( -k_p \begin{bmatrix} x_e \\ y_e \\ \phi_e \end{bmatrix} - k_i \begin{bmatrix} \int x_e \\ \int y_e \\ \int \phi_e \end{bmatrix} \right) \tag{5}$$

With $k_p$ and $k_i$ are positive because of the error position system, the robot speed takes the form:

$$\begin{bmatrix} \dot{x}_e \\ \dot{y}_e \\ \dot{\phi}_e \end{bmatrix} = \begin{bmatrix} \dot{x} \\ \dot{y} \\ \dot{\phi} \end{bmatrix} = \begin{bmatrix} V_x \\ V_y \\ \Omega \end{bmatrix} = -k_p \begin{bmatrix} x_e \\ y_e \\ \phi_e \end{bmatrix} - k_i \begin{bmatrix} \int x_e \\ \int y_e \\ \int \phi_e \end{bmatrix} \tag{6}$$

Once knowing this, the related works can be better analyzed.

## 3 Previous Works

Omni-directional robots have been extensively studied and applied to control tasks. If we talk about Robotino, we are talking about a 3-wheel omni-directional robot. However, there are robots that are used for the same tasks, which are aesthetically different, but that is not a limitation so that fuzzy logic may be necessary.

An example of that is [2], where a fuzzy controller was implemented for 4-wheel omnidirectional robots with 3° of freedom. In the test and simulation, the results met the expectations regarding the execution time. The controller provides an optimal result to reduce the differences between the reference and the output. The final results showed the validity and effectiveness applied to these types of robots.

A work that also implemented 4-wheel omnidirectional robots, and that demonstrated that the methodology used in the works with the Robotino can be applied was [3]. Where it was worked on a flat platform called RoboCup (normally used in experiments with robots for testing soccer).

The work was carried out in 2 sections, the first was to design and execute an integrated system that had communication with a server remotely linked wirelessly. It also executed the commands it received, in the second a route planner of Fuzzy-PI type and a linear-quadratic regulator (LQR) was chosen for optimum input. To avoid dynamic problems from omni-directional robots, such as slippage, filters were used for acceleration and speed. Simulated tests were performed with, and without noise, the results showed that using the new approach in the trajectory controller actually gives more precise movement in omnidirectional robots.

Now, speaking of the works in which fuzzy logic has been applied to the Robotino in order to carry out the task of path tracking, we can talk about [4], where a new control approach was designed, this for a navigation algorithm that will be implemented in omnidirectional robots, having Robotino as the main test platform. The difference between this research, and other related ones is that most of them focus on estimating the final position, instead here it is not only intended to reach that position, but it also has a specific direction angle.

In order to raise the advantages of PI controllers, a fuzzy logic PI controller was developed. Due to this, an intense study was carried out concluding that a fuzzy-PI system does not have time to control, linear and angular speeds at the same time. To put aside these disadvantages, it was proposed to implement two independent fuzzy PI controllers that adjust certain parameters ($k_p$, $k_i$). To minimize the error caused by the dynamic variations, and the navigation challenges of the Robotino. Later, another controller was added in order to avoid obstacles.

We can say that the final purpose of all the works involved with omni-directional robots, is to make man's life easier. One that makes it very clear is [5], where it focuses the objective on the automotive industry with the purpose of avoiding collisions, for this purpose a Fuzzy controller was modeled and developed in Matlab and Simulink. The controller was tested on the Robotino using the infrared sensors and their actuators. The simulation results had an approximation between Matlab and the values that were calculated giving a minimal margin of error.

Another work in which fuzzy logic was implemented in Robotino was [6], where the fuzzy controller application was designed and carried out on the omni-directional robot with the main purpose of reaching a desired position. The controller was tested remotely through the robot's main platform (Festo), using the Matlab programming language as its environment. Part of the job involves obstacle avoidance and odometry given the nature of the purpose of the work involved.

Since path tracking is one of the main and most important tasks for robots, [7] emphasizing this task in conjunction with the avoidance of obstacles that are in motion or not in motion, with the aim of reaching a destination predetermined. Due to the complications of searching for an optimal mathematical model for the Robotino, it was decided to design two fuzzy Sugeno-type controllers. One controller was implemented with 153 rules to attack the route tracking problem, on the other hand, the second controller was implemented with 27 rules to attack the obstacle avoidance problem, both controllers were designed in Matlab. According to the tests carried out in real time, their results showed good capabilities of the already mentioned controllers.

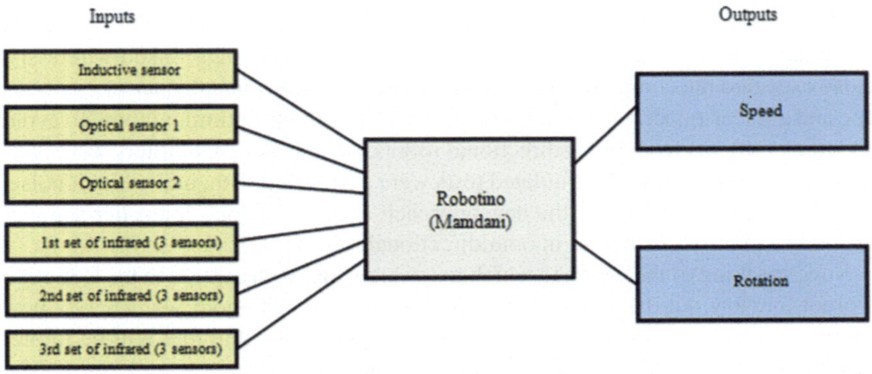


**Fig. 2** Inputs and outputs of fuzzy controller reported in [8]

A similar work to the previous one is [8], in that research a fuzzy controller using the Robotino platform, it was designed and implemented, in order to apply it to the path tracking with a high level of difficulty.

All the sensors were used, the 9 infrared sensors, 2 optical and inductive sensors, which were the inputs to the system. The outputs were the speed and rotation of the Robotino (Fig. 2). The controller was a Mamdani type with 22 rules. To obtain better results and efficiency, multiple linear regression was applied. The experiments on Robotino in real time showed optimal results.

There are some documents where the purpose is to make comparisons, one of them is [9], in that document comparisons are made between conventional proportional controllers (P, PI) and fuzzy controllers (fuzzy-P, fuzzy-PI.) Comparisons were made in Robotino focusing on route tracking, where the route was drawn by lines. The 2 optical sensors were used to define the angular velocity of the Robotino, and the inductive sensor for the linear speed. For the tests simulations were performed in Simulink and Robotino SIM (software program for the 3D simulation of Robotino in an exciting, predefined virtual experimentation environment). The results showed that the fuzzy-PI controller runs successfully, and in less time than the other controllers.

As mentioned at the beginning of this chapter, the Robotino system is a CPU, allowing it to work with a large number of hardware devices, as well as software. In [10], a new approach is presented to design and implement a fuzzy-PI controller for application in Robotino, as its navigation system using an FPGA board. The first step that was carried out was to design the controller (in Matlab), the corresponding simulations were made (in Robotino SIM) to optimize it. The controller can be run on an FPGA board having different approaches (either hardware or software). If you decide to run it by hardware you have to route it, to be able to download to an FPGA.

The implementation of fuzzy logic in omnidirectional robots is not limited to using only type-1 fuzzy logic. There are works where, the application of type-2 fuzzy logic is present. As in the case of [11], in this work the capacity offered by degrees of freedom was taken with the capabilities offered by type-2 fuzzy logic to solve real problems. To control the Robotino in environments that were dynamic and unpredictable in nature, with or without obstacles. A system was considered to plan and modify the trajectory in real time. The idea of type-2 fuzzy sets supported by zSlices was considered, which allowed Robotino to face situations with a high level of perplexity, in unstructured, and energetic closed environments. The test results proved that the Robotino was able to achieve a desired conformation by evading both, motionless and moving obstacles.

Another work that also implements type-2 fuzzy logic is [12], which aims to solve the problems faced by the navigation systems used today. This is achieved with the help of the FLS as well as the IT2FLS, which use fuzzy sets of type 2, and can take and shape such problems (such as uncertainty). In addition to that, these unlike their type 1 opposites, give better results. However, the level of difficulty and the time it takes to reduce types, is closely linked to the structure of what the membership functions are, and the number of rules restricts part of its functions in real situations.

Given the applications in the navigation of omni-directional robots, and its remarkable analysis in the mathematical field. As well as its simplicity with the APF method. In this work the IT2FLS approach is coupled with APF. The tests were carried out at Robotino proving to be efficient.

Some related works are [13, 14], where the intention is to maintain the position and specific behavior for mobile robots by means of type-2 fuzzy logic in order to control their behavior. The system was developed in Matlab and Simulink, it can successfully drive a mobile robot in both simulated and real environments. The simulations performed showed advantages of the proposed approach. It was also tested on a Mecanum wheeled robot demonstrating good maneuverability.

There are works which implement fuzzy logic but rely on other methodologies such as neural networks, this is the case of [15], who applied it to the Robotino, in which they justify the neuro-fuzzy application given recent research which has sufficient evidence to affirm that the type-2 FLS are better than the conventional, the type-1 FLS, speaking of situations that handle uncertainty. The problem in this type of application is the time factor, it takes too long to reduce the type, which limits its application in simple problems executed in real time. For this reason, a new hybrid adaptation is proposed to minimize the time it takes to try to reduce the type. As mentioned, a MIT2FNN was developed for omnidirectional robot navigation applied to environments of a dynamic and unstructured nature. Part of the process of MIT2FNN consists of a type-2 fuzzy linguistic medium as input, followed by a 2-layered neural network. To make parameter adjustments of the MIT2FNN controller. a back propagation algorithm was applied. The results of the experiments carried out were applied to the Robotino, validating the effectiveness of the new approach.

A work to be analyzed that implemented fuzzy logic in a different way is [16], where a specific prototype for uncertainty systems is created using 2 MIFREN (adaptive controller based on emulated networks of fuzzy rules with multiple inputs), with an online instruction and stability distinction algorithm. It is proposed to obtain good results and productivity of the closed circuit, as well as the consistency of the system for the parameters that can be adjustable within the MIFREN. The system is approved by an experimental type of configuration with a Robotino. The algorithm was shown to perform better than other algorithms that only have one controller added. The superiority of the preliminary configuration of MIFREN is verified by comparing the results between this, and one based on neural networks.

A large number of works take a new step in the tasks that a robot performs, we talk about optimization. As an example of this, there are [17, 18], which implemented the firefly algorithm to optimize the fuzzy controllers for omnidirectional robots, as well as to optimize the parameters of the membership functions in the fuzzy control systems respectively, both applied to the path tracking task.

A similar work where the fuzzy logic controller was optimized is [19], in this work the parameters had not been optimized before so it was considered to use the flower pollination algorithm (FPA). The reason for using this algorithm was to find the best parameters in order to minimize the error between the reference and the desired path.

As seen in previous works, there is talk of using biologically inspired algorithms for optimization. In [20], a modification of the bee colony optimization algorithm (BCO) is presented, this in order to optimize the fuzzy calibrators. The work is done in three stages, the first is to test the traditional BCO algorithm with the optimization of fuzzy controls. The second presents the modification of the original method that includes fuzzy logic with the objective of changing the parameter values during runtime. The third and last stage consists of the optimization of the reference control problems. The results of the proposed method (fuzzy BCO), exceeds the traditional one in terms of specific diffuse designs optimally. Said fuzzy controllers were applied to mobile robots.

Another work that was optimized with a bio-inspired algorithm was [21]. In this work, a new fuzzy approach is explored with the purpose of controlling diversity in the Ant Colony Optimization (ACO) algorithm. In a few words, the intention is to avoid or delay total convergence by varying a specific parameter dynamically. A fuzzy convergence controller was created to keep diversity at a certain level so that premature convergence was minimized. The results were promising in different instances, in particular in optimizing membership functions for path controls of unicycle-type mobile robots.

## 4 Analysis of Results

It has been seen in this review, most of the analyzed works have met the expectations of the researchers, which allows getting ideas to continue with future works, starting from a better point when knowing the results of the previous investigations.

If we analyze the previous section more fully, we can start with the fuzzy logic applied to 4-wheel omnidirectional robots, where the work carried out yield optimal results when executing the route tracking task.

On the other hand, focusing on the implementation of traditional fuzzy logic in Robotino. The results are not far from the previous case, in this part of the analysis results ranging from acceptable to very good were shown. Whether it will work with different programming languages, Matlab or Simulink, with a different number of fuzzy rules or with external devices (hardware), that a traditional Robotino has. The capacity and scope that traditional fuzzy logic has in this field is clear.

Reviewing the works in which type-2 fuzzy logic is implemented, some jobs even showed to be more efficient than the conventional ones in environments with a high level of uncertainty and dynamics. This shows that fuzzy logic (either type-1 or type-2) has a great scope, since you can take advantage of both, taking into account the environment. Similarly, the fuzzy systems that worked with neural networks showed effectiveness in the Robotino in highly complex environments.

Finally, when reviewing the works in which bio-inspired methods were applied either to the fuzzy controller or only to specific parameters. It is shown that improvements can be obtained, whether a certain method does not work, there are a large number of them, reason why the field of investigation as well as of future works is very wide.

## 5 Conclusions

Given the results of the works previously analyzed, we can conclude that fuzzy logic has been shown to be one of the most viable methodologies for working in the field of robotics, with omnidirectional robots benefiting the most thanks to the work and research that has been carried out. Likewise, its versatility when working with a large number of techniques, methods, devices and methodologies is clear, demonstrating that not only is it capable of working with all of these, but it is also capable of showing effective and efficient results with each and every one.

We can assure that thanks to the work carried out in the field of fuzzy logic applied to the robot, in the task of tracking routes, there is much to study and research. What definitely leads us to a need to improve is for this reason that it is necessary to continue researching in order to apply new techniques and methodologies to optimize systems and tasks that they perform.

Finally, it is proposed to study more, in particular the works that have implemented optimizations in tasks executed by omnidirectional robots or specifically with the Robotino in order to know which techniques are the most appropriate for certain tasks. We plan in the future to apply not just type-1 fuzzy logic, but also type-2 fuzzy logic in the control of omnidirectional robots, like in [22, 23, 32]. In addition, we can try other kinds of applications, like in [24–31].

## References

1. X. Li, A. Zell, Motion control of an omnidirectional mobile robot. *Lecture Notes in Electrical Engineering* (2009). https://doi.org/10.1007/978-3-540-85640-5-14
2. R.H. Abiyev, I.S. Günsel, N. Akkaya, E. Aytac, A. Çağman, S. Abizada, Fuzzy control of omnidirectional robot. Procedia Comput. Sci. (2017). https://doi.org/10.1016/j.procs.2017.

11.286
3. M.A. Al Mamun, M.T. Nasir, A. Khayyat, Embedded system for motion control of an omnidirectional mobile robot. IEEE Access (2018). https://doi.org/10.1109/ACCESS.2018.2794441
4. M.S. Masmoudi, N. Krichen, M. Masmoudi, N. Derbel, Fuzzy logic controllers design for omnidirectional mobile robot navigation. Appl. Soft Comput. J. (2016). https://doi.org/10.1016/j.asoc.2016.08.057
5. A.C. Pinto, C.W. Gomes, D.D.L.D. De Souza, D.P. Da Silva, G.V. Mattioli, Anti-collision fuzzy logic system for automobile vehicles. SAE Technical Papers (2013). https://doi.org/10.4271/2013-36-0651
6. S.E. Oltean, M. Dulau, R. Puskas, Position control of Robotino mobile robot using fuzzy logic, in *2010 IEEE International Conference on Automation, Quality and Testing, Robotics, AQTR* (2010). https://doi.org/10.1109/AQTR.2010.5520855
7. M.R.H. Al-Dahhan, M.M. Ali, Path tracking control of a mobile robot using fuzzy logic, in *13th International Multi-conference on Systems, Signals and Devices, SSD* (2016). https://doi.org/10.1109/SSD.2016.7473656
8. F. Cuevas, O. Castillo, Design and implementation of a fuzzy path optimization system for omnidirectional autonomous mobile robot control in real-time. Stud. Comput. Intell. (2018). https://doi.org/10.1007/978-3-319-71008-2_19
9. J. González-Aguilar, O. Castillo, P. Cortés-Antonio, Implementation a fuzzy system for trajectory tracking of an omnidirectional mobile autonomous robot. *Studies in Computational Intelligence* (2020). https://doi.org/10.1007/978-3-030-34135-0_23
10. M.S. Masmoudi, N. Krichen, A.B. Koesdwiady, F. Karray, M. Masmoudi, Design and FPGA implementation of a fuzzy-PI controller for omnidirectional robot system. Adv. Intell. Syst. Comput. (2017). https://doi.org/10.1007/978-3-319-31293-4_12
11. J.B. Mbede, A. Melingui, B.E. Zobo, R. Merzouki, B.O. Bouamama, zSlices based type-2 fuzzy motion control for autonomous robotino mobile robot, in *Proceedings of 2012 8th IEEE/ASME International Conference on Mechatronic and Embedded Systems and Applications, MESA* (2012). https://doi.org/10.1109/MESA.2012.6275538
12. A. Melingui, T. Chettibi, R. Merzouki, J.B. Mbede, Adaptive navigation of an omni-drive autonomous mobile robot in unstructured dynamic environments, in *2013 IEEE International Conference on Robotics and Biomimetics, ROBIO* (2013). https://doi.org/10.1109/ROBIO.2013.6739750
13. F. Cuevas, O. Castillo, P. Cortes-Antonio, Towards an adaptive control strategy based on type-2 fuzzy logic for autonomous mobile robots, in *IEEE International Conference on Fuzzy Systems* (2019). https://doi.org/10.1109/FUZZ-IEEE.2019.8858801
14. F. Cuevas, O. Castillo, P. Cortés-Antonio, Omnidirectional four wheel mobile robot control with a type-2 fuzzy logic behavior-based strategy, in *Studies in Computational Intelligence* (2020). https://doi.org/10.1007/978-3-030-35445-9_4
15. A. Melingui, R. Merzouki, J.B. Mbede, Neuro-fuzzy controller for autonomous navigation of mobile robots, in *2014 IEEE Conference on Control Applications, CCA* (2014). https://doi.org/10.1109/CCA.2014.6981474
16. C. Treesatayapun, A discrete-time stable controller for an omni-directional mobile robot based on an approximated model. Control Eng. Pract. (2011). https://doi.org/10.1016/j.conengprac.2010.11.008
17. M.L. Lagunes, O. Castillo, J. Soria, M. Garcia, F. Valdez, Optimization of granulation for fuzzy controllers of autonomous mobile robots using the Firefly algorithm. Granul. Comput. (2019). https://doi.org/10.1007/s41066-018-0121-6
18. M.L. Lagunes, O. Castillo, F. Valdez, J. Soria, Multi-metaheuristic competitive model for optimization of fuzzy controllers. Algorithms (2019). https://doi.org/10.3390/a12050090
19. O.R. Carvajal, O. Castillo, J. Soria, Optimization of membership function parameters for fuzzy controllers of an autonomous mobile robot using the flower pollination algorithm. J. Autom. Mob. Robot. Intell. Syst. (2018). https://doi.org/10.14313/JAMRIS_1-2018/6

20. C. Caraveo, F. Valdez, O. Castillo, Optimization of fuzzy controller design using a new bee colony algorithm with fuzzy dynamic parameter adaptation. Appl. Soft Comput. J. (2016). https://doi.org/10.1016/j.asoc.2016.02.033
21. O. Castillo, H. Neyoy, J. Soria, P. Melin, F. Valdez, A new approach for dynamic fuzzy logic parameter tuning in ant colony optimization and its application in fuzzy control of a mobile robot. Appl. Soft Comput. J. (2015). https://doi.org/10.1016/j.asoc.2014.12.002
22. C.I. González, P. Melin, J.R. Castro, Olivia Mendoza, Oscar Castillo: an improved Sobel edge detection method based on generalized type-2 fuzzy logic. Soft. Comput. **20**(2), 773–784 (2016)
23. E. Ontiveros, P. Melin, O. Castillo, High order α-planes integration: a new approach to computational cost reduction of general type-2 fuzzy systems. Eng. Appl. of AI **74**, 186–197 (2018)
24. P. Melin, O. Castillo, Intelligent control of complex electrochemical systems with a neuro-fuzzy-genetic approach. IEEE Tran. Ind. Electron. **48**(5), 951–955
25. L. Aguilar, P. Melin, O. Castillo, Intelligent control of a stepping motor drive using a hybrid neuro-fuzzy ANFIS approach. Appl. Soft Comput. **3**(3), 209–219 (2003)
26. P. Melin, O. Castillo, Adaptive intelligent control of aircraft systems with a hybrid approach combining neural networks, fuzzy logic and fractal theory. Appl. Soft Comput. **3**(4), 353–362 (2003)
27. P. Melin, J. Amezcua, F. Valdez, O. Castillo, A new neural network model based on the LVQ algorithm for multi-class classification of arrhythmias. Inf. Sci. **279**, 483–497 (2014)
28. P. Melin, O. Castillo, *Modelling, simulation and control of non-linear dynamical systems: an intelligent approach using soft computing and fractal theory* (CRC Press, USA and Canada, 2002)
29. P. Melin, D. Sánchez, O. Castillo, Genetic optimization of modular neural networks with fuzzy response integration for human recognition. Inf. Sci. **197**, 1–19 (2012)
30. M.A. Sanchez, O. Castillo, J.R. Castro, P. Melin, Fuzzy granular gravitational clustering algorithm for multivariate data. Inf. Sci. **279**, 498–511 (2014)
31. D. Sanchez, P. Melin, Optimization of modular granular neural networks using hierarchical genetic algorithms for human recognition using the ear biometric measure. Eng. Appl. Artif. Intell. **27**, 41–56 (2014)
32. O. Castillo, *Type-2 fuzzy logic in intelligent control applications*. Springer (2012)

# Optimization of Routes of a Robot Using Bioinspired Algorithms

**Hector M. Guajardo and Fevrier Valdez**

**Abstract** We propose in this paper the use fuzzy logic to adjust parameters in thein the particle swarm optimization and ant colony optimization.This paper describe the comparison of the results obtained in the particle swarm optimization (PSO) and the ant colony optimization (ACO) of the resolution of the traveling salesman person (TSP), the adjustment (Xu, 2019 Chinese control and decision conference (CCDC), Nanchang, China, pp 3760–3763 [1]) is performed to improve the behavior of both methods. The particle swarm method and the ant colony methods have parameters, which need to dynamically adjust to improve the performance of both algorithms.

**Keywords** Particle Swarm Optimization (PSO) and Ant Colony Optimization (ACO) · Traveling Salesman Problem (TSP) · Fuzzy logic · Dynamic adjustment

## 1 Introduction

Nowadays companies have modernized most of their development structures, with this we have in mind the following idea: the work of men is currently being developed by machines, at this time society with its speed has forced us to carry out all our activities in a faster way due to the type of life that is presented to us, taking that same speed to everyday life as well as its development, just to give an example: in the automotive industries, a car took months to complete, without, However, in these times it has been optimized until cars can be made in 20 h. With this we give ourselves an idea about the progress and the importance of progress, so we will take that idea to the metaheuristics currently used for the development and optimization of parameters. We propose in this paper the use fuzzy logic to adjust parameters in thein the particle swarm optimization and ant colony optimization. This paper focal point is the performance of the ant colony algorithm [2–5] and the particle swarm optimization [6–9] by optimizing the main parameters, with this on aim we

H. M. Guajardo · F. Valdez (✉)
Tijuana Institute of Technology, Tijuana, BC, Mexico
e-mail: fevrier@tectijuana.mx



P. Melin et al. (eds.), *Recent Advances of Hybrid Intelligent Systems Based on Soft Computing*, Studies in Computational Intelligence 915,
https://doi.org/10.1007/978-3-030-58728-4_13

can optimize a fuzzy controller [10] to obtain better results. A simple fuzzy system of one input and one output was developed to adjust the alpha parameter in the ant colony algorithm and $c_1, c_2$ and weight of inertia in particle swarm algorithm. This paper is organized as follows. Section 2 describes Particle Swarm Optimization Algorithm, Sect. 3 presents in the details of the Ant Colony Optimization, Sect. 4 describes the Traveling Salesman Problem, Sect. 5 shows the Type 1 Fuzzy Logic, Sect. 6 describes the fuzzy optimization, Sect. 7 presents in the details the results obtained in the experiments, and Sect. 8 describes the conclusions.

## 2 Particle Swarm Optimization

A PSO algorithm works with a population (called swarm) of candidate solutions (called particles). These particles move along the search space according to simple mathematical rules. The movement of each particle depends on its best position obtained, as well as the best global position found throughout the search space. As new and better positions are discovered, they move on to orient the movements of the particles. The process is repeated with the objective, not guaranteed, of finding a satisfactory solution at some point. For the global best PSO or gbest PSO, the neighborhood for each particle is the entire swarm. The social network employed by the gbest PSO reflects the star topology. For the star neighborhood topology, the social component of particle velocity update reflects information obtained from all the particles in the swarm. In this case, the social information is the best position found by the warm, referred to as $Y^t$.

The gbest PSO, the velocity of particle i is calculated as:

$$V_{ij}(t+1) = WV_{ij}(t) + C_1 r_{1j}(t)\big[Y_{ij}(t) - X_{ij}(t)\big] + C_2 r_{2j}(t)\Big[\hat{Y}_j(t) - X_{ij}(t)\Big] \quad (1)$$

| | |
|---|---|
| $V_{ij}(t)$ | Is the velocity or particle I in dimension j = 1, …, nx at time step t. |
| $X_{ij}(t)$ | The position of the particular i in the search space-time t. |
| $V_{ij}(t)$ | It is the velocity of the particular i in dimension j over time t. |
| $c_1, c_2$ | They are constants of positive acceleration used to scale the cognitive contribution and social component respectively. |
| $r_{1j}(t), r_{2j}(t)$ | They are random values between 0 and 1 sampled from a uniform distribution ∈ [0, 1]. |
| $Y_{ij}$ | The personal best position, $Y_i$, associated with particle i is the best position the particle has visited since the first time step. Consider in minimization problems, the personal best position at the next time step, t + 1, is calculated as. |
| $W \geq 1$ | Velocities increase over time, accelerating towards maximum velocity. |
| $W < 1$ | Particles decelerate until their velocities reach zero. |

*Global best position at next time step Equation*

The equation for defining the best position is:

$$Y_i(t+1) = \{Y_i(t) \quad if\ f(X_i(t+1) \geq f(Y_i(t)) \\ \{X_i(t+1) \quad if\ f(X_i(t+1)) < F(Y_i(t) \tag{2}$$

## 3 Ant Colony Optimization

The behavior of real ants looking for food inspires the ant colony optimization algorithm (ACO). Ants propagate pheromones during their search creating a trail that influences the choice of other ants' path. The algorithm uses numerical information to represent the pheromone traces and maintains information about the search experience. Each ant makes random decisions iteratively until a complete solution is achieved; These decisions follow a probability that considers all the information available at this time.

Each artificial ant in ACO builds a stochastic solution, leading to a wide variety of different routes. Construction uses the information available at this time and also the experience gained with other solutions to add components, one by one, in the solution. If it is possible to create a procedure that generates a solution for a discrete optimization problem, then, in principle, it is possible to apply the ACO.

Ant colony optimization (ACO) is a class of optimization algorithms modeled on the foraging behavior of an ant colony. In the real world, when searching for food, ants initially explore the area surrounding their nest in a random manner. As soon as an ant finds a food source, it deposits a chemical substance, called pheromone trail which will guide other ants to the food source, on the traveled path. The quantity of pheromone trail depends on the quantity and quality of the food. The indirect communication between the ants by the pheromone trail allows them to find shortest paths from their nest to the food source. In ACO, a colony of artificial ants with the artificial pheromone trails and heuristic information are used to iteratively construct candidate solutions for a problem New solutions are generated using a parameterized probabilistic model, the parameters of which are updated using previously generated solutions so as to direct the search towards promising areas of the solution space [11].

## 4 Traveling Salesman Problem

Intuitively, the TSP is the problem of a salesman who, starting from his hometown, wants to find the shortest tour that takes him through a given set of customers cities and then back home, visiting each customer city exactly once. More formally, the

TSP can be represented by complete weighted graph G = (N, A) with N being the set of nodes representing the cities, and A being the set of arcs. (Note that if the graph is not complete, one can always add arcs to obtain a new, complete graph G' with the same optimal solutions as G; this can be achieved by assigning to the weights of the additional arcs that are large enough to guarantee that they will not be used in any optimal solution.) Each arc (i, j) € A is assigned a value (length) $d_{ij}$, which is the distance between cities i and j, with i; j € N. In the general case of the asymmetric TSP, the distance between a pair of nodes i, j is dependent on the direction of traversing the arc, that is, there is at least one arc (i, j) for which $d_{ij} \neq d_{ji}$. In the symmetric TSP, $d_{ij} = d_{ji}$ holds for all the arcs in A. The goal in the TSP is to find a minimum length Hamiltonian circuit of the graph, where a Hamiltonian circuit is a closed path visiting each of the $n = |N|$ nodes of G exactly once [12].

It was already mentioned previously, a fuzzy system was developed to make the dynamic adjustment to the alpha parameter, the fuzzy system has an input that represents the iterations and an output that is the value that takes alpha for each iteration, in Figs. 2 and 3 input and output are observed of fuzzy models where each has a range [0 1] and as linguistic variable low, medium and high.

## 5 Type-1 Fuzzy Logic

Fuzzy logic [13, 14] emerged based on the principle of incompatibility between accuracy and complexity, since normal computational procedures are so exact that they are incompatible with the complexity of human reasoning, and for this reason fuzzy logic uses linguistic variables that are words like those humans use and not exact numbers as they use machines.

In Fig. 1 type-1 triangular membership functions are illustrated.

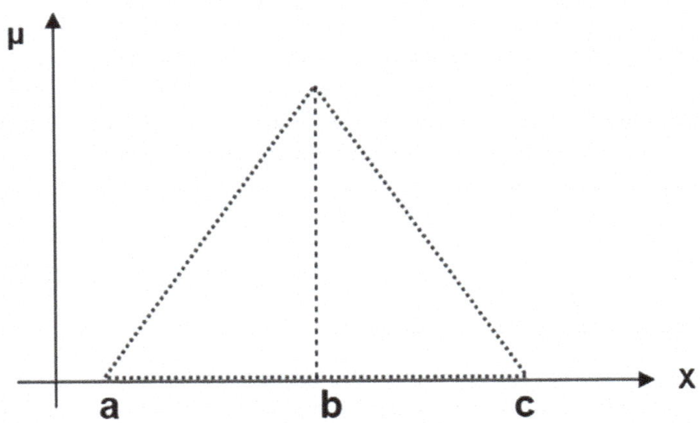


**Fig. 1** Type-1 triangular membership functions

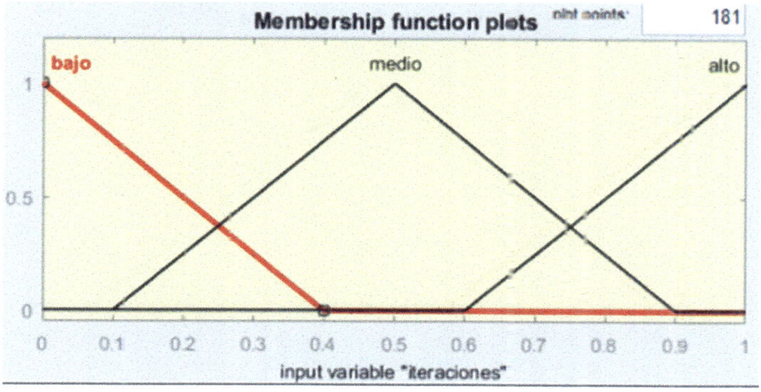


**Fig. 2** Type-1 fuzzy system input iterations with 3 triangular membership functions for $c_1$ in PSO

$$triangular(u; a, b, c) = \begin{cases} 0, & u \leq a \\ \frac{u-a}{b-c}, & a \leq x \leq b \\ \frac{c-x}{c-b} & b \leq x \leq c \\ 0, & c \leq x \end{cases} \tag{3}$$

## 6 Fuzzy Optimization

We start to describe the fuzzy system. These types of systems (fuzzy systems) are expert systems that are based on linguistic variables, that depend on membership functions (uncertainty), and the calculus that is done internally, produces one or more output variables, course, and this depends on the problem. For ACO the optimized parameters were: alpha considerably improved the results shown, in the case of PSO the parameters that were optimized were $c_1$, $c_2$ and insertion weight, with this the results obtained also improved compared to their results without optimization, the ant colony optimization algorithm was optimized using fuzzy dynamic adjustment in the algorithm, is of Mamdani [15, 16] type, which has 2 inputs and 2 outputs as shown in Figs. 2 and 3.

## 7 Simulation Results

At the beginning of the experiments, random values were taken for the $c_1$ and $c_2$ from PSO, cognitive and social parameters with a range of 0–2 as suggested by the author of the algorithm, just as the population and the iterations were randomized.

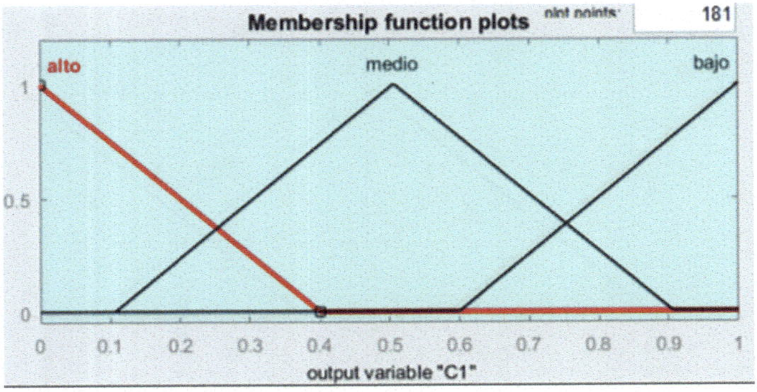


**Fig. 3** Type-1 fuzzy systems output $c_1$ PSO

These experiments performed manually trial and error were performed unfairly when evaluating the minor cost possible, the number of iterations was 30, 40, 50 and 100 nodes with each one of the algorithms searching the minor cost on 500 iterations in each experiment. We show in the following Tables 1 and 2 the results obtained from the results of the algorithms without optimization and optimization.

This Table 1 describes the values the $c_1$, $c_2$ and weight of inertia parameters from the Particle Swarm Optimization Algorithm.

This Table 2 describes the values the $c_1$, $c_2$ and weight of inertia parameters optimized from the Particle Swarm Optimization Algorithm.

In the case of ACO, the experiments were started with random parameters, after the experiments alpha was optimized, improving the results obtained. The number of iterations was 30, 40, 50 and 100 nodes with each one of the algorithms searching the minor cost on 500 iterations in each experiment. We show in Tables 3 and 4 the results obtained from the results of the algorithms without optimization and optimization.

We show in Table 5 the comparison of the first 30 experiments of the results obtained from the PSO and ACO algorithms with optimized parameters. In this case, the ACO algorithm shows better results.

As we can see the results of the experiment doing the fuzzy adjustment by iteration are better than changing the value manually. The results show that fuzzy parameters help more the behavior of the method in this particular case of decreasing the cost.

## 8 Conclusions

The algorithms help us speed up the result of the experiments we carry out and we find that the random values within the range that the author mentions work correctly, however, by attaching the fuzzy adaptation to various parameters of the PSO and the ACO, we show that the result of the obtained experiments was improved. In order to evade local spikes and obtain the opportune global result, we have used a combination

**Table 1** Particle swarm optimization algorithm random parameters

| Experiment | PSO nodes | Population | Iterations | Weigh inertia | $C_1$ | $C_2$ | Lower cost obtained |
|---|---|---|---|---|---|---|---|
| 1 | 30 | 500 | 200 | 0.5 | 1.5 | 1.5 | 664.787 |
| 2 | 30 | 500 | 200 | 0.5 | 1.5 | 1.5 | 644.443 |
| 3 | 30 | 500 | 200 | 0.5 | 1.5 | 1.5 | 812.345 |
| 4 | 30 | 500 | 200 | 0.5 | 1.5 | 1.5 | 772.205 |
| 5 | 30 | 500 | 200 | 0.5 | 1.5 | 1.5 | 808.712 |
| 6 | 30 | 500 | 200 | 0.5 | 1.5 | 1.5 | 765.971 |
| 7 | 30 | 500 | 200 | 0.5 | 1.5 | 1.5 | 780.983 |
| 8 | 30 | 500 | 200 | 0.5 | 1.5 | 1.5 | 780.921 |
| 9 | 30 | 500 | 200 | 0.5 | 1.5 | 1.5 | 705.453 |
| 10 | 30 | 500 | 200 | 0.5 | 1.5 | 1.5 | 759.215 |
| 11 | 30 | 500 | 200 | 0.5 | 1.5 | 1.5 | 801.231 |
| 12 | 30 | 500 | 200 | 0.5 | 1.5 | 1.5 | 793.548 |
| 13 | 30 | 500 | 200 | 0.5 | 1.5 | 1.5 | 872.211 |
| 14 | 30 | 500 | 200 | 0.5 | 1.5 | 1.5 | 656.582 |
| 15 | 30 | 500 | 200 | 0.5 | 1.5 | 1.5 | 706.805 |
| 16 | 30 | 500 | 200 | 0.5 | 1.5 | 1.5 | 783.938 |
| 17 | 30 | 500 | 200 | 0.5 | 1.5 | 1.5 | 711.359 |
| 18 | 30 | 500 | 200 | 0.5 | 1.5 | 1.5 | 854.112 |
| 19 | 30 | 500 | 200 | 0.5 | 1.5 | 1.5 | 759.192 |
| 20 | 30 | 500 | 200 | 0.5 | 1.5 | 1.5 | 750.813 |
| 21 | 30 | 500 | 200 | 0.5 | 1.5 | 1.5 | 708.619 |
| 22 | 30 | 500 | 200 | 0.5 | 1.5 | 1.5 | 695.276 |
| 23 | 30 | 500 | 200 | 0.5 | 1.5 | 1.5 | 824.671 |
| 24 | 30 | 500 | 200 | 0.5 | 1.5 | 1.5 | 902.247 |
| 25 | 30 | 500 | 200 | 0.5 | 1.5 | 1.5 | 700.287 |
| 26 | 30 | 500 | 200 | 0.5 | 1.5 | 1.5 | 778.975 |
| 27 | 30 | 500 | 200 | 0.5 | 1.5 | 1.5 | 738.047 |
| 28 | 30 | 500 | 200 | 0.5 | 1.5 | 1.5 | 702.125 |
| 29 | 30 | 500 | 200 | 0.5 | 1.5 | 1.5 | 834.812 |
| 30 | 30 | 500 | 200 | 0.5 | 1.5 | 1.5 | 751.249 |

of both algorithms with fuzzy logic being these ACO and PSO, in addition to that, we use adaptations with fuzzy developments of triangular membership functions, with these adaptations generated to the algorithms we have obtained in all the experiments better ACO results than PSO in TSP resolution in all the experiments obtained from 30, 40, 50 and up to 100 nodes. More research can be done on this to improve this implementation. We envision as future work the adaptation of trapezoidal and

**Table 2** Particle swarm optimization algorithm optimized parameters

| Experiment | PSO nodes | Population | Iterations | Weight inertia | $C_1$ | $C_2$ | Lower cost obtained |
|---|---|---|---|---|---|---|---|
| 1 | 30 | 500 | 200 | 0.87 | 1.74 | 1.3 | 744.4945 |
| 2 | 30 | 500 | 200 | 0.87 | 1.74 | 1.3 | 691.7225 |
| 3 | 30 | 500 | 200 | 0.87 | 1.74 | 1.3 | 662.3489 |
| 4 | 30 | 500 | 200 | 0.87 | 1.74 | 1.3 | 649.7018 |
| 5 | 30 | 500 | 200 | 0.87 | 1.74 | 1.3 | 710.994 |
| 6 | 30 | 500 | 200 | 0.87 | 1.74 | 1.3 | 824.098 |
| 7 | 30 | 500 | 200 | 0.87 | 1.74 | 1.3 | 660.6467 |
| 8 | 30 | 500 | 200 | 0.87 | 1.74 | 1.3 | 769.9469 |
| 9 | 30 | 500 | 200 | 0.87 | 1.74 | 1.3 | 636.9222 |
| 10 | 30 | 500 | 200 | 0.87 | 1.74 | 1.3 | 892.2515 |
| 11 | 30 | 500 | 200 | 0.87 | 1.74 | 1.3 | 941.5559 |
| 12 | 30 | 500 | 200 | 0.87 | 1.74 | 1.3 | 814.8211 |
| 13 | 30 | 500 | 200 | 0.87 | 1.74 | 1.3 | 820.2219 |
| 14 | 30 | 500 | 200 | 0.87 | 1.74 | 1.3 | 1032.7783 |
| 15 | 30 | 500 | 200 | 0.87 | 1.74 | 1.3 | 872.1368 |
| 16 | 30 | 500 | 200 | 0.87 | 1.74 | 1.3 | 719.906 |
| 17 | 30 | 500 | 200 | 0.87 | 1.74 | 1.3 | 1029.4988 |
| 18 | 30 | 500 | 200 | 0.87 | 1.74 | 1.3 | 678.4458 |
| 19 | 30 | 500 | 200 | 0.87 | 1.74 | 1.3 | 735.5286 |
| 20 | 30 | 500 | 200 | 0.87 | 1.74 | 1.3 | 679.765 |
| 21 | 30 | 500 | 200 | 0.87 | 1.74 | 1.3 | 732.884 |
| 22 | 30 | 500 | 200 | 0.87 | 1.74 | 1.3 | 785.9776 |
| 23 | 30 | 500 | 200 | 0.87 | 1.74 | 1.3 | 713.3184 |
| 24 | 30 | 500 | 200 | 0.87 | 1.74 | 1.3 | 652.2732 |
| 25 | 30 | 500 | 200 | 0.87 | 1.74 | 1.3 | 705.6665 |
| 26 | 30 | 500 | 200 | 0.87 | 1.74 | 1.3 | 791.1604 |
| 27 | 30 | 500 | 200 | 0.87 | 1.74 | 1.3 | 779.7708 |
| 28 | 30 | 500 | 200 | 0.87 | 1.74 | 1.3 | 814.2845 |
| 29 | 30 | 500 | 200 | 0.87 | 1.74 | 1.3 | 602.5625 |
| 30 | 30 | 500 | 200 | 0.87 | 1.74 | 1.3 | 679.1898 |

sinusoidal membership functions to the aforementioned parameters to optimize the results and have more comparisons with other applications.

**Table 3** Ant colony optimization algorithm random parameters

| Experiment | ACO nodes | Population | Iterations | Evaporation coefficient | Alpha | COMUN cost disposal | Lower cost obtained |
|---|---|---|---|---|---|---|---|
| 1 | 30 | 500 | 200 | 0.15 | 0.2 | 0.97 | 304.4906 |
| 2 | 30 | 500 | 200 | 0.15 | 0.2 | 0.97 | 258.151 |
| 3 | 30 | 500 | 200 | 0.15 | 0.2 | 0.97 | 267.0012 |
| 4 | 30 | 500 | 200 | 0.15 | 0.2 | 0.97 | 267.8232 |
| 5 | 30 | 500 | 200 | 0.15 | 0.2 | 0.97 | 261.9851 |
| 6 | 30 | 500 | 200 | 0.15 | 0.2 | 0.97 | 255.407 |
| 7 | 30 | 500 | 200 | 0.15 | 0.2 | 0.97 | 251.852 |
| 8 | 30 | 500 | 200 | 0.15 | 0.2 | 0.97 | 256.1221 |
| 9 | 30 | 500 | 200 | 0.15 | 0.2 | 0.97 | 271.481 |
| 10 | 30 | 500 | 200 | 0.15 | 0.2 | 0.97 | 273.11 |
| 11 | 30 | 500 | 200 | 0.15 | 0.2 | 0.97 | 279.493 |
| 12 | 30 | 500 | 200 | 0.15 | 0.2 | 0.97 | 268.891 |
| 13 | 30 | 500 | 200 | 0.15 | 0.2 | 0.97 | 254.301 |
| 14 | 30 | 500 | 200 | 0.15 | 0.2 | 0.97 | 260.606 |
| 15 | 30 | 500 | 200 | 0.15 | 0.2 | 0.97 | 268.4786 |
| 16 | 30 | 500 | 200 | 0.15 | 0.2 | 0.97 | 275.722 |
| 17 | 30 | 500 | 200 | 0.15 | 0.2 | 0.97 | 268.691 |
| 18 | 30 | 500 | 200 | 0.15 | 0.2 | 0.97 | 234.726 |
| 19 | 30 | 500 | 200 | 0.15 | 0.2 | 0.97 | 240.47 |
| 20 | 30 | 500 | 200 | 0.15 | 0.2 | 0.97 | 279.729 |
| 21 | 30 | 500 | 200 | 0.15 | 0.2 | 0.97 | 262.159 |
| 22 | 30 | 500 | 200 | 0.15 | 0.2 | 0.97 | 248.0383 |
| 23 | 30 | 500 | 200 | 0.15 | 0.2 | 0.97 | 258.729 |
| 24 | 30 | 500 | 200 | 0.15 | 0.2 | 0.97 | 250.521 |
| 25 | 30 | 500 | 200 | 0.15 | 0.2 | 0.97 | 241.618 |
| 26 | 30 | 500 | 200 | 0.15 | 0.2 | 0.97 | 268.328 |
| 27 | 30 | 500 | 200 | 0.15 | 0.2 | 0.97 | 256.002 |
| 28 | 30 | 500 | 200 | 0.15 | 0.2 | 0.97 | 270.584 |
| 29 | 30 | 500 | 200 | 0.15 | 0.2 | 0.97 | 235.257 |
| 30 | 30 | 500 | 200 | 0.15 | 0.2 | 0.97 | 279.729 |

**Table 4** Ant colony optimization algorithm optimized parameters

| Experiment | ACO nodes | Population | Iterations | Evaporation coefficient | Alpha | COMUN cost disposal | Lower cost obtained |
|---|---|---|---|---|---|---|---|
| 1 | 30 | 500 | 200 | 0.15 | 0.2 | 0.97 | 160.3544 |
| 2 | 30 | 500 | 200 | 0.15 | 0.2 | 0.97 | 137.3128 |
| 3 | 30 | 500 | 200 | 0.15 | 0.2 | 0.97 | 151.3842 |
| 4 | 30 | 500 | 200 | 0.15 | 0.2 | 0.97 | 149.2125 |
| 5 | 30 | 500 | 200 | 0.15 | 0.2 | 0.97 | 142.121 |
| 6 | 30 | 500 | 200 | 0.15 | 0.2 | 0.97 | 141.2545 |
| 7 | 30 | 500 | 200 | 0.15 | 0.2 | 0.97 | 160.2511 |
| 8 | 30 | 500 | 200 | 0.15 | 0.2 | 0.97 | 163.2114 |
| 9 | 30 | 500 | 200 | 0.15 | 0.2 | 0.97 | 165.2144 |
| 10 | 30 | 500 | 200 | 0.15 | 0.2 | 0.97 | 146.7415 |
| 11 | 30 | 500 | 200 | 0.15 | 0.2 | 0.97 | 166.2541 |
| 12 | 30 | 500 | 200 | 0.15 | 0.2 | 0.97 | 140.1895 |
| 13 | 30 | 500 | 200 | 0.15 | 0.2 | 0.97 | 156.1478 |
| 14 | 30 | 500 | 200 | 0.15 | 0.2 | 0.97 | 147.6574 |
| 15 | 30 | 500 | 200 | 0.15 | 0.2 | 0.97 | 151.944 |
| 16 | 30 | 500 | 200 | 0.15 | 0.2 | 0.97 | 140.6536 |
| 17 | 30 | 500 | 200 | 0.15 | 0.2 | 0.97 | 154.1236 |
| 18 | 30 | 500 | 200 | 0.15 | 0.2 | 0.97 | 167.2141 |
| 19 | 30 | 500 | 200 | 0.15 | 0.2 | 0.97 | 159.1244 |
| 20 | 30 | 500 | 200 | 0.15 | 0.2 | 0.97 | 156.3701 |
| 21 | 30 | 500 | 200 | 0.15 | 0.2 | 0.97 | 174.1214 |
| 22 | 30 | 500 | 200 | 0.15 | 0.2 | 0.97 | 133.0451 |
| 23 | 30 | 500 | 200 | 0.15 | 0.2 | 0.97 | 160.8558 |
| 24 | 30 | 500 | 200 | 0.15 | 0.2 | 0.97 | 160.2252 |
| 25 | 30 | 500 | 200 | 0.15 | 0.2 | 0.97 | 169.3854 |
| 26 | 30 | 500 | 200 | 0.15 | 0.2 | 0.97 | 158.9195 |
| 27 | 30 | 500 | 200 | 0.15 | 0.2 | 0.97 | 154.1634 |
| 28 | 30 | 500 | 200 | 0.15 | 0.2 | 0.97 | 164.2297 |
| 29 | 30 | 500 | 200 | 0.15 | 0.2 | 0.97 | 170.4214 |
| 30 | 30 | 500 | 200 | 0.15 | 0.2 | 0.97 | 161.3722 |

**Table 5** Comparison of best results between ACO and PSO in the resolution of the TSP

| Experiment | Shorter distance obtained PSO | Shorter distance obtained ACO |
|---|---|---|
| 1 | 744.4945 | 160.3544 |
| 2 | 691.7225 | 137.3128 |
| 3 | 662.3489 | 151.3842 |
| 4 | 649.7018 | 149.2125 |
| 5 | 710.994 | 142.121 |
| 6 | 824.098 | 141.2545 |
| 7 | 660.6467 | 160.2511 |
| 8 | 769.9469 | 163.2114 |
| 9 | 636.9222 | 165.2144 |
| 10 | 892.2515 | 146.7415 |
| 11 | 941.5559 | 166.2541 |
| 12 | 814.8211 | 140.1895 |
| 13 | 820.2219 | 156.1478 |
| 14 | 1032.7783 | 147.6574 |
| 15 | 872.1368 | 151.944 |
| 16 | 719.906 | 140.6536 |
| 17 | 1029.4988 | 154.1236 |
| 18 | 678.4458 | 167.2141 |
| 19 | 735.5286 | 159.1244 |
| 20 | 679.765 | 156.3701 |
| 21 | 732.884 | 174.1214 |
| 22 | 785.9776 | 133.0451 |
| 23 | 713.3184 | 160.8558 |
| 24 | 652.2732 | 160.2252 |
| 25 | 705.6665 | 169.3854 |
| 26 | 791.1604 | 158.9195 |
| 27 | 779.7708 | 154.1634 |
| 28 | 814.2845 | 164.2297 |
| 29 | 602.5625 | 170.4214 |
| 30 | 679.1898 | 161.3722 |

## References


1. L.A. Zadeh, Fuzzy logic. Computer **21**(4), 83–93 (1988)
2. L.A. Zadeh, *Fuzzy Algorithms*, vol. 12 (Department of Electrical Engineering and Project MAC, 1968), pp. 94–102
3. C. Caraveo, F. Valdez, O. Castillo, Optimization mathematical functions for multiple variables using the algorithm of self-defense of the plants, in *2018 33rd Youth Academic Annual Conference of Chinese Association of Automation (YAC)* (2018). Accessed 17 Dec 2019

4. D. Tian, N. Li, Fuzzy particle swarm optimization algorithm, in *2009 International Joint Conference on Artificial Intelligence* (2009). Accessed 17 Dec 2019
5. M. Dorigo, T. Stützle, Ant colony optimization algorithms for the traveling salesman problem, in *Ant Colony Optimization*, vol. 547 (A Bradford Book, Tijuana, BC, 2004), pp. 65–119
6. Z. Zhang, K. Zou, Simple ant colony algorithm for combinatorial optimization problems, in *2017 36th Chinese Control Conference (CCC)* (2017). Accessed 20 Jan 2020
7. Y. Zhang J. Chen, C. Bingham, M. Mahfouf, A new adaptive Mamdani-type fuzzy modeling strategy for industrial gas turbines, *IEEE International Conference on Fuzzy Systems (FUZZ-IEEE)* (2014). Accessed 15 Jan 2020
8. A.K. Sharma, V. Singh, N.K. Verma, J. Liu, Condition based monitoring of machine using mamdani fuzzy network, in *Prognostics and System Health Management Conference (PHM-Chongqing)* (2018), Accessed 15 Jan 2020
9. Y. Xu, Some new operations on triangular fuzzy number intuitionistic fuzzy set, in *Chinese Control And Decision Conference (CCDC)* (2019). Accessed 15 Jan 2020
10. Z.F. Hao, R.C. Cai, H. Huang, An adaptive parameter control strategy for ACO, in *2006 International Conference on Machine Learning and Cybernetics* (2006). Accessed 15 Sept 2019
11. A. Shetty, K.S. Puthusseri, R. Shankaramani, A. Shetty, An improved ant colony optimization algorithm: minion ant (MAnt) and its application on TSP, in *2018 IEEE Symposium Series on Computational Intelligence (SSCI)* (2018). https://ieeexplore.ieee.org/document/8628805. Accessed 15 Nov 2019
12. R.S. Jadon, U. Dutta, Modified ant colony optimization algorithm with uniform mutation using self-adaptive approach for travelling salesman problem, in *2013 Fourth International Conference on Computing, Communications and Networking Technologies (ICCCNT), Tiruchengode, 2013* (2013). Accessed 16 Dec 2019
13. M. Dutta, P. Pragya, TSP solution using dimensional ant colony optimization, in *2015 Fifth International Conference on Advanced Computing & Communication Technologies* (2015). Accessed 16 Dec 2019
14. Y. Wang, Improving artificial bee colony and particle swarm optimization to solve TSP problem, in *2018 International Conference on Virtual Reality and Intelligent Systems (ICVRIS)* (2018). Accessed 16 Dec 2019
15. H. Qian, T. Su, Hybrid algorithm based on max and min ant system and particle swarm optimization for solving TSP problem, in *2018 33rd Youth Academic Annual Conference of Chinese Association of Automation (YAC)* (2018). https://ieeexplore.ieee.org/document/8406459. Accessed 16 Dec 2019
16. S. Akter, S.S. Rahman, M.H. Rahman, M.A. Akhand, Particle swarm optimization with partial search to solve traveling salesman problem, in *2012 International Conference on Computer and Communication Engineering (ICCCE)* (2012). https://ieeexplore.ieee.org/document/6271164. Accessed 16 Dec 2019

# Optimal Design of Fuzzy Logic Systems Through a Chicken Search Optimization Algorithm Applied to a Benchmark Problem

**Leticia Amador-Angulo and Oscar Castillo**

**Abstract** In this paper an implementation of the CSO (Chicken Search Optimization) algorithm in benchmark problems is presented. CSO algorithm is used on solving the problem to find the optimal distribution on the Membership Functions (MFs) in the Type-1 Fuzzy Logic System (T1FLS) applied to fuzzy controller specifically for the water tank problem. Optimization in the structure and parameters for designing for a fuzzy tracking benchmark controller is presented. An efficiently CSO algorithm for the optimization in Fuzzy Logic Controllers (FLC) is presented. When level of noise is added in the model the CSO shows an excellent development. CSO algorithm shows better results when is compared with others metaheuristic in the simulation results for this benchmark control problem.

**Keywords** Chicken search optimization · Fuzzy controller · Fuzzy logic system · Bio-inspired algorithms · Benchmark problem

## 1 Introduction

Currently meta-heuristic algorithms are used to solve various optimization problems, and this work is based on CSO algorithms as a tool in the stabilization of one benchmark problem apply to controlling the filling of a water tank [1–4].

The problem for stabilization in control systems is of recent interest for many authors. However, most of the methods require to use the mathematical models in the control design methods, which, is an important problem in the practice for complex plants; this problem is a reason for interest in the scientific community, it is therefore that various research are focused in the hybridization of algorithms inspired by nature which have demonstrated an excellent result to find the optimal

L. Amador-Angulo · O. Castillo (✉)
Division of Graduate Studies and Research, Tijuana Institute of Technology, Tijuana, Mexico
e-mail: ocastillo@tectijuana.mx

L. Amador-Angulo
e-mail: gloria.amador@tectijuana.edu.mx



P. Melin et al. (eds.), *Recent Advances of Hybrid Intelligent Systems Based on Soft Computing*, Studies in Computational Intelligence 915,
https://doi.org/10.1007/978-3-030-58728-4_14

design in the T1FLS. On the other hand, fuzzy control has demonstrated to solve many nonlinear control problems successfully, for mentioning some works; which are presented in [1–3, 5–7].

In this paper the main contribution consists in the implementation of the CSO algorithm to optimize design and implementation of fuzzy controllers especially in benchmark problems, the goal is to find the optimal design of parameters in the type-1 fuzzy logic system that allow minimizer the error in modeling complex problems in control.

The content of this paper is divided as follows. Section 2 shows an overview of Related Works important for this work. Section 3 outlines the general concepts of the theory of the type-1 fuzzy logic system. Section 4 describes the bio-inspired algorithm used. Section 5 presents the problem statement. Various simulation results are presented in Sect. 6. Section 7 shows a comparative results with others meta-heuristics. Finally, Sect. 8 offers the conclusion and future works.

## 2 Related Works

Several searches are interested in metaheuristics algorithms as a technique in solving problems in the field of computing. CSO algorithm has implementation to solve various problems, for mentioning some; in [8] an Ultrasonic Machining Process Optimization by Cuckoo Search and Chicken Swarm Optimization Algorithms is presented, in [9] an Optimization of Laser Transmission Welding Parameters Using Chicken Swarm Optimization Algorithm: Chicken Swarm Algorithm Optimization is presented, [10] a Chicken S-BP: An Efficient Chicken Swarm Based Back-Propagation Algorithm is presented, in [11] an Improved chicken swarm optimization to classify dementia MRI images using a novel controlled randomness optimization algorithm is presented, in [12] a Parameter estimation for chaotic systems based on improved boundary chicken swarm optimization is presented, in [13] a New Teaching–Learning-based Chicken Swarm Optimization Algorithm is presented, in [14] a Switched Capacitor-Coupled Inductor DC–DC Converter for Grid-Connected PV System using LFCSO-Based Adaptive Neuro-Fuzzy Inference System is presented, in [5] a Chicken Swarm–Based Method for Ascent Trajectory Optimization of Hypersonic Vehicles is presented, in [7] a Comparative Study of Bio-inspired Algorithms Applied in the Optimization of Fuzzy Systems is presented, in [15] a CSOCA: Chicken Swarm Optimization based Clustering Algorithm for Wireless Sensor Networks is presented, in [16] a Chicken Swarm Optimization Based on Elite Opposition-Based Learning is presented, and [17] a Assessment of water resource carrying capacity based on the chicken swarm optimization-projection pursuit model is presented.

Other important field in which CSO algorithm has been used efficiently is in medicine, for example; in [18] a CSO-based algorithm with support vector machine for brain tumor's disease diagnosis is presented, in [19] a Deadlock-free migration for virtual machine consolidation using Chicken Swarm Optimization algorithm is presented, and in [20] a Application of Chicken Swarm Optimization in Detection of Cancer and Virtual Reality is presented. This research is focused on applying the CSO algorithm in the implementation of fuzzy controllers specifically to solve a benchmark problem.

## 3 Fuzzy Logic System

Zadeh in 1965 was the first one to introduce the main idea of fuzzy logic systems [21–23], and Mamdani was the first one to introduce the concepts of the control theory in 1974 [3, 24]. Both ideas allow that the fuzzy controllers have successfully been used for different applications [2, 6, 7, 25]. Figure 1 illustrates the general description of T1FLS.

A type-1 fuzzy set in the universe X is characterized by a membership function $u_A(x)$ taking values on the interval [0, 1] and can be represented as a set of ordered pairs of an element and the membership degree of an element to the set are defined by the following Eq. (1) [22, 23]:

$$A = \{(x, \mu_A(x))|x \in X\} \tag{1}$$

where $\mu_A : X \to [0, 1]$.

In this definition $\mu_A(x)$ represents the membership degree of the element $x \in X$ to the set $A$. In this work we are going the use the following notation: $A(x) = \mu_A(x)$ for all $x \in X$.

### 3.1 Fuzzy Logic Controllers

Based on ideas the Mamdani in the representation of the fuzzy controllers [3, 24]. The representation of a generic fuzzy controller is illustrated in Fig. 2.

Many researchers have shown an interest in the study and stabilization of fuzzy controllers, just to mention a few in [1, 3, 4, 25, 26].

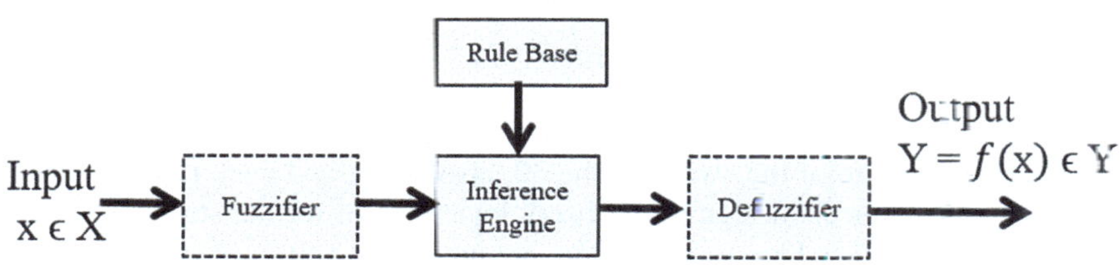


**Fig. 1** Architecture of a Type-1 fuzzy logic sets

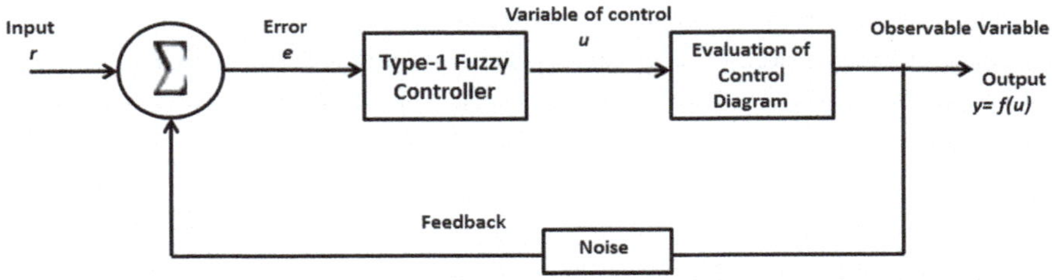


**Fig. 2** Graphical representation of fuzzy controller

## 4 Chicken Search Optimization

Chicken swarm optimization (CSO) is an interesting bio-inspired algorithm based on the behaviors of cocks, hens, and chicks to find food. This algorithm presents chicken swarm in the searching space which is mapped a specific particle individual.

Cock swarm, chick swarm and hen swarm are the types of chicken are used and the whole chicken swarm. The main function and methodology in this algorithm is that the cock swarms represents a highest fitness and chick swarm presents lowest fitness. The role (cock, hen and chick) for each chicken are sorted according to fitness value. The mechanism of the search for each subswarm is different. An important characteristic in this algorithm is that a chicken in the swarm represents a solution in the problem [27]. The goal of the CSO algorithm is this paper is to solve a minimization problem.

In the CSO algorithm, for all the particles, those with the best fitness can be selected as the cock particle swarm, and is represented by Eq. 2:

$$x_{i,j}^{t+1} = x_{i,j}^{t} + randn(0, \sigma^2) \cdot x_{i,j}^{t} \tag{2}$$

where $t$ represents the iterations, $x_{i,j}^{t+1}$ and $x_{i,j}^{t}$ indicate the position of $j$th dimension of a chicken in t + 1. The function randn(0, $\sigma^2$) is used as a random number of Gaussian distribution and the variance is found by Eq. 3:

$$\sigma^2 = \left\{ \begin{array}{l} 1, ------------------ > fit_i < fit_k \\ \exp\left(\dfrac{(fit_k - fit_i)}{(|fit_i| + \xi)}\right), ------- > fit_i \geq fit_k, \end{array} \right\} \tag{3}$$

where $i, k \in [1, \text{rsize}]$ and $i \neq k$. In this case study, rsize indicates the number of cock swarms, the size of the swarm is the 30, also, a bee and cock represents the number of values for the design of the fuzzy logic system for the studied case 30 values are required (see Fig. 4). $fit_i$ and $fit_k$ are the fitness values of cock particle i and k, respectively; $\xi$ represents a number which is used to avoid zero-division-error, is the smallest constant in the computer. Otherwise, for CSO algorithm, hen swarm was selected with good fitness. Its random search is performed via cocks of hen population and that of others, which can be represented by Eq. 4:

$$x_{i,j}^{t+1} = x_{i,j}^{t} + S1 \cdot rand \cdot \left(x_{r1,j}^{t} - x_{i,j}^{t}\right) + S2 \cdot rand \cdot \left(x_{r2,j}^{t} - x_{i,j}^{t}\right) \tag{4}$$

where $x_{i,j}^{t}$ and $x_{r2,j}^{t}$ indicates the position of the cock individual 1 in the population of hen and cock individual 2 in the other population, respectively. An uniform random number with the range of [0, 1] is represented by function *rand*. Equation 5 denote the weight calculated which to calculate the values of 1 and 2.

$$S1 = \exp\left(\frac{\left(fit_i - fit_{r1}\right)}{\left(\left|fit_{r2}\right| + \xi\right)}\right),\ S2 = \exp\left(fit_{r2} - fit_i\right), \tag{5}$$

The fitness value of the cock individual is represented by r1 ($fit_{r1}$) in the population of hen, and the cock individual in the other population is represented by r2 ($fit_{r2}$). All individuals, except for the cock swarm and hen swarm, are defined as chick swarm. Its search mode follows that of the hen swarm, which is given by Eq. 6:

$$x_{i,j}^{t+1} = x_{i,j}^{t} + FL \cdot \left(x_{m,j}^{t} - x_{i,j}^{t}\right),\ FL \in [0, 2] \tag{6}$$

In CSO algorithm, the chick would follow its mother to forage for food, therefore $x_{m,j}^{t}$, is the position of *j*th dimension of particle *m* in $t + 1$, t indicate the iteration, then, FL represents the parameter where this characteristic is evaluated. Hence, *m* represents the position of the th chick's mother (m ∈ [1 N]), where N indicates the size of swarm(population). The Pseudocode of the CSO algorithm is illustrated in Table 1.

A total of 9 step has the CSO algorithm. Figure 3 represents the Flowchart that illustrates the mechanism in this algorithm.

## 5 Study Case

The benchmark control problem is the water level control in a tank and is illustrated in Fig. 4.

The process of filling the water tank is presented as a differential equation for the height of water in the tank, H, and is given by Eq. (7).

$$\frac{d}{dt}Vol = A\frac{dH}{dt} = bV - a\sqrt{H} \tag{7}$$

where **Vol** is the volume of water in the tank, **A** is the cross-sectional area of the tank, **b** is a constant related to the flow rate into the tank and **a** is a constant related to the flow rate out of the tank. The equation describes the height of water, H, as a function of time, due to the difference between flow rates into and out of the tank

**Table 1** Pseudocode CSO Algorithm

| |
|---|
| 1. Initialize a population of $N$ chickens and define the relation parameters |
| 2. Evaluate the $N$ chickens' fitness values, $t$=0 |
| 3. While ($t$ < Max_Generation) |
| 4. If ($t$% G = = 0) |
| 5. Rank the chickens' fitness values and establish a hierarchal order in the swarm |
| 6. Divide the swarm into different groups, and determine the relationship between the chicks and mother hens in a group |
| 7. End if |
| 8. For i = 1; $N$ |
| 9. If (i == rooster) |
| 10. Update its solution/location using Eq. (2) |
| 11. End if |
| 12. If (i == hen) |
| 13. Update its solution/location using Eq. (4) |
| 14. End if |
| 15. If (i == chick) |
| 16. Update its solution/location using Eq. (6) |
| 17. End if |
| 18. Evaluate the new solution; |
| 19. If the new solution is better its previous one; update it |
| 20. End For |
| 21. End While |

[25]. The fuzzy system for this problem is of Mamdani type, which is composed of two inputs which has three triangular membership functions and one output which has five triangular membership functions, as represented in Fig. 5, Table 2 shows the fuzzy rules that are used, and Fig. 5 shows the representation as IF-Then rules for this benchmark problem.

The fuzzy rules that are outlined in Fig. 6 are those that maintain control of the opening of the valve and with this have leveled the water tank. The 5 rules were obtained in Table 2 are based on analyzing the filling of a water tank [25]. The model is shown in Fig. 7.

## 6 Simulations Results

Previously, the study case of the Water tank was presented, which was used to validate the CSO algorithm. Table 3 shows the parameters that were used in the CSO algorithm.

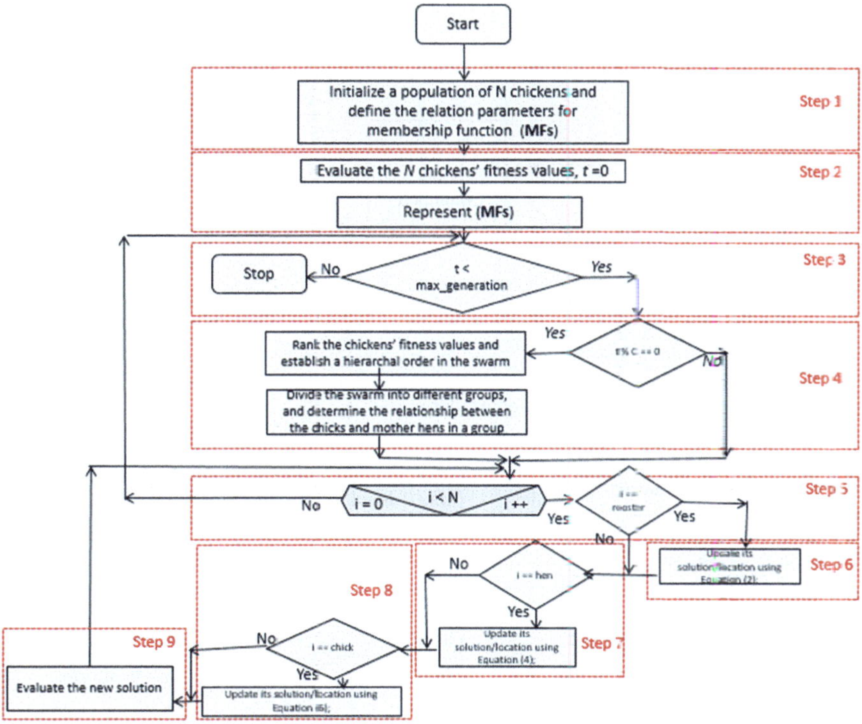


**Fig. 3** Flowchart of the step in the CSO algorithm

**Fig. 4** Water tank system

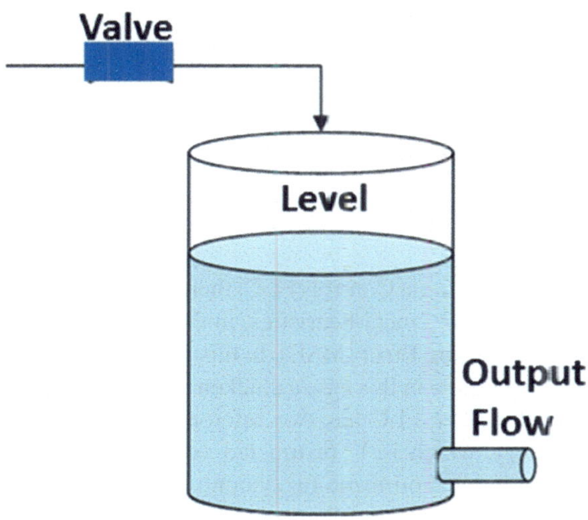

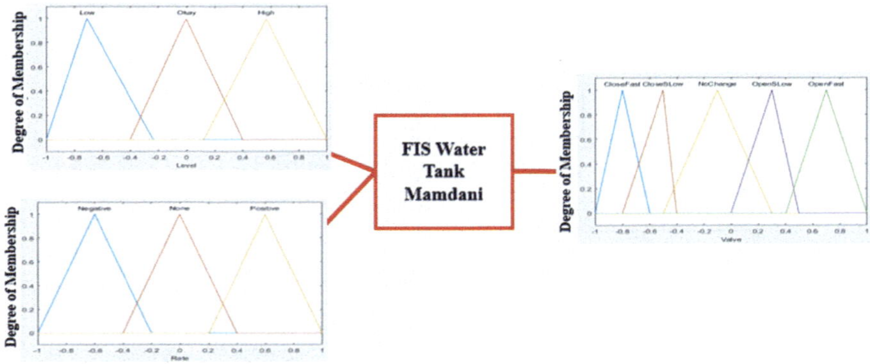


**Fig. 5** Fuzzy system for the water tank controller

**Table 2** Rules for the studied case

| # Rules | Input 1 level | Input 2 rate | Output valve |
|---|---|---|---|
| 1 | Okay | – | NoChange |
| 2 | Low | – | OpenFast |
| 3 | High | – | CloseFast |
| 4 | Okay | Positive | CloseSlow |
| 5 | Okay | Negative | OpenSlow |

```
. If (level is Okay) then (valve is no_change) (1)
. If (level is High) then (valve is open_fast) (1)
. If (level is Low) then (valve is close_fast) (1)
. If (level is Okay) and (rate is Positive) then (valve is close_slow) (1
. If (level is Okay) and (rate is Negative) then (valve is open_slow) (1)
```

**Fig. 6** Fuzzy rules for the water tank controller

In Fuzzy Logic Control (FLC) there exists a metric to evaluate the performance of the bio-inspired meta-heuristics, in this case CSO algorithms. This metric is known as Mean Square Error (MSE). In this paper, this metric is used as the function fitness that the particle in this case a chicken that should to can find in the execution for each algorithm. The FLC has two references, the first represents the desired reference trajectory, which is $Y_i$ (estimated value) and the second represents the trajectory created by the optimized fuzzy controller, in this case, is the observed value, which is $\overline{Y}_i$ (signal value) and $n$ represents the total number of observed samples. The metric to use in this paper is represented by Eq. 7.

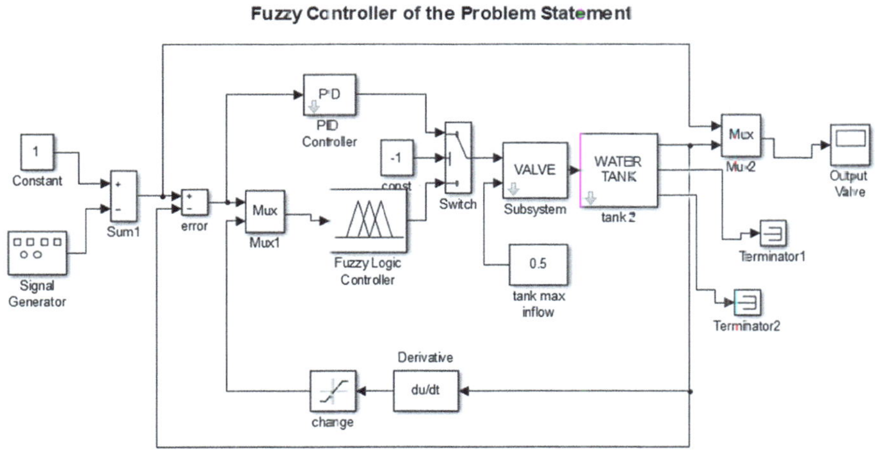


**Fig. 7** Model of benchmark problem water tank

**Table 3** Parameters for the CSO algorithm

| Parameters | Values |
|---|---|
| N (Population) | 50 |
| RN | 0.15 * N |
| HN | 0.7 * N |
| MN | 0.15 * N |
| G | 10 |
| FL | [0.5, 0.9] |
| Generations | 100 |

$$MSE = \frac{1}{n}\sum_{i=1}^{n}(\bar{Y}_i - Y_i)^2 \tag{7}$$

A total of 30 experiments in the model is executed, also the model is simulated 1000 unit of time for each experiment. Table 4 shows the best, worst average and standard deviation of the fitness function in the CSO algorithm the metric MSE explicated in Eq. 7.

Table 4 shows the best MSE results with a value of ***1.56E−04*** with the experiment number 8, and the worst MSE with value of **1.17E−01** with the experiment number 3, respect to BEST MSE in each experiment. The worst value to found was of 1.21E+00. The best result is the experiment number 8 the convergence is shown in Fig. 8.

To observe the behavior of CSO algorithm Fig. 9 shows the Best MSE (a), and Average MSE (b) results for the 30 experiments executed in the model.

The Best experiments was the number 8 (see Table 4). The distribution of the Membership Functions for the best experiment is shown in Fig. 10.

**Table 4** Results of simulation for the case study

| No. | Best | Worst | Average | Standard deviation ($\sigma$) |
|---|---|---|---|---|
| 1 | 6.35E−04 | 2.25E+00 | 1.41E+00 | 5.27E−02 |
| 2 | 7.78E−02 | 2.65E−01 | 1.43E−01 | 4.51E−02 |
| 3 | **1.17E−01** | 1.31E+00 | 8.01E−01 | 2.70E−02 |
| 4 | 7.27E−02 | 1.31E+00 | 7.52E−01 | 2.83E−02 |
| 5 | 8.17E−02 | 1.31E+00 | 7.89E−01 | 2.84E−02 |
| 6 | 5.70E−02 | 1.31E+00 | 8.12E−01 | 2.82E−02 |
| 7 | 8.00E−02 | 1.31E+00 | 7.83E−01 | 2.98E−02 |
| 8 | ***1.56E−04*** | 2.25E+00 | 1.35E+00 | 5.81E−02 |
| 9 | 7.86E−02 | 1.31E+00 | 7.04E−01 | 2.49E−02 |
| 10 | 8.46E−02 | 1.31E+00 | 6.98E−01 | 3.18E−02 |
| 11 | 9.16E−02 | 1.31E+00 | 8.31E−01 | 2.61E−02 |
| 12 | 7.96E−02 | 1.31E+00 | 7.72E−01 | 2.59E−02 |
| 13 | 8.46E−02 | 1.31E+00 | 8.25E−01 | 2.43E−02 |
| 14 | 8.64E−02 | 1.31E+00 | 7.61E−01 | 2.67E−02 |
| 15 | 7.51E−02 | 1.31E+00 | 7.20E−01 | 2.62E−02 |
| 16 | 7.78E−02 | 1.31E+00 | 7.41E−01 | 2.92E−02 |
| 17 | 7.68E−02 | 1.31E+00 | 7.93E−01 | 2.53E−02 |
| 18 | 7.77E−02 | 1.31E+00 | 8.57E−01 | 2.61E−02 |
| 19 | 7.19E−02 | 1.31E+00 | 7.89E−01 | 2.65E−02 |
| 20 | 7.50E−02 | 1.31E+00 | 1.31E+00 | 2.70E−02 |
| 21 | 1.10E−01 | 1.31E+00 | 8.23E−01 | 2.69E−02 |
| 22 | 7.74E−02 | 1.31E+00 | 7.64E−01 | 2.61E−02 |
| 23 | 7.37E−02 | 1.31E+00 | 7.86E−01 | 2.61E−02 |
| 24 | 8.45E−02 | 1.31E+00 | 7.94E−01 | 2.35E−02 |
| 25 | 8.17E−02 | 1.31E+00 | 7.84E−01 | 2.69E−02 |
| 26 | 7.40E−02 | 1.31E+00 | 8.13E−01 | 2.69E−02 |
| 27 | 7.34E−02 | 1.31E+00 | 7.90E−01 | 2.44E−02 |
| 28 | 7.59E−02 | 1.31E+00 | 7.73E−01 | 2.40E−02 |
| 29 | 7.48E−02 | 1.31E+00 | 8.27E−01 | 2.82E−02 |
| 30 | 7.33E−02 | 1.31E+00 | 8.07E−01 | 2.67E−02 |

One adaptation in the model was added, with the goal to observe the efficiently of the CSO algorithm. The experimentation was performed with the external perturbations, one pulse generator was added. The setting for noise was an amplitude of 1, Period (seconds) of 10, and Pulse Width (% of period) of 5. Figure 11 shows the block of the pulse generator added in the model.

Table 5 show the results for the 30 experiments executed with noise in the model.

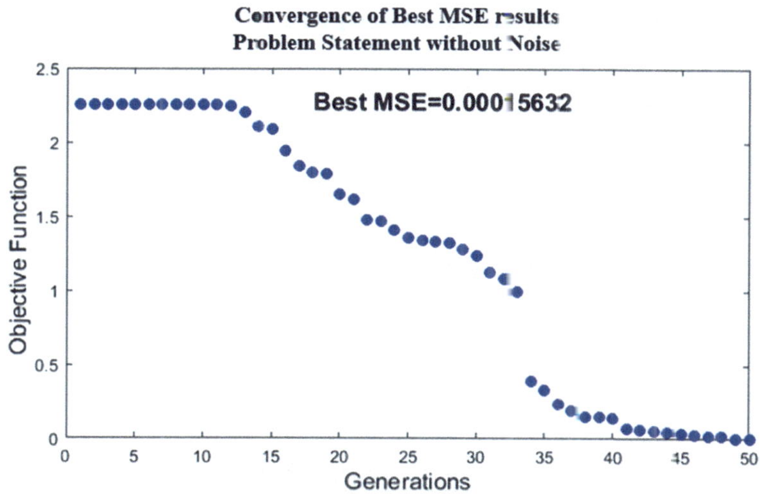


**Fig. 8** Best MSE result of problem statement without Noise in the model

Table 5 shows the best MSE results with a value of ***2.18E−08*** with the experiment number 4, and the worst MSE with value of **2.25E+00** with the experiment number 16, respect to BEST MSE in each experiment. The worst value to found was of 2.25E+00. The best result is the experiment number 4 is shown in the convergence in Fig. 12.

The Best experiments was the number 4 (See Table 5). The distribution of the Membership Functions for the best experiment is shown in Fig. 13.

## 7 Comparative Results

To analyze the behavior of the CSO algorithm, several comparison are shown. First Table 5 shows some performance index such as; average, standard Deviation (σ), Best and Worst MSE to found by CSO algorithm without Noise and with Noise in the model.

Table 5 shows that when noise is added in the model the Best MSE to found by CSO algorithm is of **2.18E−08.** The average has similarity in the results for CSO algorithm without noise the value is of 7.48E−02 with respect to CSO algorithm with noise the value is of 7.54E−02. This results allows to observe that when the disturbance is added in the model then the CSO algorithm to explore in local minima and the behavior is more stable in the model, with acceptable values in the average. Figure 14 illustrates a comparative with the Average MSE without Noise and with Noise in the model.

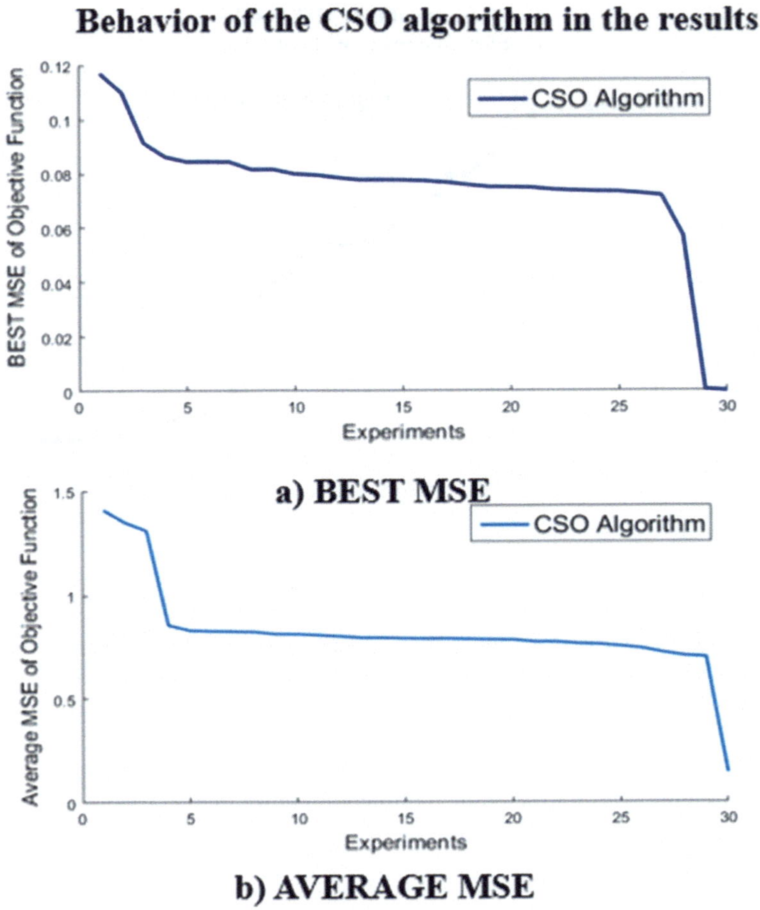


**Fig. 9** Behavior of the CSO algorithm in the results

Figure 14 illustrates the results of average in the fitness function of CSO algorithm. The behavior is similar when noise is added in the model represents by the red line, which indicates a major stabilization in the model.

Tables 7 and 8 show a comparative of results with other meta-heuristics such as; Harmony search algorithm (HS), Differential Evolutional Algorithm (DE), Bee Colony Optimization (BCO) [2], without and with noise in the model, respectively.

The results shown in Table 8 indicate that Chicken Search Optimization (CSO) is a efficiently algorithm in the optimization of fuzzy controller, for example, the average MSE without noise in the model is of **7.48E−02,** instead for the Bee Colony Optimization (BCO) the average MSE is of *5.21E−01*, for Harmony Search (HS) is of *5.66E−01* and for Differential Evolutional (DE) the average MSE is of *9.37E−02*.

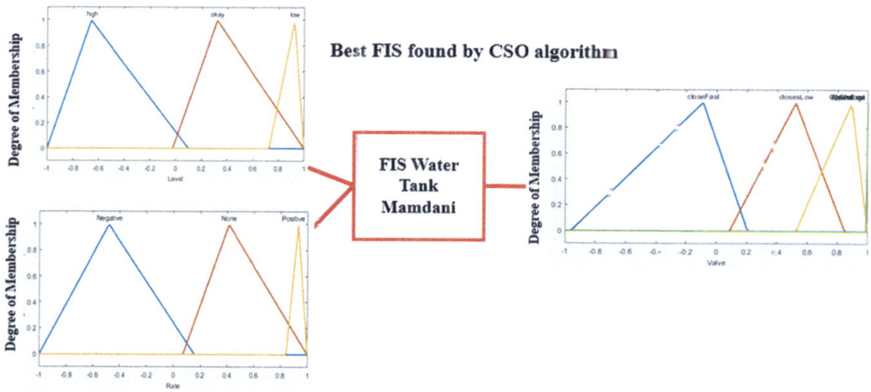


**Fig. 10** Best fuzzy logic system found by CSO algorithm

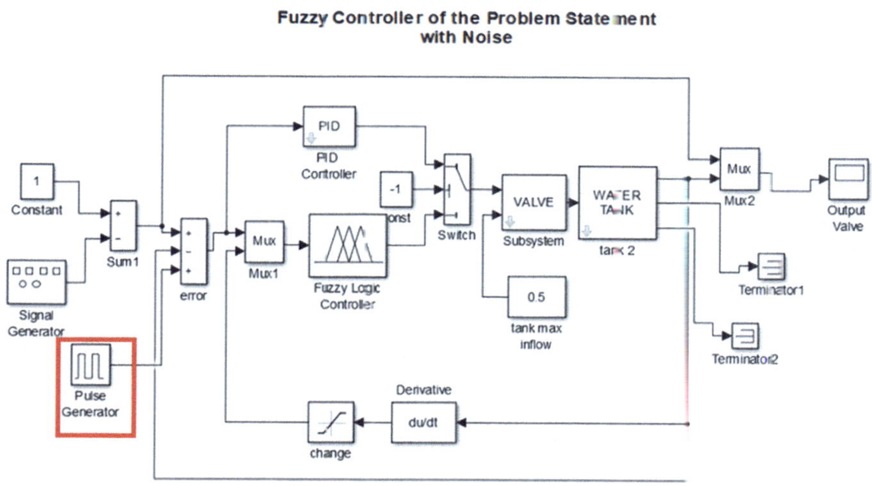


**Fig. 11** Model of benchmark problem water tank with noise

The nearest value is to found by DE. Thus, CSO algorithm to obtain the best Average MSE.

To analyze the results when noise is added in the model, the CSO obtained a value of average MSE of **7.54E−02**, BCO to obtain of value of 4.96E−01, HS with a value of 5.54E−01 and finally DE to obtain a value of 9.44E−02. In this case, the nearest value is to found by DE algorithm.

**Table 5** Results of simulation for the case study with noised add in the model

| No. | Best | Worst | Average | Standard deviation ($\sigma$) |
|---|---|---|---|---|
| 1 | 2.78E−04 | 2.25E+00 | 8.59E−01 | 4.61E−02 |
| 2 | 4.06E−04 | 3.79E−02 | 1.85E−02 | 8.17E−03 |
| 3 | 5.36E−05 | 2.25E+00 | 1.02E+00 | 7.76E−02 |
| 4 | ***2.18E−08*** | 2.25E+00 | 1.13E+00 | 6.65E−02 |
| 5 | 9.88E−05 | 2.25E+00 | 9.94E−01 | 3.83E−02 |
| 6 | 3.78E−04 | 2.25E+00 | 8.30E−01 | 4.80E−02 |
| 7 | 2.19E−04 | 2.25E+00 | 1.07E+00 | 2.98E−02 |
| 8 | 5.78E−04 | 2.25E+00 | 9.56E−01 | 4.63E−02 |
| 9 | 8.55E−04 | 2.25E+00 | 1.01E+00 | 4.12E−02 |
| 10 | 3.21E−04 | 2.25E+00 | 9.63E−01 | 2.86E−02 |
| 11 | 4.08E−04 | 2.25E+00 | 9.34E−01 | 3.73E−02 |
| 12 | 5.27E−05 | 2.25E+00 | 1.03E+00 | 3.48E−02 |
| 13 | 1.26E−04 | 2.25E+00 | 1.08E+00 | 3.52E−02 |
| 14 | 8.91E−04 | 2.25E+00 | 9.50E−01 | 3.98E−02 |
| 15 | 7.37E−04 | 2.25E+00 | 8.94E−01 | 4.07E−02 |
| 16 | **2.25E+00** | 1.27E+00 | 9.12E−01 | 3.22E−02 |
| 17 | 8.00E−05 | 2.25E+00 | 9.93E−01 | 3.48E−02 |
| 18 | 8.98E−05 | 2.25E+00 | 1.04E+00 | 3.36E−02 |
| 19 | 9.24E−04 | 2.25E+00 | 9.99E−01 | 3.71E−02 |
| 20 | 2.17E−04 | 2.25E+00 | 9.32E−01 | 4.39E−02 |
| 21 | 1.85E−05 | 2.25E+00 | 1.08E+00 | 3.12E−02 |
| 22 | 3.16E−04 | 2.25E+00 | 1.00E+00 | 3.35E−02 |
| 23 | 4.19E−04 | 2.25E+00 | 9.81E−01 | 1.32E−01 |
| 24 | 1.26E−03 | 2.25E+00 | 9.20E−01 | 3.77E−02 |
| 25 | 3.95E−04 | 2.25E+00 | 1.07E+00 | 3.01E−02 |
| 26 | 4.14E−04 | 2.25E+00 | 8.46E−01 | 4.51E−02 |
| 27 | 1.12E−03 | 2.25E+00 | 8.36E−01 | 4.16E−02 |
| 28 | 1.68E−04 | 2.25E+00 | 1.24E+00 | 3.33E−02 |
| 29 | 5.33E−04 | 2.25E+00 | 9.68E−01 | 3.32E−02 |
| 30 | 4.35E−06 | 2.25E+00 | 9.05E−01 | 4.47E−02 |

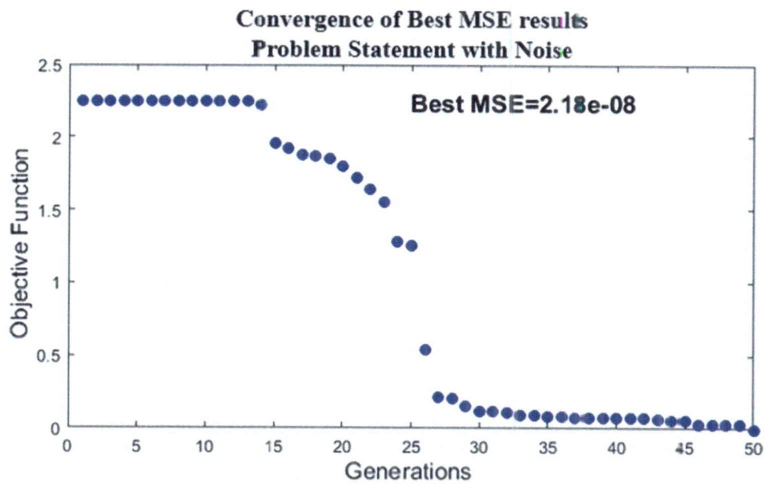


**Fig. 12** Best MSE result of problem statement with Noise in the model

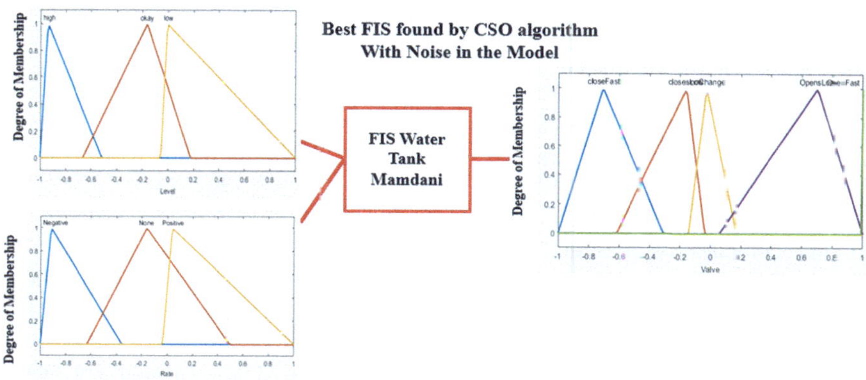


**Fig. 13** Best Fuzzy logic system found by CSO algorithm with noise in the model

## 8 Conclusions

It is important to mention the efficiency the CSO algorithm in the results obtained. For Fuzzy Controller CSO algorithm shows a stabilization for the problems statement implemented in this paper. A comparative with noise in the model allows to observer that CSO algorithm is an excellent meta-heuristics for apply in Fuzzy Controller (see Table 5 and 6). When CSO is compared with noise in the model the results indicates that this algorithm can to find good average errors (see Fig. 14).

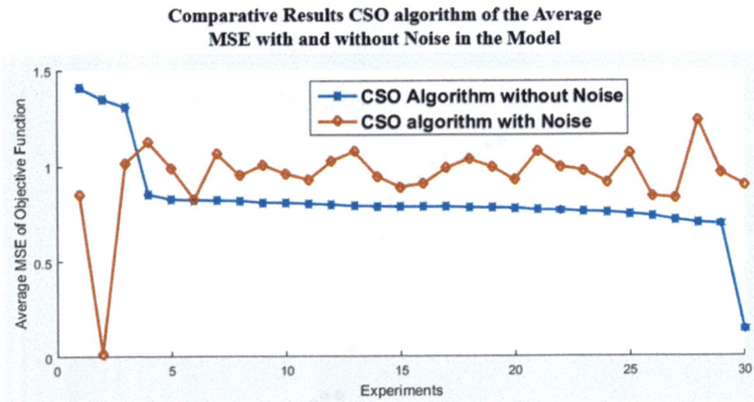


**Fig. 14** Comparative results of the best MSE without noise and with noise in the model

**Table 6** Comparative results without and with noise in the model

| Performance index | Output (Valve) | |
|---|---|---|
| | CSO algorithm without noise | CSO algorithm with noise |
| Average | 7.48E−02 | 7.54E−02 |
| Standard deviation (σ) | 2.29E−02 | 4.11E−01 |
| Best | 1.56E−04 | **2.18E−08** |
| Worst | 1.17E−01 | 2.25E+00 |

Other comparative presents in this paper is with others meta-heuristics for example, BCO, HS and DE algorithms. The results to obtain for CSO algorithms indicates that the excellent results to improve to other meta-heuristics algorithms (see Tables 7 and 8).

For the future works, an idea interesting is to apply the CSO algorithm in a problem no benchmark for example, in the automatization of trajectories in an autonomous mobile robot, also in the design of the optimal Interval Type-2 Fuzzy Logic System,

**Table 7** Comparative results with meta-heuristics without noise in the model

| Performance index | Meta-heuristics algorithms | | | |
|---|---|---|---|---|
| | CSO | BCO [2] | HS [2] | DE [2] |
| Average | **7.48E−02** | 5.21E−01 | 5.66E−01 | 9.37E−02 |
| Standard deviation (σ) | **2.29E−02** | 3.47E−01 | 3.26E−02 | 4.81E−02 |
| Best | **1.56E−04** | 4.50E−01 | 4.98E−01 | 1.65E−02 |
| Worst | 1.17E−01 | 5.91E−01 | **6.24E−01** | 2.36E−01 |

**Table 8** Comparative results with meta-heuristics with noise in the model

| Performance index | Meta-heuristics algorithms | | | |
|---|---|---|---|---|
| | CSO | BCO [2] | HS [2] | DE [2] |
| Average | **7.54E−02** | 4.96E−01 | 5.54E−01 | 9.44E−02 |
| Standard deviation (σ) | 4.11E−01 | 3.81E−01 | **2.92E−02** | 3.34E−02 |
| Best | **2.18E−08** | 4.24E−01 | 4.36E−01 | 4.82E−02 |
| Worst | 2.25E+00 | 5.69E−01 | **6.25E−01** | 1.82E−01 |

in this type of future works is considered that CSO algorithm to obtain excellent results. In addition, we can try the proposed method in type-2 fuzzy optimization [28–32] or in other kinds of applications [33–41].

## References


1. L. Amador-Angulo, O. Castillo, Comparative analysis of designing different types of membership functions using bee colony optimization in the stabilization of fuzzy controllers, in *Nature-Inspired Design of Hybrid Intelligent Systems* (Springer, Cham, 2017), pp. 551–571
2. O. Castillo, et al., Comparative study in fuzzy controller optimization using bee colony, differential evolution, and harmony search algorithms. Algorithms **12**(1), 9 (2019)
3. E.H. Mamdani, Application of fuzzy algorithms for control of simple dynamic plant, in *Proceedings of the Institution of Electrical Engineers*, vol. 121, no. 12 (IET, 1974), pp. 1585–1588
4. D. Qiao, et al., Improved evolutionary algorithm and its application in PID controller optimization. Inf. Sci. **63**, 1–199205 (2020)
5. Y. Li, Y. Wu, X. Qu, Chicken swarm-based method for ascent trajectory optimization of hypersonic vehicles. J. Aerosp. Eng. **30**(5), 04017043 (2017)
6. X. Liang, D. Kou, L. Wen, An improved chicken swarm optimization algorithm and its application in robot path planning. IEEE Access **8**, 49543–49550 (2020)
7. I. Miramontes, P. Melin, G. Prado-Arechiga, Comparative study of bio-inspired algorithms applied in the optimization of fuzzy systems, in *Hybrid Intelligent Systems in Control, Pattern Recognition and Medicine* (Springer, Cham, 2020), 219–231
8. B. Acherjee, D. Maity, A.S. Kuar, Ultrasonic machining process optimization by cuckoo search and chicken swarm optimization algorithms. Int. J. Appl. Metaheur. Comput. (IJAMC) **11**(2) 1–26 (2020)
9. B. Acherjee, D. Maity, A.S. Kuar, S. Mitra, D. Misra, Optimization of laser transmission welding parameters using chicken swarm optimization algorithm: chicken swarm algorithm optimization, in *Handbook of Research on Manufacturing Process Modeling and Optimization Strategies* (IGI Global, 2017), pp. 142–161
10. N. Akhter, et al., Chicken S-BP: an efficient chicken swarm based back-propagation algorithm, in *Recent Advances on Soft Computing and Data Mining: The Second International Conference on Soft Computing and Data Mining (SCDM-2016), Bandung, Indonesia, August 18–20, 2016 Proceedings*, vol. 549 (Springer, 2016)
11. N. Bharanidharan, R. Harikumar Rajaguru, Improved chicken swarm optimization to classify dementia MRI images using a novel controlled randomness optimization algorithm. Int. J. Imaging Syst. Technol. (2020)

12. S. Chen, R. Yan, Parameter estimation for chaotic systems based on improved boundary chicken swarm optimization, in *International Symposium on Optoelectronic Technology and Application 2016* (International Society for Optics and Photonics, 2016), pp. 101571K–101571K
13. S. Deb, et al., A new teaching–learning-based chicken swarm optimization algorithm. Soft Comput. **24**(7), 5313–5331 (2020)
14. J.K.M. Kumar, H. Abdul Rauf, R. Umamaheswari, switched capacitor-coupled inductor DC–DC converter for grid-connected PV system using LFCSO-based adaptive neuro-fuzzy inference system. J. Circ. Syst. Comput. 2050201 (2020)
15. W. Osamy, A.A. El-Sawy, A. Salim, CSOCA: chicken swarm optimization based clustering algorithm for wireless sensor networks. IEEE Access (2020)
16. C. Qu, S.A. Zhao, Y. Fu, W. He, Chicken swarm optimization based on elite opposition-based learning. Math. Prob. Eng. (2017)
17. X. Yu, et al., Assessment of water resource carrying capacity based on the chicken swarm optimization-projection pursuit model. Arab. J. Geosci. **13**(1), 39 (2020)
18. S.A. Taie, W. Ghonaim, CSO-based algorithm with support vector machine for brain tumor's disease diagnosis, in *2017 IEEE International Conference on Pervasive Computing and Communications Workshops (PerCom Workshops)* (IEEE 2017), pp. 183–187
19. F. Tian, R. Zhang, J. Lewandowski, K.M. Chao, L. Li, B. Dong, Deadlock-free migration for virtual machine consolidation using chicken swarm optimization algorithm. J. Intell. Fuzzy Syst. **32**(2), 1389–1400 (2017)
20. A.K. Tripathi, et al., Application of chicken swarm optimization in detection of cancer and virtual reality, in *Advanced Computational Intelligence Techniques for Virtual Reality in Healthcare* (Springer, Cham, 2020), pp. 165–192
21. L. Zadeh, Fuzzy sets. Inf. Control **8**(338) (1965)
22. L.A. Zadeh, Fuzzy sets as a basis for a theory of possibility. Fuzzy Sets Syst. **1**(1), 3–28 (1978)
23. L.A. Zadeh, The concept of a lingüistic variable and its application to approximate reasoning, Part II. Information Sciences **8**, 301–357 (1975)
24. E.H. Mamdani, N. Baaklini, Prescriptive method for deriving control policy in a fuzzy-logic controller. Electron. Lett. **11**(25), 625–626 (1975)
25. O. Castillo, et al., A comparative study of type-1 fuzzy logic systems, interval type-2 fuzzy logic systems and generalized type-2 fuzzy logic systems in control problems. Inf. Sci. **354**, 257–274 (2016)
26. C.C. Lee, Fuzzy logic in control systems: fuzzy logic controller. I. IEEE Trans. Syst. Man Cybern. **20**(2), 404–418 (1990)
27. X. Meng, Y. Liu, X. Gao, H. Zhang, A new bio-inspired algorithm: chicken swarm optimization, in *International Conference in Swarm Intelligence* (Springer, Cham, 2014), pp. 86–94
28. G.M. Mendez, O. Castillo, Interval type-2 TSK fuzzy logic systems using hybrid learning algorithm, in *The 14th IEEE International Conference on Fuzzy Systems. FUZZ'05* (2005), pp. 230–235
29. P. Melin, C.I. González, J.R. Castro, O. Mendoza, O. Castillo, Edge-detection method for image processing based on generalized type-2 fuzzy logic. IEEE Trans. Fuzzy Syst. **22**(6), 1515–1525 (2014)
30. C.I. González, P. Melin, J.R. Castro, O. Castillo, O. Mendoza, Optimization of interval type-2 fuzzy systems for image edge detection. Appl. Soft Comput. **47**, 631–643 (2016)
31. C.I. González, P. Melin, J.R. Castro, O. Mendoza, O. Castillo, An improved Sobel edge detection method based on generalized type-2 fuzzy logic. Soft. Comput. **20**(2), 773–784 (2016)
32. E. Ontiveros, P. Melin, O. Castillo, High order α-planes integration: a new approach to computational cost reduction of general type-2 fuzzy systems. Eng. Appl. of AI **74**, 186–197 (2018)
33. P. Melin, O. Castillo, Intelligent control of complex electrochemical systems with a neuro-fuzzy-genetic approach. IEEE Trans. Ind. Electron. **48**(5), 951–955
34. L. Aguilar, P. Melin, O. Castillo, Intelligent control of a stepping motor drive using a hybrid neuro-fuzzy ANFIS approach. Appl. Soft Comput. **3**(3), 209–219 (2003)

35. P. Melin, O. Castillo, Adaptive intelligent control of aircraft systems with a hybrid approach combining neural networks, fuzzy logic and fractal theory. Appl. Soft Comput. **3**(4), 353–362 (2003)
36. P. Melin, J. Amezcua, F. Valdez, O. Castillo, A new neural network model based on the LVQ algorithm for multi-class classification of arrhythmias. Inf. Sci. **279**, 483–497 (2014)
37. P. Melin, O. Castillo, *Modelling, simulation and control of non-linear dynamical systems: an intelligent approach using soft computing and fractal theory* (CRC Press, USA and Canada, 2002)
38. P. Melin, D. Sánchez, O. Castillo, Genetic optimization of modular neural networks with fuzzy response integration for human recognition. Inf. Sci. **197**, 1–19 (2012)
39. M.A. Sanchez, O. Castillo, J.R. Castro, P. Melin, Fuzzy granular gravitational clustering algorithm for multivariate data. Inf. Sci. **279**, 498–511 (2014)
40. D. Sanchez, P. Melin, Optimization of modular granular neural networks using hierarchical genetic algorithms for human recognition using the ear biometric measure. Eng. Appl. Artif. Intell. **27**, 41–56 (2014)
41. O. Castillo, *Type-2 Fuzzy Logic in Intelligent Control Applications* (Springer, 2012)

# Optimization of Fuzzy Trajectory Tracking in Autonomous Mobile Robots Based on Bio-inspired Algorithms

**Leticia Amador-Angulo and Oscar Castillo**

**Abstract** In this paper a comparison between the Bee Colony Optimization (BCO) and the Chicken Search Optimization (CSO) algorithms in optimal fuzzy control design is presented. The comparison is based on solving the problem to find the optimal distribution on the Membership Functions (MFs) of fuzzy controllers for mobile robots. Optimization in the structure and parameters for designing for a fuzzy tracking controller is presented. A new CSO algorithm for the optimization in Fuzzy Logic Controllers (FLC) is presented. The main aim is to use the two bio-inspired algorithms individually to improve of trajectories of an Autonomous Mobile Robot. CSO algorithm shows better results when compared to BCO in the simulation results for this control problem.

**Keywords** Bio-inspired algorithms · Type-1 fuzzy logic system · Chicken search optimization · Bee colony optimization · Autonomous mobile robot

## 1 Introduction

The problem for stabilization in control systems is of recent interest for many authors. However, most of the methods require to use the mathematical models in the control design methods, which, is an important problem in the practice for complex plants; due to the processing time, computational cost and other factors. On the other hand, fuzzy control has demonstrated to solve many nonlinear control problems successfully, for mentioning some works; which are presented in [1–4].

L. Amador-Angulo · O. Castillo (✉)
Division of Graduate Studies, Tijuana Institute of Technology, Tijuana, Mexico
e-mail: ocastillo@tectijuana.mx

L. Amador-Angulo
e-mail: gloria.amador@tectijuana.edu.mx



P. Melin et al. (eds.), *Recent Advances of Hybrid Intelligent Systems Based on Soft Computing*, Studies in Computational Intelligence 915,
https://doi.org/10.1007/978-3-030-58728-4_15

Actually, when optimization traditional algorithms are implemented the computing time increases exponentially. To solve these problems, the algorithms to realize a simulation of the behavior that the individuals o insects in nature, and these behaviors are represented in the swarm optimization algorithms, including the artificial bee colony (ABC) [5], ant colony optimization (ACO) [6], a new improved krill herd algorithm [7], bee colony optimization algorithm (BCO) [8], artificial bee colony (ABC) algorithm (ABC) [9], particle swarm optimization (PSO) [10], cuckoo search (CS) algorithm [11], firefly algorithm (FA) [12], fish swarm algorithm (FSA) [13], and a moth search algorithm [14], all these methods have achieved favorable results in diverse applications. Some important characteristics that the intelligent swarm optimization presents are; high precision, excellent stability, can obtain the exact solution, fast convergence rate and an approximate solution for large-scale optimum problems within limited time.

All bio-inspired algorithms have shown excellent results in the optimization and stabilization of fuzzy controllers. The CSO algorithm is a good technique to solve complex problems [15]. In addition, there are several works in which the CSO algorithm has been used to solve different kinds of problems. The BCO is a good technique for solving problems [1, 16, 17]. In fuzzy controller design, it has demonstrated that an excellent design of the Type-1 FLC can be achieved and is evaluated as more stable [2, 12, 16].

Bioinspired optimization algorithms exhibit the main characteristics of a good technique at numerical optimization and finding excellent results from a defined search space [18]. In this paper, the main motivation consists in the implementation of a new bio-inspired algorithm namely the Chicken Search Optimization for tuning the membership functions in fuzzy controllers. Also, a comparison with the Bee Colony Optimization algorithm is performed to be able to observe the robustness of each algorithm especially in achieving the stabilization of the studied case in this paper.

The first part in this paper presents a review of algorithms applied to solve complex optimization problems is presented. In Sect. 3 the basic concepts of Type-1 Fuzzy Logic system is presented. In Sect. 4 the case study is presented. In Sect. 5 both Bio-Inspired Algorithms are explained. In Sect. 6 the experiments results are presented for each bio-inspired algorithm, individually. In Sect. 7 a comparative analysis for results is presented. In Sect. 8 the statistical test is presented, and finally Sect. 9 describes some conclusions and future works.

## 2 Related Works

There have been recent work on algorithms inspired by nature for solving complex optimization problems, i.e.; in [19] a Chicken Swarm Optimization (CSO) is studied, in [5] an Artificial Bee Colony Optimization is implemented, in [20] a Bat Algorithm

(BA) is presented, in [6] an Ant System (AS) is studied, in [11] a Cuckoo Search algorithm (CSA) is presented to solve structural optimization problems, in [12] a Firefly Algorithms (FA) is presented to optimize PID controllers, in [21] a Flower Pollination Algorithm (FPA) is presented, in [13] an Artificial Fish Swarm (AFS) is presented, in [22] and a Spider Monkey Optimization (SMO) algorithm is presented.

The CSO is a new technique that has been implemented for various authors, like in [15, 19, 23–27]. Various problems have been solved with the implementation of the algorithms based on the behavior the species, animals, insects, etc., like in [12, 18].

Recently various authors have shown interest in performing comparisons with several bio-inspired algorithms in solving complex problems, like in [2, 4, 16, 18, 20, 21, 28–37].

## 3 Basic Concepts of Fuzzy Systems

### 3.1 Fuzzy Logic System

Type-1 fuzzy sets is the main concept in the design of a Fuzzy Logic System (FLS). The essential elements are shown in Fig. 1, which is known as Type-1 FLS [38].

The general representation of fuzzy sets in the universe $u$ is shown by Eq. 1 [38]:

$$A = \{(x, u_A(x))\}|x \in X\} \quad (1)$$

where $\mu_A : X \to [0, 1]$, in Eq. 1 $\mu_A(x)$ represents the membership degree of the element x $\in$ X to the set A. For this paper the following notation for $A(x)$ is given by $\mu_A(x)$ for all $x \in$ X. The main blocks that composed a Type-1 FLS are; a Fuzzifier which to require crisp input and maps them into Fuzzy Sets, the Inference engine that

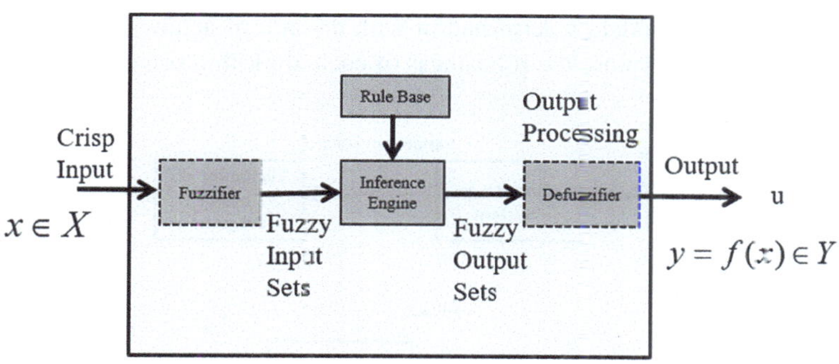


**Fig. 1** Structure of the Type-1 FLS

needs a system of Rules Base, and a Defuzzifier, which is the output of the process to obtaining the final crisp value.

## 3.2 *Fuzzy Logic Controllers*

Input–output uncertainties are the main objective of a Fuzzy Logic Controller (FLC) which are represented through type-1 fuzzy sets. The theory proposed by Ebrahim Mamdani in 1975 is essential for control system. The Mamdani's method is to create with the goal to control a steam engine and boiler combination based on linguistic control rules through of the experienced human operators [39]. Lotfi Zadeh's on the paper called fuzzy algorithms for complex systems and decision process of was the inspiration for Mamdani, therefore, various works have used this technique, such as in [1, 2, 11, 12, 16, 22, 40, 41]. Based on the inference process, each bio-inspired algorithm in a separate way is evaluated in the model with the main goal of finding a fuzzy controller design with better results in the minimization of the error. In Fig. 2 the generic representation of the FLC is illustrated [1, 16].

Figure 2 shows the representation of a Fuzzy Logic Controller (FLC), the model is used for each bio-inspired algorithm with the goal of obtaining a comparison in their performance. A robot composed of one passive wheel and two wheel motors is studied. The goal consists in controlling the two wheel motors with the hybridization of two techniques: the FLS and the bio-inspired algorithms. The evaluation is realized in a FLC, which is detailed as follows; the input (r) is represented by the values of linear and angular error (see Fig. 3), and the Error (e) is the MSE that is found in the simulation. The fuzzy logic system (FLS) is designed as follows, the variable of control (u) indicates the behavior in the outputs in this case the torque 1 (wheel right) and torque 2 (wheel left) for both outputs the MSE is added, the FLS found for each bio-inspired algorithm is evaluated in the model, a new observable value of the variable is found, and it becomes a cycle until we find the minimum error. Noise level is added in the model as perturbation with the aim of achieving stabilization, thus in each iteration explore the robustness of each algorithm used.

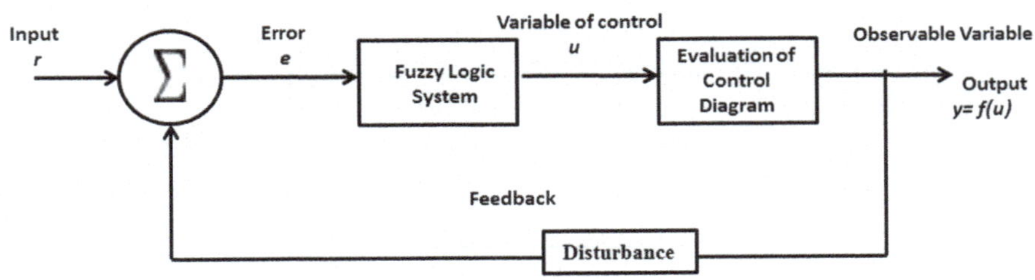


**Fig. 2** General representation of a FLC

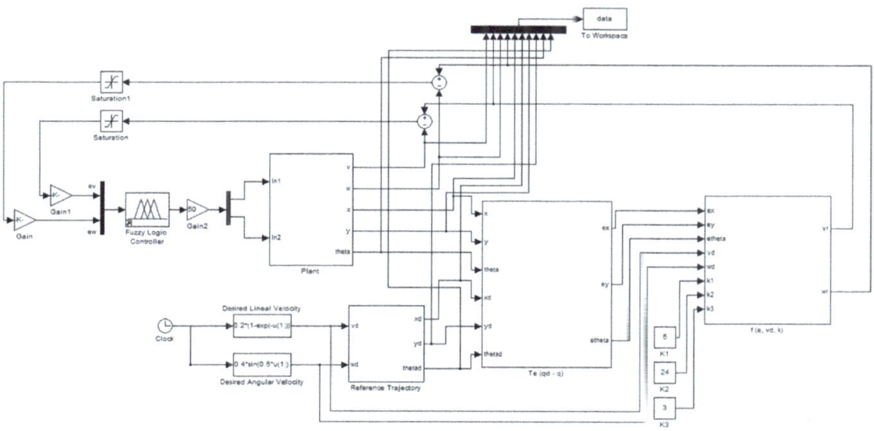


**Fig. 3** Complex plant for the AMR

## 4 Case Study

To test the two bio-inspired algorithms in this paper a complex problem was selected; an autonomous mobile robot (AMR) which task is to minimize the trajectory error through the optimal design of the FLS. Figure 3 illustrates the complex plant for the case study.

The robot body illustrated in Fig. 3 has two wheel motors and one passive wheel for stabilization. The motion of the passive wheel can be ignored in the dynamics represented by the following set of equations from [32].

$$M(q)\dot{v} + C(q, \dot{q})v + Dv = \tau + P(t) \tag{2}$$

$$\dot{q} = \begin{bmatrix} \cos\theta & 0 \\ \sin\theta & 0 \\ 0 & 1 \end{bmatrix} \begin{bmatrix} v \\ w \end{bmatrix} \tag{3}$$

where $q = (x, y, \theta)^T$ is the vector of the configuration coordinates; $v = (v, w)^T$ is the vector of velocities; $\tau = (\tau_1, \tau_2)$ is the vector of torques applied to the wheels of the robot, where $\tau_1$ and $\tau_2$ denote the torques of the right and left wheel, respectively; $P \in R^2$ is the uniformly bounded disturbance vector; $M(q) \in R^{2\times 2}$ is the positive-definite inertia matrix; $C(q, \dot{q})v$ is the vector of centripetal and Coriolis forces; and $D \in R^{2\times 2}$ is a diagonal positive-definite damping matrix. Equation (3) represents the kinematics of the system, where $(x, y)$ is the position in the $X$–$Y$ (world) reference frame; $\theta$ is the angle between the heading direction and the *x-axis*; and $v$ and $w$ are the linear and angular velocities, respectively.

In Fig. 3 a fuzzy logic system (FLS) is implemented which main task to minimize the error in the trajectory to control the two motors of the robot. Figure 4 shows the

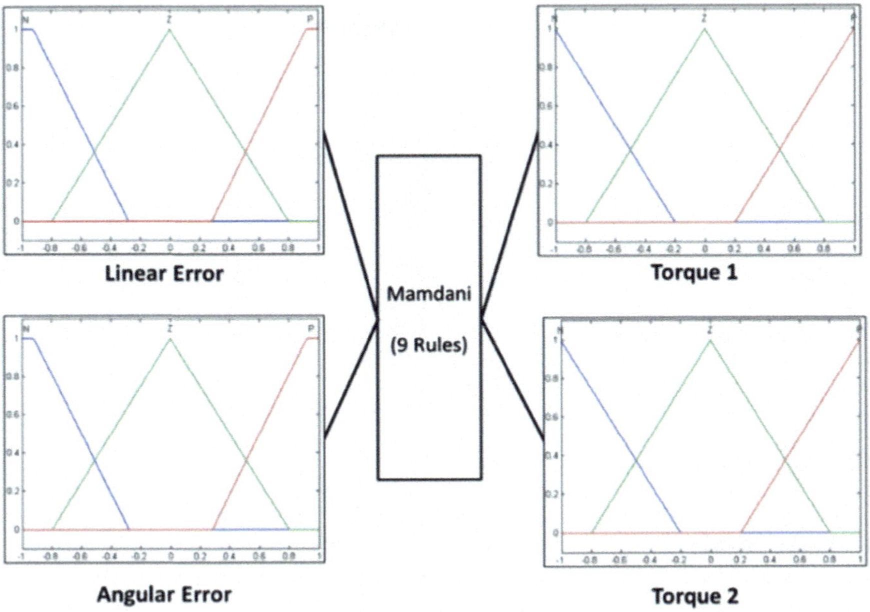


**Fig. 4** Fuzzy system for control

FS; this is the FLS that CSO and BCO algorithms is optimizing, in this case study only the points (parameters) of the membership functions (MF) optimized.

Table 1 shows the linguistic values for each input and output to evaluate the fuzzy controller in the model, these values are represented as follows; Negative is represented by "N", Zero by "Z" and Positive by "P". The linguistics values are

**Table 1** Fuzzy rules for control

| No. | Inputs | | Outputs | |
|---|---|---|---|---|
| | Linear error | Angular error | Torque 1 | Torque 2 |
| 1 | N | N | N | N |
| 2 | N | Z | N | Z |
| 3 | N | P | N | P |
| 4 | Z | N | Z | N |
| 5 | Z | Z | Z | Z |
| 6 | Z | P | Z | P |
| 7 | P | N | P | N |
| 8 | P | Z | P | Z |
| 9 | P | P | P | P |

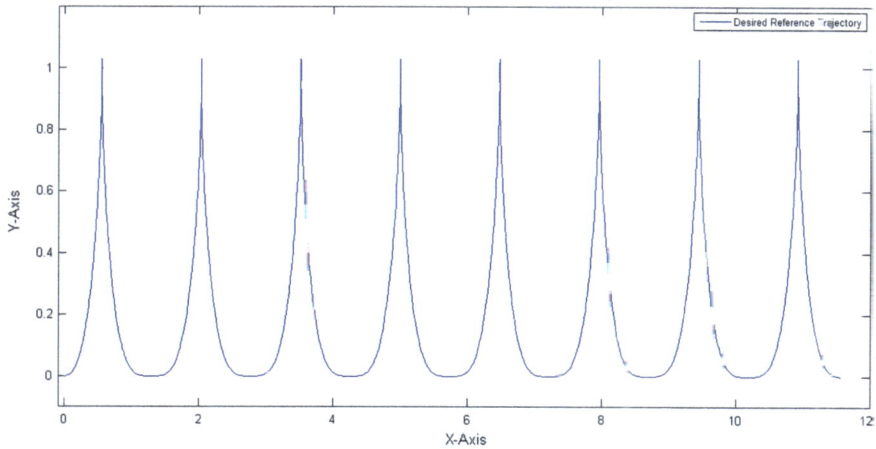


**Fig. 5** Reference trajectory for the AMR

obtained with the observation in the behavior the some test and simulations of the fuzzy controller for the study case.

The fuzzy controller from Fig. 4 shows the linear error and angular error as inputs, and Torque 1 and Torque 2 for each motorized wheel of the robot as outputs. The inputs are granulated into one triangular MF in the middle and two trapezoidal MFs in the ends, the outputs are granulated into three triangular MFs with a range of [−1 1] in each MF. The search space in each algorithm is in the range of [−1, 1]. Figure 4 shows that the FS for control has 2 inputs and 2 outputs; the inputs have trapezoidal MFs in the extremes and triangular MF in the middle and the triangular MFs in outputs, each triangular MF requires 3 values and each trapezoidal MF requires 4 values. Thus, a bee and cock have a size of 40 values. In the order to create a complex problem for the algorithms a trajectory in shape of "U" is created. Figure 5 shows the desired reference trajectory, where it starts from the point (0, 0) [4].

# 5 Bio-inspired Algorithm

## *5.1 Chicken Search Algorithm*

Chicken swarm optimization (CSO) is a new intelligent bio-inspired algorithm based on the behaviors of cocks, hens, and chicks to find food. In CSO algorithm, chicken swarm in searching space is mapped a specific particle individual. Three type of chicken are used and the whole chicken swarm has cock swarm, chick swarm and hen swarm. The mechanism is that the cock swarms represents a highest fitness and chick swarm presents lowest fitness. The role (cock, hen and chick) for each chicken are sorted according to fitness value. The methodology of the search for each subswarm is different. An important characteristic in this algorithm is that a chicken in the swarm represents a solution in the problem [26]. The goal of the CSO algorithm is this paper is to solve a minimization problem.

In the CSO algorithm, for all the particles, those with the best fitness can be selected as the cock particle swarm, and is represented by Eq. 4:

$$x_{i,j}^{t+1} = x_{i,j}^{t} + randn(0, \sigma^2) \cdot x_{i,j}^{t} \tag{4}$$

where $t$ represents the iterations, $x_{i,j}^{t+1}$ and $x_{i,j}^{t}$ indicate the position of $j$th dimension of a chicken in t + 1. The function randn(0, $\sigma^2$) is used as a random number of Gaussian distribution and the variance is found by Eq. 5:

$$\sigma^2 = \left\{ \begin{array}{l} 1, ------------------> fit_i < fit_k \\ \exp\left(\frac{(fit_k - fit_i)}{(|fit_i| + \xi)}\right), -------> fit_i \geq fit_k, \end{array} \right\} \tag{5}$$

where $i$, $k \in$ [1, rsize] and $i \neq k$. In this case study, rsize indicates the number of cock swarms, the size of the swarm is the 40, also, a bee and cock represents the number of values for the design of the fuzzy logic system for the studied case 40 values are required (see Fig. 4). $fit_i$ and $fit_k$ are the fitness values of cock particle i and k, respectively; $\xi$ represents a number which is used to avoid zero-division-error, is the smallest constant in the computer. Otherwise, for CSO algorithm, hen swarm was selected with good fitness. Its random search is performed via cocks of hen population and that of others, which can be represented by Eq. 6:

$$x_{i,j}^{t+1} = x_{i,j}^{t} + S1 \cdot rand \cdot \left(x_{r1.j}^{t} - x_{i,j}^{t}\right) + S2 \cdot rand \cdot \left(x_{r2.j}^{t} - x_{i,j}^{t}\right) \tag{6}$$

where $x_{i,j}^{t}$ and $x_{r2.j}^{t}$ indicates the position of the cock individual 1 in the population of hen and cock individual 2 in the other population, respectively. An uniform random number with the range of [0, 1] is represented by function *rand*. Equation 7 denote the weight calculated which to calculate the values of 1 and 2.

$$S1 = \exp\left(\frac{(fit_i - fit_{r1})}{(|fit_{r2}| + \xi)}\right),\ S2 = \exp(fit_{r2} - fit_i), \tag{7}$$

The fitness value of the cock individual is represented by r1 ($fit_{r1}$) in the population of hen , and the cock individual in the other population is represented by r2 ($fit_{r2}$). All individuals, except for the cock swarm and hen swarm, are defined as chick swarm. Its search mode follows that of the hen swarm, which is given by Eq. 8:

$$x_{i,j}^{t+1} = x_{i,j}^{t} + FL \cdot \left(x_{m,j}^{t} - x_{i,j}^{t}\right),\ FL \in [0, 2] \tag{8}$$

**Table 2** Pseudocode CSO algorithm

| |
|---|
| 1. Initialize a population of *N* chickens and define the relation parameters |
| 2. Evaluate the *N* chickens fitness values, $t=0$ |
| 3. While ($t$ < Max_Generation) |
| 4. If ($t$% G = = 0) |
| 5. Rank the chickens' fitness values and establish a hierarchal order in the swarm |
| 6. Divide the swarm into different groups, and determine the relationship between the chicks and mother hens in a group |
| 7. End if |
| 8. For i = 1; *N* |
| 9. If (i == rooster) |
| 10. Update its solution/location using Eq. (4) |
| 11. End if |
| 12. If (i == hen) |
| 13. Update its solution/location using Eq. (6) |
| 14. End if |
| 15. If (i == chick) |
| 16. Update its solution/location using Eq. (8) |
| 17. End if |
| 18. Evaluate the new solution |
| 19. If the new solution is better its previous one; update it |
| 20. End For |
| 21. End While |

In CSO algorithm, the chick would follow its mother to forage for food, therefore $x^{t}_{m,j}$, is the position of *j*th dimension of particle *m* in $t + 1$, t indicate the iteration, *then, FL represents the parameter where this characteristic is evaluated. Hence, m* represents the position of the ith chick's mother (m ∈ [1, N]), where N indicates the size of swarm (population). The Pseudocode of the CSO algorithm is illustrated in Table 2.

## 5.2 *Bee Colony Optimization*

Dušan B. Teodorović was the first in developer the meta-heuristics called Bee Colony, bioinspired in the method that explores collective intelligence applied by the honey bees to improve the nectar collecting process [1, 8, 16, 42, 43]. Some interesting and efficient characteristics has this algorithm, for example: ***backward pass*** and ***forward pass,*** and assign roles for each follower bee, scout bee and bee [8]. Equations 9–12 allow to define the dynamics of BCO algorithm; Eq. 7 indicates the probability that a bee has located the best solution, represented in the following equations:

$$Pij, n = \frac{[\rho ij, n]^{\alpha} . \left[\frac{1}{dij}\right]^{\beta}}{\sum_{j \in Ai,n} [\rho ij, n]^{\alpha} . \left[\frac{1}{dij}\right]^{\beta}} \tag{9}$$

$$D_i = K . \frac{Pf_i}{Pf_{colony}} \tag{10}$$

$$Pf_i = \frac{1}{L_I}, L_i = \text{Tour Length} \tag{11}$$

$$Pf_{colony} = \frac{1}{N_{Bee}} \sum_{i=1}^{N_{Bee}} Pf_i \tag{12}$$

Equation 9 describes the probability of bee *k* located on a node *i* and the next node is selected by j, therefore, $N_i^k$ indicates all nodes in a neighborhood, where the node i is to connect with the bee *k*, also, in this equation, $\rho_{\mathrm{ij}}$ is used to find the rating of the edge between node *i* and node *j*, the probability to visit the following node is represented by value of $\beta$, respect to, $d_{ij}$ is used by to found the heuristic distance of node *i* to node *j,* and in the special for this algorithm, this value indicates the total dance that a bee has in the moment, and finally, $\alpha$ represents the better solution to find for the BCO algorithm, this variable is in the range of [0, 1]. Equation 10 is used to find the duration that a bee has as waggle dance, determined by a linear function, where *K* denotes the waggle dance scaling factor. The probability scores of bee *i* is denoted by $Pf_i$ represented by Eq. 11, and $Pf_{colony}$ is found by Eq. 12 that represents the bee colony's average profitability, in this case, when a bee to complete its tour, this equation is executed, and each bee update its value in the colony. The pseudocode of the BCO algorithm is illustrated in Table 3.

## 6 Simulation Results

Both bio-inspired algorithms were applied to the optimization of the FLS (from Fig. 2) for control of the case study, with the following setting in each algorithm which is described in Tables 4 and 5, respectively.

In Fuzzy Logic Control (FLC) there exists a metric to evaluate the performance of the bio-inspired meta-heuristics, in this case CSO and BCO algorithms. This metric is known as Mean Square Error (MSE). In this paper, this metric is used as the function fitness that the particle (bee and chicken) that should to can find in the execution for each algorithm. The FLC has two references, the first represents the desired reference trajectory, which is $Y_i$ (estimated value) and the second represents the trajectory created by the optimized fuzzy controller, in this case, is the observed

**Table 3** Pseudocode BCO algorithm

| |
|---|
| 1. Initialization; an empty solution is assigned to every bee |
| 2. For every bee do the forward pass |
| a) Set *k*=; // counter for constructive moves in the forward pass |
| b) Evaluate all possible constructive moves |
| c) According to the evaluation, choose on move using the roulette wheel |
| d) *k*=*k*+1; if k ≤ NC goto step b |
| 3. All bees are back to the hive; // backward pass stars. |
| 4. Evaluate (partial) objective function value for each bee |
| 5. Every bee decides randomly whether to continue its own exploration and become a recruiter, or to become a follower (bees with higher objective function value have greater chance to continue their own exploration) |
| 6. For every bee follower, choose a new solution from recruiters by roulette wheel |
| 7. If solutions have not completed goto step 2 |
| 8. Evaluate all solutions and find the best one |
| 9. Output the best result |

**Table 4** Parameters for the BCO algorithm

| Parameter | BCO |
|---|---|
| Population | 30 |
| Iterations | 15 |
| Alpha | 0.5 |
| Beta | 2.5 |
| Food number | 15 |
| Limit | 100 |

**Table 5** Parameters for the CSO algorithm

| Parameter | CSO |
|---|---|
| N (Population) | 30 |
| RN | 0.15 * N |
| HN | 0.7 * N |
| MN | 0.15 * N |
| G | 10 |
| FL | [0.4, 0.5] |
| Generations | 15 |

value, which is $\overline{Y}_i$ (signal value) and $n$ represents the total number of observed samples. The metric to use in this paper is represented by Eq. 13.

**Table 6** Comparison of results of the plant for the two bio-inspired algorithms

| MSE | BCO | CSO |
|---|---|---|
| Average | 0.1410 | 0.0786 |
| Best | 0.0055 | 3.625E-05 |
| Worst | 1.2903 | 0.9164 |
| Standard deviation | 0.2720 | 0.1758 |

$$MSE = \frac{1}{n}\sum_{i=1}^{n}(\bar{Y}_i - Y_i)^2 \tag{13}$$

With the objective to realize a comparison in the performance for the two bio-inspired algorithms, a total of 30 times were applied for each algorithm. For each simulation for both algorithms the average, best, worst and standard deviation are calculated and these results are summarized in Table 6.

Table 6 shows that the CSO algorithm allows the stabilization of the trajectory in the autonomous mobile robot and it is better when compared to the BCO algorithm. The best results for the 30 experiments for CSO is of **3.625E−05** and for the BCO is of **0.0055.** Also, the standard deviation for the CSO is better with a value of **0.1758**. The best trajectories from each bio-inspired algorithm is illustrated in Fig. 6, with the objective to realize a visual comparison; is important to note that the results in the simulation for each algorithm are very similar to the desired trajectory.

Figure 6 shows that the BCO algorithm allows for the better stabilization in the trajectories when compared to the CSO algorithm. To observe the behavior in the fitness function (MSE) for each algorithm, Figs. 7 and 8 illustrate the best convergence for each bio-inspired algorithm. The convergence indicates all the MSE errors in the simulation, where the error that was found is the smallest, which represents the experiment with minimal error in the performed simulations.

Figure 7 illustrates the BCO algorithm (a) to find the minimal errors compared to CSO algorithm (b).

## 7 Comparative Analysis

For the comparison, Fig. 8 illustrates the Fitness Function (best MSE error) in the 30 experiments for CSO and BCO algorithms.

The worst results for CSO and BCO algorithms are illustrated in Fig. 9.

The simulation results show that both bio-inspired algorithm allow the stabilization in the case study.

The noise in problems of fuzzy controllers represents a mechanism to be able to better analyze the optimization by bio-inspired algorithms, in this work, an adaptation in the model is added. For each input (linear error (ew) and angular error (ev)), a

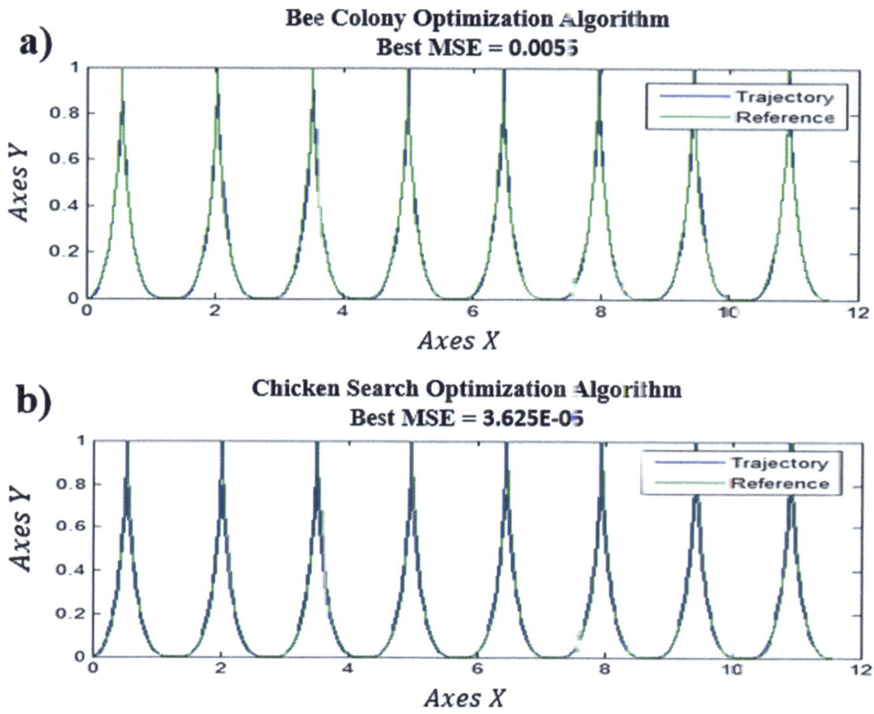


**Fig. 6** Best simulation of **a** BCO and **b** CSO algorithms

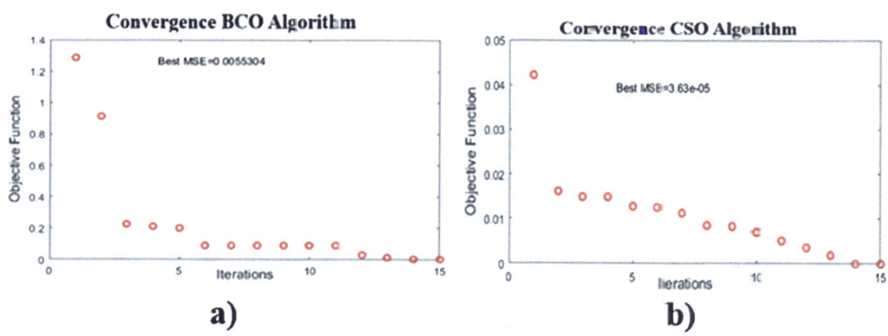


**Fig. 7** **a** Convergence for the best results by BCO algorithm, **b** convergence for the best results by CSO algorithm

random number generator is added in the model to simulate noise with a value of 5%. Figure 10 shows the modification in the model.

Figure 11 shows the new desired reference trajectory with perturbation in the model. It's important to mention that the noise allows that both bio-inspired algorithm shows a better performance.

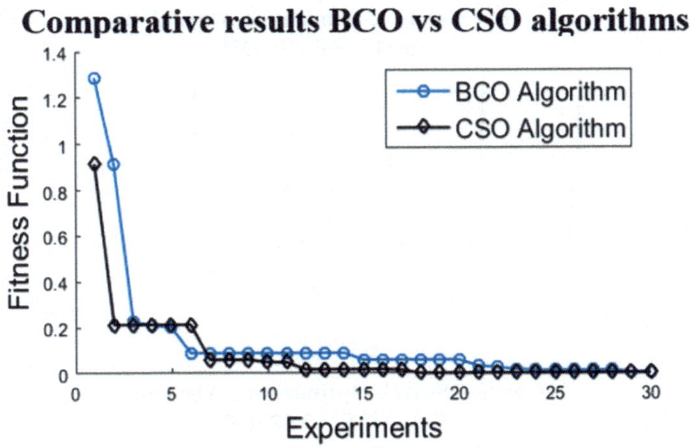


**Fig. 8** Comparison results based on the MSE for CSO and BCO algorithms

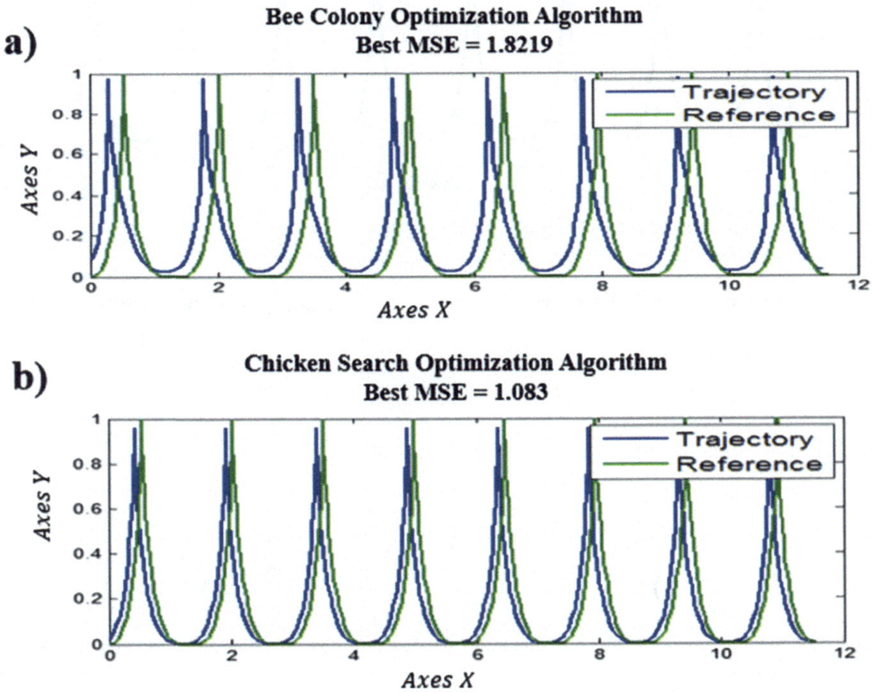


**Fig. 9** **a** Worst simulation of BCO and **b** worst simulation of CSO algorithm

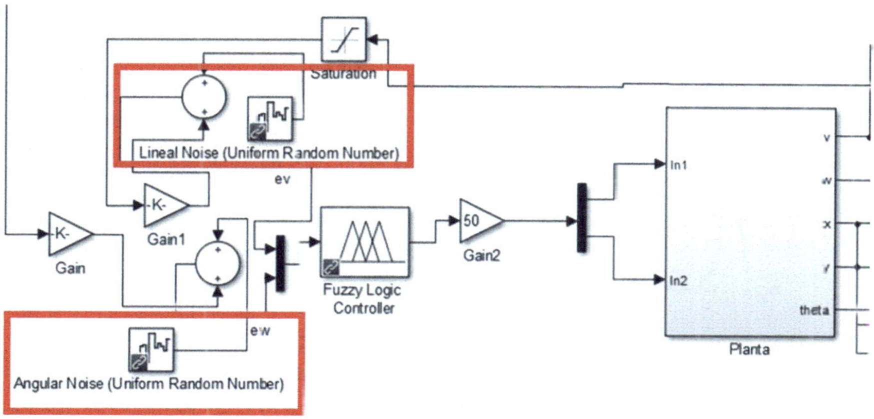


**Fig. 10** Complex plant for an autonomous mobile robot with noise

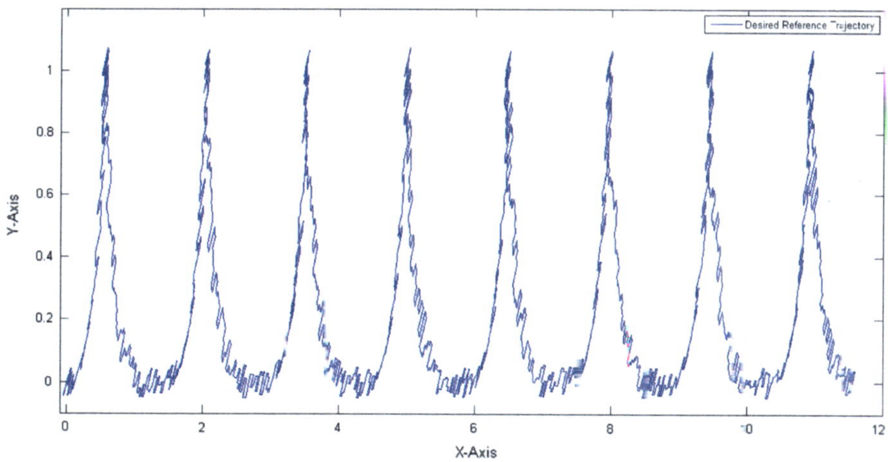


**Fig. 11** Reference trajectory with noise for the autonomous mobile robot

With the references trajectories in the model without noise (See Fig. 3) and with noise (see Fig. 10), both bio-inspired algorithms were compared, and evaluate the performance by CSO and BCO algorithms, the best fitness functions (MSE) is compared for each algorithm simulated in the plant and the comparison is presented in Fig. 12.

Figure 13 shows the stability with noise in the model for both bio-inspired algorithms.

Both bio-inspired algorithms show a stable trajectory, therefore, BCO and CSO algorithms are excellent techniques in the optimization of FLCs, in this case, the trajectory in an autonomous mobile robot. Another important analysis is to observe

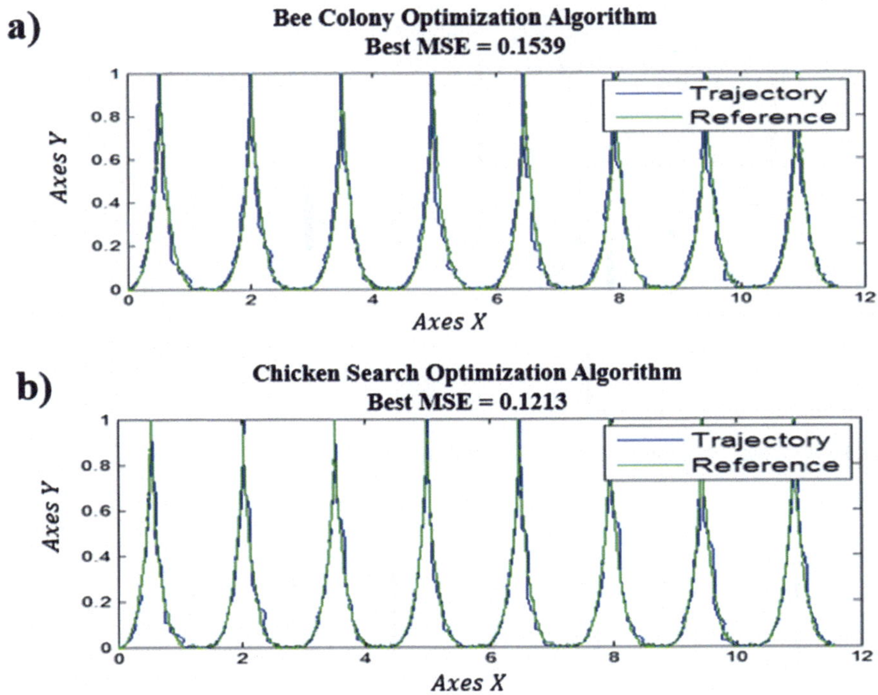


**Fig. 12** Simulation results with noise in the model. **a** BCO and **b** CSO algorithm

the behavior of the averages in each bio-inspired algorithm in the iterations. Figure 14 shows a comparative with the averages in each experiment.

Figure 14 shows that the CSO algorithm finds the better averages during the execution. The best and worst average errors for the CSO algorithm are **0.0106** and **0.0507**, respectively. And for the BCO algorithm is of **0.0128** for the best average and **1.2903** for the worst average. Therefore, the stabilization of the trajectory in the AMR is better implemented the CSO algorithm. The BCO algorithm allows to find minimal errors in short iterations. Figure 15 shows a comparative of the average MSE for the first experiment by BCO and CSO algorithms.

Figures 16 and 17 show the distribution for the Type-1 FLS that the BCO and CSO algorithms to find with the best MSE.

From the results in the comparison with BCO and CSO algorithms, when noise is added to the plant the stabilization in the trajectory is excellent and better results are found with CSO. The average in the errors in the Fitness Functions are lower when CSO is used in the tuning of the parameters in a fuzzy controller with a value of **0.0786** compared to BCO with a value of **0.1410**. With respect to the worst errors in CSO was of **0.9164** and **1.2903** for BCO, respectively. Both bio-inspired algorithms show that they are efficient techniques in the optimization of fuzzy controllers, in this case in controlling the trajectory in an autonomous mobile robot. Figure 8 shows

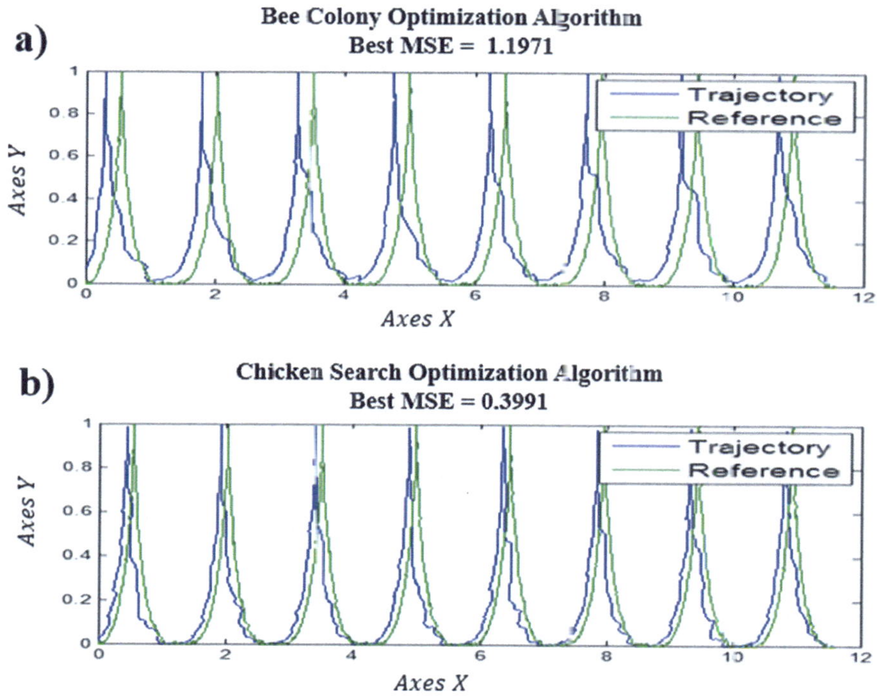


**Fig. 13** Simulations results with noise in the model. **a** BCO and **b** CSO algorithm

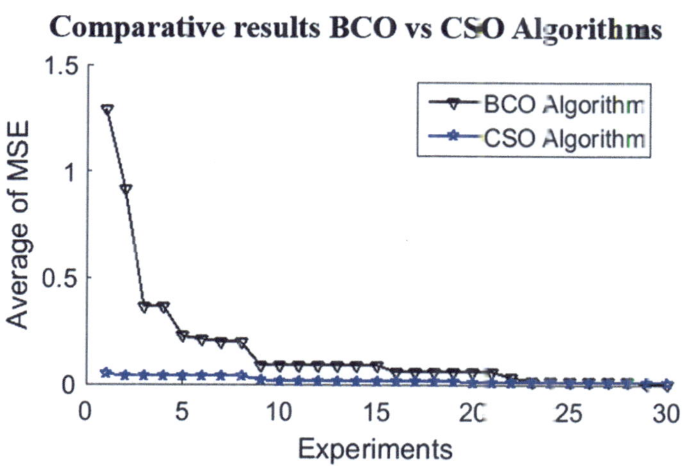


**Fig. 14** Average MSE for each bio-inspired algorithms

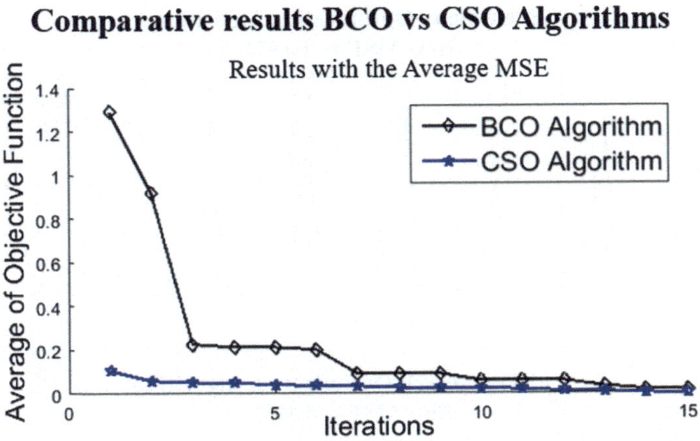


**Fig. 15** Average MSE for the first experiment

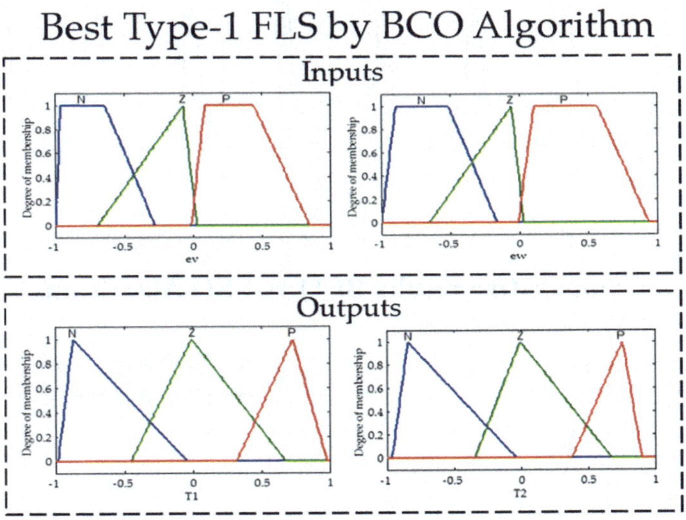


**Fig. 16** Best Type-1 FLS found by BCO algorithm

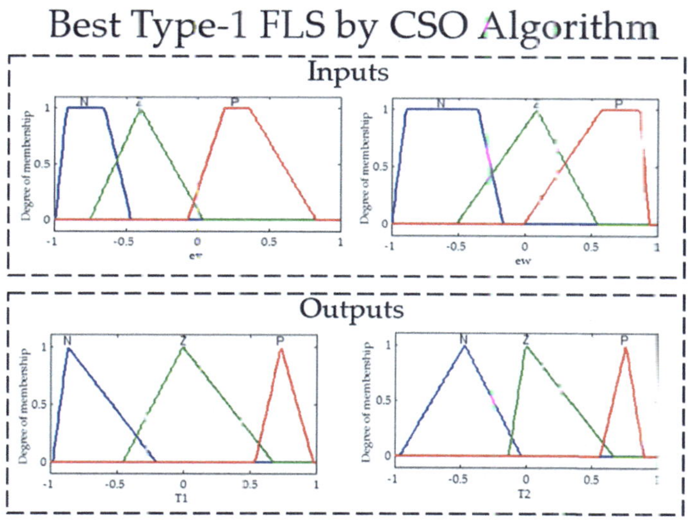


**Fig. 17** Best Type-1 FLS found by CSO algorithm

that the BCO algorithm begins with lower errors, also presents rapid convergence in the solution of the analyzed problem. Figure 12 illustrates that when noise is added in the robot model the CSO algorithm is more stable than the BCO algorithm. Based on the experiments, the BCO algorithm starts with a high error (see Fig. 14), thus, requires of a higher number the generations or iterations to found minimum errors in the simulation of the model, instead; in the CSO algorithm the errors remain minimal during the execution (see Fig. 15).

## 8 Statistical Test

The selected statistical test is Z test and the parameters used are described in Table 7.

The hypotheses used for the statistical test indicate the following: $H_o$: CSO algorithm is greater or equal to the BCO algorithm, and $H_a$: CSO algorithm is smaller than the BCO algorithm. A sample of 30 experiments was randomly chosen from

**Table 7** Parameters for the statistical z-test

| Parameter | Value |
|---|---|
| Level of confidence | 95% |
| Alpha | 5% |
| $H_a$ | $\mu_1 < \mu_2$ |
| $H_0$ | $\mu_1 \geq \mu_2$ |
| Critical value | −1.645 |

**Table 8** Results of applying the statistical z-test for problem statement for the statistical

| Algorithm | Compared with | Z value | Evidence |
|---|---|---|---|
| CSO | BCO | −2.8396 | S |

each bio-inspired algorithm, and according to the values used in Table 6, there is a rejection zone for values lower than −1.64. The equation of the z-test is shown as follows:

$$Z = \frac{\left(_1 - \overline{X}_2\right) - (\mu_1 - \mu_2)}{\sigma_{\overline{X}_1 - \overline{X}_2}} \tag{14}$$

Table 8 show the values of Z, and "S" means that is found evidence of significance.

Table 8 shows that when the Z value is smaller than −1.645 the evidence is "Significant" to reject the null hypothesis, in other words, the alternative hypothesis is accepted, in this case the Z value is **−2.8396** then CSO algorithm is better than the BCO algorithm, then the alternative hypothesis is accepted.

## 9 Conclusions

Chicken Search Optimization is an excellent bio-inspired optimization algorithm, capable of handling complex problems and obtaining very good results, in this case the optimization of membership functions of a fuzzy controller for autonomous mobile robots.

One way to check the robustness and efficiency of the CSO is to use a type-2 fuzzy logic system, and so you can analyze the performance of CSO as a technique in optimization of fuzzy controllers, when the noise is added in the model the meta-heuristic CSO allows to demonstrate better execution. As future work we envision using type-2 fuzzy logic in other forms, like in [44], or consider other types of applications, like in [45–51].

## References

1. L. Amador-Angulo, O. Mendoza, J.R. Castro, A. Rodríguez-Díaz, P. Melin, O. Castillo, Fuzzy sets in dynamic adaptation of parameters of a bee colony optimization for controlling the trajectory of an autonomous mobile robot. Sensors **16**(9), 1458 (2016)
2. O. Castillo, R. Martinez-Marroquin, P. Melin, F. Valdez, J. Soria, Comparative study of bio inspired algorithms applied to the optimization of type-1 and type-2 fuzzy controllers for an autonomous mobile robot. Inf. Sci. **192**, 19–38 (2012)
3. E. Cuevas, H. Sossa, A comparison of nature inspired algorithms for multi-threshold image segmentation. Expert Syst. Appl. **40**(4), 1213–1219 (2013)

4. F. Olivas, et al., Comparative study of type-2 fuzzy particle swarm, bee colony and bat algorithms in optimization of fuzzy controllers. Algorithms **10**(3), 101 (2017)
5. A. Chatterjee, M. Biswas, D. Maji, D. Jana, S. Brojabasi, G. Sarkar, S. Das, Discrete wavelet transform based vi image fusion with artificial bee colony optimization, in *Computing and Communication Workshop and Conference (CCWC), 2017 IEEE 7th Annual* (IEEE, 2017), pp. 1–4
6. M. Doringo, V. Maniezzo, A. Colorni, Ant system: optimization by a colony of cooperating agents. IEEE Trans. Syst. Man Cybern. B Cybern. **26**(1), 29–41 (1996)
7. L. Guo, G.G. Wang, A.H. Gandomi, A.H. Alavi, H. Duan, A new improved krill herd algorithm for global numerical optimization. Neurocomputing **138**, 392–402 (2014)
8. A. Jovanovic, M. Nikolic, D. Teodorovic, Area-wide urban traffic control: a bee colony optimization approach. Transp. Res. Part C: Emerging Technol. **77**, 329–350 (2017)
9. D. Karaboga, B. Basturk, A powerful and efficient algorithm for numerical function optimization: artificial bee colony (ABC) algorithm. J. Global Optim. **39**(3), 459–471 (2007)
10. J. Kennedy, R. Eberhart, Particle swarm optimization, in *Proceedings of the IEEE International Conference on Neural Networks* (December 1995), pp. 1942–1948
11. A.H. Gandomi, X.-S. Yang, A.H. Alavi, Cuckoo search algorithm: a metaheuristic approach to solve structural optimization problems. Eng. Comput. **29**(1), 17–35 (2013)
12. T.S. Gorripotu, R.K. Sahu, S. Panda, Firefly algorithm optimised PID controller for automatic generation control with redox flow battery. Int. J. Comput. Syst. Eng. **3**(1–2), 48–57 (2017)
13. M. Mao, L. Zhang, B.V.P. Chong, M. Musembi, Q. Duan, Artificial fish swarm algorithm based-maximum power generation for grid-connected PV panels, in *2017 UKSim-AMSS 19th International Conference on Modelling & Simulation* (Institute of Electrical and Electronics Engineers (IEEE), 2017), pp. 130–135
14. G.G. Wang, Moth search algorithm: a bio-inspired metaheuristic algorithm for global optimization problems. Memetic Comput. 1–14 (2016)
15. F. Tian, R. Zhang, J. Lewandowski, K.M. Chao, L. Li, B. Dong, Deadlock-free migration for virtual machine consolidation using Chicken Swarm Optimization algorithm. J. Intell. Fuzzy Syst. **32**(2), 1389–1400 (2017)
16. L. Amador-Angulo, O. Castillo, Comparative study of bio-inspired algorithms applied in the design of fuzzy controller for the water tank, in *Recent Developments and New Direction in Soft-Computing Foundations and Applications* (Springer, 2016)
17. T. Chaiyatham, I. Ngamroo, A bee colony optimization based fuzzy logic-PID control design of electrolyzer for microgrid stabilization. **8**(9), 6049–6066 (2012)
18. B. Ervural, B.C. Ervural, C. Kahraman, A comprehensive literature review on nature-inspired soft computing and algorithms: tabular and graphical analyses, in *Handbook of Research on Soft Computing and Nature-Inspired Algorithms* (IGI Global, 2017), pp. 34–68
19. B. Acherjee, D. Maity, A.S. Kuar, S. Mitra, D. Misra, D. Optimization of laser transmission welding parameters using chicken swarm optimization algorithm: chicken swarm algorithm optimization, in *Handbook of Research on Manufacturing Process Modeling and Optimization Strategies* (IGI Global, 2017), pp. 142–161
20. A. Chakri, R. Khelif, M. Benouaret, X.S. Yang, New directional bat algorithm for continuous optimization problems. Expert Syst. Appl. **69**, 159–175 (2017)
21. X. He, X.S. Yang, M. Karamanoglu, Y. Zhao, Global convergence analysis of the flower pollination algorithm: a discrete-time Markov chain approach. Procedia Comput. Sci **108**, 1354–1363 (2017)
22. K. Selvam, D.V. Kumar, Frequency control of micro grid with wind perturbations using levy walks with spider monkey optimization algorithm. Int. J. Renew. Energy Res. (IJRER) **7**(1), 146–156 (2017)
23. N. Akhter, A. Ullah, M.Z. Rehman, N. Abdul Hamid, H. Chiroma, Chicken S-BP: an efficient chicken swarm based back-propagation algorithm, in *Recent Advances on Soft Computing and Data Mining: The Second International Conference on Soft Computing and Data Mining (SCDM-2016), Bandung, Indonesia, August 18th–20th, 2016 Proceedings*, vol. 549 (Springer, 2017), p. 122

24. Chen, S., R. Yan, Parameter estimation for chaotic systems based on improved boundary chicken swarm optimization, in *International Symposium on Optoelectronic Technology and Application 2016* (International Society for Optics and Photonics, 2016), pp. 101571K–101571K
25. Y. Li, Y. Wu, X. Qu, Chicken swarm-based method for ascent trajectory optimization of hypersonic vehicles. J. Aerosp. Eng. **30**(5), 04017043 (2017)
26. X. Meng, Y. Liu, X. Gao, H. Zhang, A new bio-inspired algorithm: chicken swarm optimization, in *International Conference in Swarm Intelligence* (Springer, Cham, 2014), pp. 86–94
27. C. Qu, S.A. Zhao, Y. Fu, W. He, Chicken swarm optimization based on elite opposition-based learning. Math. Prob. Eng. (2017)
28. S. Agounad, E.H. Aassif, Y. Khandouch, G. Maze, D. Décultot, Characterization and prediction of the backscattered form function of an immersed cylindrical shell using hybrid fuzzy clustering and bio-inspired algorithms. Ultrasonics (2017)
29. D. Cica, B. Sredanovic, S. Borojevic, D. Kramar, An integration of bio-inspired algorithms and fuzzy logic for tool wear estimation in hard turning, in *International Conference on Advanced Manufacturing Engineering and Technologies* (Springer, 2017), pp. 1–12
30. Y. Feng, G.G. Wang, S. Deb, M. Lu, X.J. Zhao, Solving 0–1 knapsack problem by a novel binary monarch butterfly optimization. Neural Comput. Appl. **28**(7), 1619–1634 (2017)
31. J. Galan-Jimenez, A. Gazo-Cervero, Using bio-inspired algorithms for energy levels assessment in energy efficient wired communication networks. J. Netw. Comput. Appl. **37**, 171–185 (2014)
32. T.C. Lee, K.T. Song, C.H. Lee, C.C. Teng, Tracking control of unicycle-modeled mobile robots using a saturation feedback controller. IEEE Trans. Control Syst. Technol. **9**, 305–318 (2001)
33. J. Ma, Z. Bi, T.O. Ting, S. Hao, W. Hao, Comparative performance on photovoltaic model parameter identification via bio-inspired algorithms. Sol. Energy **132**, 606–616 (2016)
34. S.M. Nigdeli, G. Bekdaş, X.S. Yang, Optimum tuning of mass dampers by using a hybrid method using harmony search and flower pollination algorithm, in *International Conference on Harmony Search Algorithm* (Springer, Singapore, 2017), pp. 222–231
35. N. Palmieri, X.S. Yang, F. De Rango, S. Marano, Comparison of bio-inspired algorithms applied to the coordination of mobile robots considering the energy consumption. Neural Comput. Appl. 1–24 (2017)
36. A. Pandey, S. Banerjee, Bio-inspired computational intelligence and its application to software testing, in *Handbook of Research on Soft Computing and Nature-Inspired Algorithms* (IGI Global, 2017), pp. 429–444
37. S. Swayamsiddha, S. Parija, S.S. Singh, P.K. Sahu, Bio-inspired algorithms for mobile location management—a new paradigm, in *Proceedings of the 5th International Conference on Frontiers in Intelligent Computing: Theory and Applications* (Springer, Singapore, 2017), pp. 35–44
38. L.A. Zadeh, The concept of a lingüistic variable and its application to approximate reasoning, Part II. Inf. Sci. **8**, 301–357 (1975)
39. E.H. Mamdani, N. Baaklini, Prescriptive method for deriving control policy in a fuzzy-logic controller. Electron. Lett. **11**(25), 625–626 (1975)
40. F. Chao, et al.: Type-2 fuzzy hybrid controller network for robotic systems. IEEE Trans. Cybern. (2019)
41. S.K. Injeti, Optimal tuning of PID controller for LFC of two area power system (PV-diesel) using bio-inspired optimization algorithms. World J. Modell. Simul. **12**(2), 112–124 (2016)
42. J.C. Biesmeijer, T.D. Seeley, The use of waggle dance information by honey bees throughout their foraging careers. Behav. Ecol. Sociobiol. **59**(1), 133–142 (2005)
43. F.C. Dyler, The biology of the dance language. Annu. Rev. Entomol. **47**, 917–949 (2002)
44. O. Castillo, *Type-2 Fuzzy Logic in Intelligent Control Applications* (Springer, 2012)
45. L. Aguilar, P. Melin, O. Castillo, Intelligent control of a stepping motor drive using a hybrid neuro-fuzzy ANFIS approach. Appl. Soft Comput. **3**(3), 209–219 (2003)
46. P. Melin, O. Castillo, Adaptive intelligent control of aircraft systems with a hybrid approach combining neural networks, fuzzy logic and fractal theory. Appl. Soft Comput. **3**(4), 353–362 (2003)

47. P. Melin, J. Amezcua, F. Valdez, O. Castillo, A new neural network model based on the LVQ algorithm for multi-class classification of arrhythmias. Inf. Sci. **279**, 483–497 (2014)
48. P. Melin, O. Castillo, *Modelling, Simulation and Control of Non-linear Dynamical Systems: An Intelligent Approach Using Soft Computing and Fractal Theory* (CRC Press, USA and Canada, 2002)
49. P. Melin, D. Sánchez, O. Castillo, Genetic optimization of modular neural networks with fuzzy response integration for human recognition. Inf. Sci. **197**, 1–19 (2012)
50. M.A. Sanchez, O. Castillo, J.R. Castro, P. Melin, Fuzzy granular gravitational clustering algorithm for multivariate data. Inf. Sci. **279**, 498–511 (2014)
51. D. Sanchez, P. Melin, Optimization of modular granular neural networks using hierarchical genetic algorithms for human recognition using the ear biometric measure. Eng. Appl. Artif. Intell. **27**, 41–56 (2014)

# Swarm Intelligence: A Review of Optimization Algorithms Based on Animal Behavior

**Fevrier Valdez**

**Abstract** In this paper a review about the optimization algorithms based on swarm intelligence (SI) with animal behavior is presented. In this review, are analyzed the most popular algorithms such as Firefly Algorithm (FA), Particle Swarm Optimization (PSO), Ant Colony Optimization (ACO), Bee Colony Optimization (BCO) and the Bat algorithm (BA). These algorithms are mentioned in the paper because they inspired on animal behavior demonstrating be useful for solving optimization problems in several applications, and also the algorithms are inspired in swarm intelligence and share similarities in some aspects.

**Keywords** Optimization · Swarm · Intelligence · Animal · Collective · Review · ACO · PSO · BCO · FA · BA

## 1 Introduction

The aim of this chapter is to review the algorithms inspired on collective intelligent based on animal behavior and their applications. Therefore, the literature review about these methods is analyzed. The main contribution of this paper, is review the importance in the last years about the methods using animal behavior as the main motivation and how the authors have adopted these algorithms to solve optimization problems and improve their results. We use the Web of Science Core Collection (WoS) to search the more recent works about the these methods. Also, the VOS viewer software was used to calculate the network, clusters and important information that is presented in this research.

Metaheuristic algorithms are often nature-inspired and they are now among the most widely used algorithms for optimization [92]. They have many advantages over conventional algorithms. Metaheuristic algorithms are very diverse, including

F. Valdez (✉)
Computer Science, Tijuana Institute of Technology, Calzada Tecnologico S/N, Tijuana 22450, BC, Mexico
e-mail: fevrier@tectijuana.mx



P. Melin et al. (eds.), *Recent Advances of Hybrid Intelligent Systems Based on Soft Computing*, Studies in Computational Intelligence 915,
https://doi.org/10.1007/978-3-030-58728-4_16

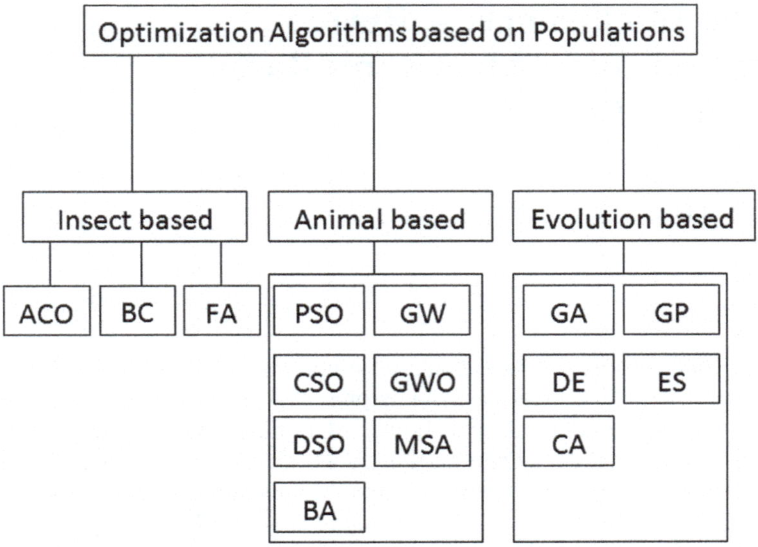


**Fig. 1** Optimization algorithms based on populations

genetic algorithms (GA), programming genetic (GP), evolutionary programming (EP), simulated annealing (SA), differential evolution (DE), ant colony and bee algorithms, bat algorithm, particle swarm optimization, harmony search, firefly algorithm, flower pollination algorithm, cuckoo search (CS), dolphin swarm optimization (DSO), monkey search algorithm (MSA), glow-worm (GW), cat swarm optimization (CSO), and others [1, 9, 32, 38, 39, 45, 48, 78, 91, 94, 104]. Here, we will review about the algorithms based on animals and insects. However, Fig. 1 shows a general classification about the optimization methods.

Swarm intelligence originated from the study of colonies, or swarms of social organisms. Studies of the social behavior of organisms in swarms prompted the design of very efficient optimization and clustering algorithms. For example, simulation studies of the graceful, but unpredictable, choreography of bird flocks led to the design of the particle swarm optimization algorithm, and studies of the foraging behavior of ants resulted in ant colony optimization algorithms. A major thrust in algorithmic development is the design of algorithmic models to solve increasingly complex problems. Enormous successes have been achieved through the modeling of biological and natural intelligence, resulting in so-called "intelligent systems". These intelligent algorithms include artificial neural networks, evolutionary computation, swarm intelligence, artificial immune systems, and fuzzy systems. Together with logic, deductive reasoning, expert systems, case-based reasoning and symbolic machine learning systems, these intelligent algorithms form part of the field of Artificial Intelligence (AI). Just looking at this wide variety of AI techniques, AI can be seen as a combination of several research disciplines, for example, computer science, physiology, philosophy, sociology and biology.

Studies of social animals and social insects have resulted in a number of computational models of swarm intelligence. Biological swarm systems that have inspired computational models include ants, termites, bees, spiders, fish schools, and bird flocks. Within these swarms, individuals are relatively simple in structure, but their collective behavior is usually very complex. The complex behavior of a swarm is a result of the pattern of interactions between the individuals of the swarm over time. This complex behavior is not a property of any single individual, and is usually not easily predicted or deduced from the simple behaviors of the individuals. This chapter is organized as follows: in Sect. 1 a brief introduction about of the swarm intelligence is presented, Sect. 2 describe the literature review and the analyzed methods in the chapter are presented, Sect. 3 shows the applications that have been used these methods with fuzzy logic, neural networks and other techniques, in Sect. 4 the conclusion are presented and finally the references in the last section are included.

## 2 Literature Review

Nowadays application of swarm intelligence optimization algorithms has been widely used because they are able to solve several complex optimization problems. The term "Swarm intelligence", that denotes this "collective intelligence" has come into use [11–13, 15]. Swarm Intelligence [15] is the part of Artificial Intelligence based on study of actions of individuals in various decentralized systems. This chapter describes some heuristic techniques based on animal behavior and populations throughout history have been used by researchers as optimization methods to solve problems of everyday life and industry. This chapter relates the history and description of the stochastic algorithms PSO, ACO, BCO and BA. However, in this section a table with the most popular optimization algorithms based in populations in chronological order is presented in Table 1.

### *2.1 Firefly Algorithm*

FA was first proposed by Yang [101–103], which was based on the flashing patterns and behavior of fireflies. Nowadays FA works has expanded significantly in the last years with many papers published about this algorithm as can be seen in [27] where was made an extensive review of this method. In essence, FA uses the following three idealized rules:

- Fireflies are unisex so that one firefly will be attracted to other fireflies regardless of their sex.
- The attractiveness is proportional to the brightness, and they both decrease as their distance increases. Thus, for any two flashing fireflies, the less bright one

**Table 1** Popular optimization algorithms inspired in populations in chronological order

| Year | Optimization algorithms inspired in populations |
|---|---|
| 2018 | Monkey search algorithm [55] |
| 2016 | Dolphin swarm algorithm [96] |
| 2015 | Ant lion optimizer [69] |
| 2012 | Krill Herd algorithm [40] |
| 2010 | Bat algorithm [100] |
| 2009 | Cuckoo search [99] |
| 2010 | Artificial bee algorithm [49] |
| 2010 | Firefly algorithm [102] |
| 2009 | Gravitational search algorithm [75] |
| 2007 | Intelligent water drops [47] |
| 2006 | Cat swarm optimization [25] |
| 2006 | Glow-worm [54] |
| 2005 | Harmony search algorithm [42] |
| 2005 | Honey bee algorithm [86] |
| 2002 | Bacterial foraging algorithm [85] |
| 2001 | Bee colony optimization [62] |
| 2002 | Estimation of distribution algorithm [71] |
| 1995 | Particle swarm optimization [35] |
| 1992 | Ant colony optimization [26] |
| 1989 | Tabu search [43] |
| 1983 | Simulated annealing [52] |
| 1979 | Cultural algorithms [77] |
| 1975 | Genetic algorithms [46] |
| 1966 | Evolutionary programming [68] |
| 1965 | Evolution strategies [50] |

will move towards the brighter one. If there is no brighter one than a particular firefly, it will move randomly.

## 2.2 *Ant Colony Optimization*

One of the first behaviors studied by researchers was the ability of ants to find the shortest path between their nest and a food source. From these studies and observations followed the first algorithmic models of the foraging behavior of ants, as developed by Dorigo [27, 28]. Since then, research in the development of Algorithms based on Ants have been very popular, resulting in a large number of algorithms and applications. These algorithms that were developed as a result of studies of ant foraging behavior are referred to as instances of the ant colony optimization meta-heuristic [29, 33].

The first Ant Colony Optimization (ACO) algorithm has been applied to the Traveling Salesman Problem (TSP) [27–30, 33]. Given a set of cities and the distances between them, the TSP is the problem of finding the shortest possible path which visits every city exactly once. Possible path which visits every city exactly once, it can be represented by a complete weighted graph G = (N, E) N is the set of nodes representing the cities and E is the set of edges. Each edge is assigned a value $d_{ij}$ which is the distance between cities i and j. When applying the ACO algorithm to the TSP, a pheromone strength $\tau_{ij}$ (t) is associated to each edge (i, j), where $\tau_{ij}$ (t) is a numerical value which is modified during the execution of the algorithm and t is the iteration counter.

The ACO metaheuristic is inspired by observing the behavior of real ant colonies, which presented an interesting feature, how to find the shortest paths between the nest and food, on its way the ants deposit a substance called pheromone, this trail allows the ants back to their nest from the food; it uses the evaporation of pheromone to avoid an unlimited increase of pheromone trails and allow to forget the bad decisions [31, 34]. Where is the probability of ant k goes from node $i$ to node $j$, is the heuristic information, say, prior knowledge, is the possible neighborhood of an ant k when in the city $i$, is the pheromone trail, $\alpha$ and $\beta$ are parameters that determine the influence of the pheromone and prior knowledge, respectively.

### 2.2.1 Ant System

Initially, three different versions of AS were proposed [27–29, 33]. These were called ant-density, ant quantity, and ant-cycle. Whereas in the ant-density and antquantity versions the ants updated the pheromone directly after a move from one city to an adjacent city, in the ant-cycle version the pheromone update was only done after all the ants had constructed the tours and the amount of pheromone deposited by each ant was set to be a function of the tour quality. Nowadays, when referring to AS, one actually refers to ant-cycle since the two other variants were abandoned because of their inferior performance.

The two main phases of the AS algorithm constitute the ants' solution construction and the pheromone update. In AS a good heuristic to initialize the pheromone trails is to set them to a value slightly higher than the expected amount of pheromone deposited by the ants in one iteration; a rough estimate of this value can be obtained by setting, $(i, j)$, $\tau_{ij} = \tau_0 = m/C^{nn}$, where $m$ is the number of ants, and $C^{nn}$ is the length of a tour generated by the nearest-neighbor heuristic. The reason for this choice is that if the initial pheromone values $\tau_0$'s are too low, then the search is quickly biased by the first tours generated by the ants, which in general leads toward the exploration of inferior zones of the search space. On the other side, if the initial pheromone values are too high, then many iterations are lost waiting until pheromone evaporation reduces enough pheromone values, so that pheromone added by ants can start to bias the search. In AS, $m$ ants build a tour of the TSP. Initially, ants are put on randomly chosen cities. At each construction step, ant $k$ applies a probabilistic action choice rule, called random proportional rule, to decide which city to visit next. In particular, the probability with which ant $k$, currently at city $i$, chooses to go to city $j$ is

$$p_{ij}^{k} = \frac{[\tau_{ij}]^{\alpha}[\eta_{ij}]^{\beta}}{\sum l \in N_i^k [\tau_{ij}]^{\alpha}[\eta_{ij}]^{\beta}}, if \quad j \in N_i^k, \tag{1}$$

where $\eta_{ij}= 1/d_{ij}$ is a heuristic value that is available a priori, $\alpha$ and $\beta$ are two parameters which determine the relative influence of the pheromone trail and the heuristic information, and $N_i^k$ is the feasible neighborhood of ant k when being at city i, that is, the set of cities that ant k has not visited yet. The update of pheromone trails is as follows.

After all the ants have constructed their tours, the pheromone trails are updated. This is done by first lowering the pheromone value on all arcs by a constant factor, and the adding pheromone on the arcs the ants have crossed in their tours, Pheromone evaporation is made by

$$\tau_{ij} \leftarrow (1-\rho)\tau_{ij}, \forall (i, j) \in L, \tag{2}$$

where $0 < \rho \leq 1$ The parameter $\rho$ is used to avoiding unlimited accumulation of the pheromone trails and it enables the algorithm to delete bad decisions previously taken by ants. In fact, if an arc is not chosen by the ants, its associated pheromone value decreases exponentially in the number of iterations. After evaporation, all ants deposit pheromone on the arcs they have crossed in their tour:

$$\tau_{ij} = \tau_{ij} + \sum_{k=1}^{m} \Delta\tau_{ij}^{k} \forall (i, j) \in L \tag{3}$$

### 2.2.2 Elitist Ant System

A first improvement on the initial AS, called the elitist strategy for Ant System (EAS), was introduced in Dorigo [30, 31, 34]. The idea is to provide strong additional reinforcement to the arcs belonging to the best tour found since the start of the algorithm; this tour is denoted as $T^{bs}$ (best-so-far tour) in the following. Note that this additional feedback to the best-so-far tour (which can be viewed as additional pheromone deposited by an additional ant called best-so-far ant) is another example of a daemon action of the ACO metaheuristic. The update of pheromone trails is very similar as in AS only differs in the additional reinforcement of tour $T^{bs}$ is achieved by adding a quantity $e/C^{bs}$ to its arcs, where $e$ is a parameter that defines the weight given to the best-so-far tour $T^{bs}$, and $C^{bs}$ is its length. Thus, the follow equation for the pheromone deposit becomes

$$\tau_{ij} = \tau_{ij} + \sum_{k=1}^{m} \Delta\tau_{ij}^{k} + e\Delta\tau_{ij}^{bs}, \tag{4}$$

In EAS, the pheromone evaporation is implemented as in AS. Computational results presented in Dorigo [27, 28, 31, 34]. Suggest that the use of the elitist strategy with an appropriate value for parameter $e$ allows AS to both find better tours and find them in a lower number of iterations.

### 2.2.3 Rank-Based Ant System

Bullnheimer et al. [16] proposed a modification of AS to: (1) allow only the best ant to update pheromone concentrations on the links of the global-best path, (2) to use elitist ants, and (3) to let ants update pheromone on the basis of a ranking of the ants. In ASrank the ants are sorted by increasing tour length and the quantity of pheromone an ant deposits is weighted according to the rank $r$ of the ant. In the iteration only the ($w$-best-ranked ants and the ant that produced the best-so-fat tour are allowed to deposit pheromone. The best-so-far tour gives the strongest feedback, with $w$. For ASrank, the global update rule is defined as follow:

$$\tau_{ij} \leftarrow \tau_{ij} + \sum_{r=1}^{w-1} (w-r)\Delta\tau_{ij}^{r} + w\Delta\tau_{ij}^{bs}, \tag{5}$$

The results of an experimental evaluation by Bullnheimer et al. (suggest that ASrank performs slightly better than EAS and significantly better than AS).

### 2.2.4 MAX-MIN Ant System

MAX–MIN Ant System (MMAS) [81, 82] introduces four main modifications with respect to AS. First, it strongly exploits the best tours found: only either the iteration—best ant, that is, the ant that produced the best tour in the current iteration, or the best-so-far ant is allowed to deposit pheromone. Unfortunately, such a strategy may lead to a stagnation situation in which all the ants follow the same tour, because of the excessive growth of pheromone trails on arcs of a good, although suboptimal, tour. To counteract this effect, a second modification introduced by MMAS is that it limits the possible range of pheromone trail values to the interval $[\tau_{\min}, \tau_{\max}]$. Third, the pheromone trails are initialized to the upper pheromone trail limit, which, together with a small pheromone evaporation rate, increases the exploration of tours at the start of the search. Finally, in MMAS, pheromone trails are reinitialized each time the system approaches stagnation or when no improved tour has been generated for a certain number of consecutive iterations. The update of pheromone trail is as follows:

$$\tau_{ij} \leftarrow \tau_{ij} + \Delta\tau_{ij}^{best} \tag{6}$$

## 2.3 Particle Swarm Optimization

Particle swarm optimization (PSO) is a population based stochastic optimization technique developed by Eberhart and Kennedy in 1995, inspired by the social behavior of bird flocking or fish schooling [50].

PSO shares many similarities with evolutionary computation techniques such as Genetic Algorithms (GA) [17]. The system is initialized with a population of random solutions and searches for optima by updating generations. However, unlike the GA, the PSO has no evolution operators such as crossover and mutation. In PSO, the potential solutions, called particles, fly through the problem space by following the current optimum particles [6].

Each particle keeps track of its coordinates in the problem space, which are associated with the best solution (fitness) it has achieved so far (The fitness value is also stored). This value is called pbest. Another "best" value that is tracked by the particle swarm optimizer is the best value, obtained so far by any particle in the neighbors of the particle. This location is called lbest. When a particle takes all the population as its topological neighbors, the best value is a global best and is called gbest.

The particle swarm optimization concept consists of, at each time step, changing the velocity of (accelerating) each particle toward its pbest and lbest locations (local version of PSO). Acceleration is weighted by a random term, with separate random numbers being generated for acceleration toward pbest and lbest locations.

In the past several years, PSO has been successfully applied in many research and application areas. It is demonstrated that PSO gets better results in a faster, cheaper way compared with other methods [5].

Another reason that PSO is attractive is that there are few parameters to adjust. One version, with slight variations, works well in a wide variety of applications. Particle swarm optimization has been used for approaches that can be used across a wide range of applications, as well as for specific applications focused on a specific requirement.

## 2.4 Bee Colony Optimization

The Bee Colony Optimization (BCO) metaheuristic [62–65, 87] was proposed recently by Lučić and Teodorović as a new direction method in the field of Swarm Intelligence. The BCO has been successfully applied to various engineering and management problems by Teodorović et al. [67, 88–90]. The BCO approach is a "bottom-up" approach to modeling where special kinds of artificial agents are created by analogy with bees. Artificial bees represent agents, which collaboratively solve complex combinatorial optimization problem.

Lučić and Teodorović [62–65] were among first who used basic principles of collective bee intelligence in solving combinatorial optimization problems. The BCO is a population-based algorithm. Population of artificial bees searches for the optimal solution. Artificial bees represent agents, which collaboratively solve complex combinatorial optimization problems. Every artificial bee generates one solution to the problem. The algorithm consists of two alternating phases: forward pass and backward pass. In each forward pass, every artificial bee is exploring the search space. It applies a predefined number of moves, which construct and/or improve the solution, yielding to a new solution. Having obtained new partial solutions, the bees go again to the nest and start the second phase, the so-called backward pass. In the backward pass, all artificial bees share information about their solutions.

In nature, bees would perform a dancing ceremony, which would notify other bees about the quantity of food they have collected, and the closeness of the patch to the nest. In the BCO, the artificial bees publicize the quality of the solution, i.e. the objective function value. During the backward pass, every bee decides with a certain probability whether to abandon the created partial solution and become again uncommitted follower, or dance and thus recruit the nest mates before returning to the created partial solution. During the second forward pass bees expand previously created partial solutions, by a predefined number of nodes, and after that perform again the backward pass and return to the hive. In the hive, bees again participate in a decision-making process, make a decision, perform third forward pass, etc. The two phases of the search algorithm, forward and backward pass, are performed iteratively, until a stopping condition is met. The possible stopping conditions could be, for example, the maximum total number of forward/backward passes, the maximum total number of forward/backward passes without the improvement of the objective function, etc.

## 2.5 Bat Algorithm

Bat algorithm is a bio-inspired algorithm developed by Yang in 2010 and BA has been found to be very efficient. If we idealize some of the echolocation characteristics of microbats, we can develop various bat-inspired algorithms or bat algorithms. For simplicity, we now use the following approximate or idealized rules [97]:

1. All bats use echolocation to sense distance, ant they also 'know' the difference between food/prey and background barriers in some magical way.
2. Bats fly randomly witch velocity $\mathbf{v}_i$ at position $\mathbf{x}_i$ witch a fixed frequency $\mathbf{f}_{min}$, varying wavelength $\lambda$ and loudness $\mathbf{A}_0$ to search for prey They can automatically depending on the proximity of their target.
3. Although loudness can vary in many ways, we assume that the loudness varies from a large (positive) $\mathbf{A}_0$ to a minimum constant value $\mathbf{A}_{min}$.

For simplicity, the frequency f $\varepsilon$ [0, $f_{max}$], the new solutions $\mathbf{x}_i^t$ and velocity $\mathbf{v}_i^t$ at a specific time step t are represented by a random vector drawn from a uniform distribution [44].

# 3 Applications

In this chapter a review about of the applications with the algorithms above mentioned is presented. In this case, we describe only the applications that have used swarm computing. In particular, in this paper, we have analyzed with detail only four algorithms based in swarm such ACO, PSO, BCO and BA algorithms. However, in this section is shown some applications with these algorithms.

## *3.1 Firefly Applications*

Here, we present some application with the firefly algorithm in several areas with different techniques. FA combined with fuzzy logic has been used in the last years to improve the results. Here, according to the search in web of science (WoS) 'firefly with fuzzy logic' we found the following results. Only 48 journal papers were found with the search above described. However, we are presenting a brief description about the first 5 papers, but the complete updated references can be seen with the same search in WoS. In [66] were compared the different intelligent controllers such as Fuzzy, ANFIS and Hybrid Fuzzy and Firefly Algorithm (HFFA) for Maximum Power Point Tracking (MPPT) of 100 W PV system using a Super Lift Boost Converter (SLBC). The proposed intelligent controller was designed and simulated in MATLAB environment under various weather conditions. By other hand, in [4] was presented an Adaptive Fuzzy Map Approach that Relies on FA for Accruing Velocity of Big Data and Decentralized Decision Making. An advantage of the Firefly algorithm is the providing of a small number of iterations comparing to the other methods, which minimize the execution time. Furthermore, the Firefly algorithm is significant to the fuzzy logic system to get its inputs. In addition to the Firefly algorithm, Kalman filter was used to scale the distances of Big data sets, where it generates output by assigning the match and mismatch. In [37] was proposed a Fuzzy Logic and Bio-Inspired Firefly Algorithm Based Routing Scheme in Intrabody Nanonetworks, where the authors proposed a fuzzy logic-based correlation region selection and bio-inspired FA based nano biosensors (NBSs) nomination jointly contribute to energy conservation by minimizing transmission of correlated spatial data. Their proposed fuzzy logic-based correlation region selection mechanism aims at selecting those correlated regions for data aggregation that are enriched in terms of energy and detected information. In [84] was proposed a study of the semi-active control schemes, namely fuzzy logic based controllers tuned using a novel optimization algorithm called advanced firefly algorithm (AFA) which was proposed to regulate the body of the vehicle's suspension from any disturbances acted to the system.

The AFA is to be introduced based on the improvement of the original FA to enhance the solution quality of the FA. Also, in [74] was developed a Hybrid improved firefly pattern search optimized fuzzy aided PID controller for automatic generation control of power systems with multi-type generations, where the performance of original FA was improved by adding memory, newborn fireflies and using a new updating formula of fireflies which eliminate the wandering movement of the fireflies during the iteration process. Figure 2 shows the total of papers by authors with at least 2 papers.

Also, using the software VosViewer [72] an analysis with this search was calculated. Here, we found a network with the works above mentioned. Figure 3 shows the

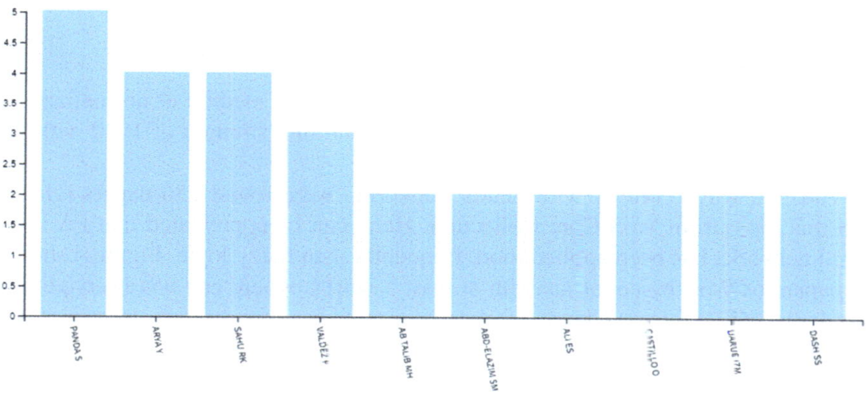


**Fig. 2** Firefly with fuzzy logic by authors

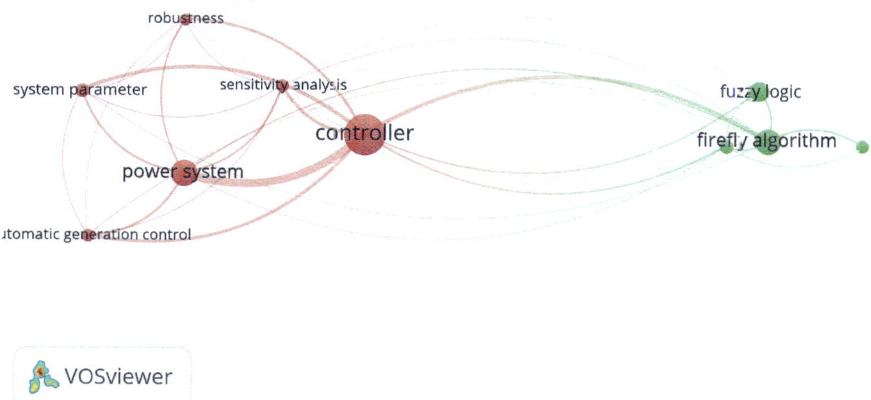


**Fig. 3** Network visualization with the search' firefly with fuzzy logic'

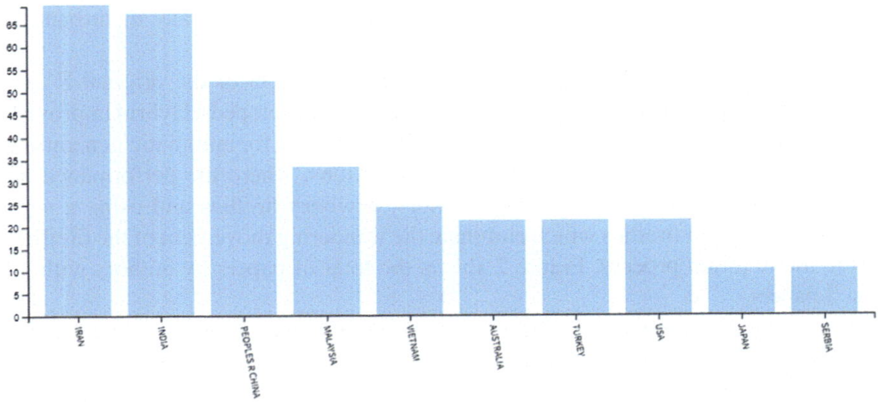


**Fig. 4** Firefly with neural networks by authors

network visualization obtaining two clusters applying the method of normalization called association strength, with 37 links and a total link strength of 1040 with 10 items described in Fig. 3.

With the search 'Firefly with neural networks' were found 256 papers related with this topic from WoS Core Collection. Here, can be appreciated that FA with neural networks has been applied more frequently than fuzzy logic. Figure 4 shows the papers of WoS by countries with at least 5 works, where can be observed that Iran, India and Peoples R China, has the major number of works with this topic. We cited only the first 5 works in this research [10, 23, 76, 79, 80], but the complete references can be reviewed in WoS with the topic above described.

Also, of the same way and using the software VosViewer [72] an analysis with this search was calculated taking as reference the title and abstract fields and the main weight was by occurrences, we found a network with the works above mentioned.

Figure 5 shows the network visualization obtaining four clusters applying the method of normalization called association strength, with 1926 links and a total link strength of 1926 with 70 items shown in Fig. 5. Figure 6, shows the cluster density of this search.

## 3.2 ACO Applications

In this section, are presented some applications with the ACO algorithm in fuzzy logic, neural networks and other works. ACO combined with fuzzy logic has been used for parameter adaptation to improve the performance. Also, according to the topic in web of science (WoS) 'ACO with fuzzy logic' were found the following results. In total, 41 journal papers were found with the search above described. However, we are presenting a brief description about the first 5 papers, but the

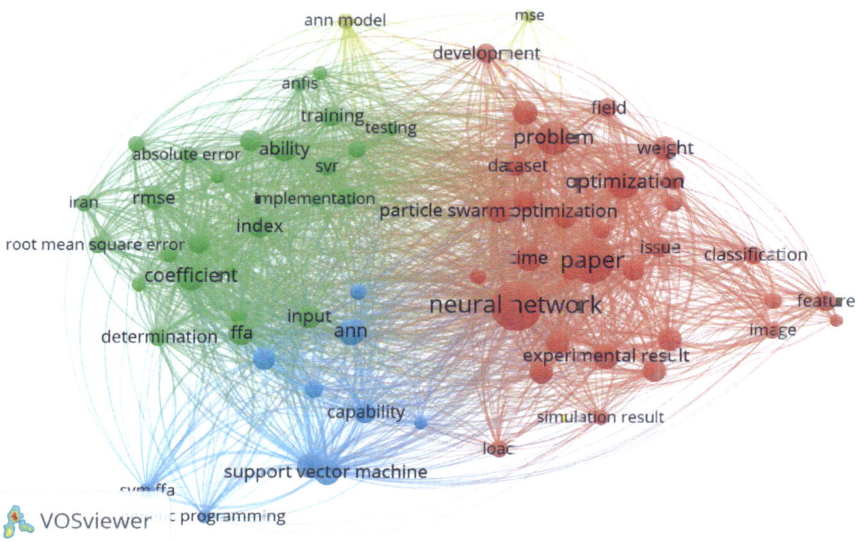


**Fig. 5** Network visualization with the search' firefly with neural networks'

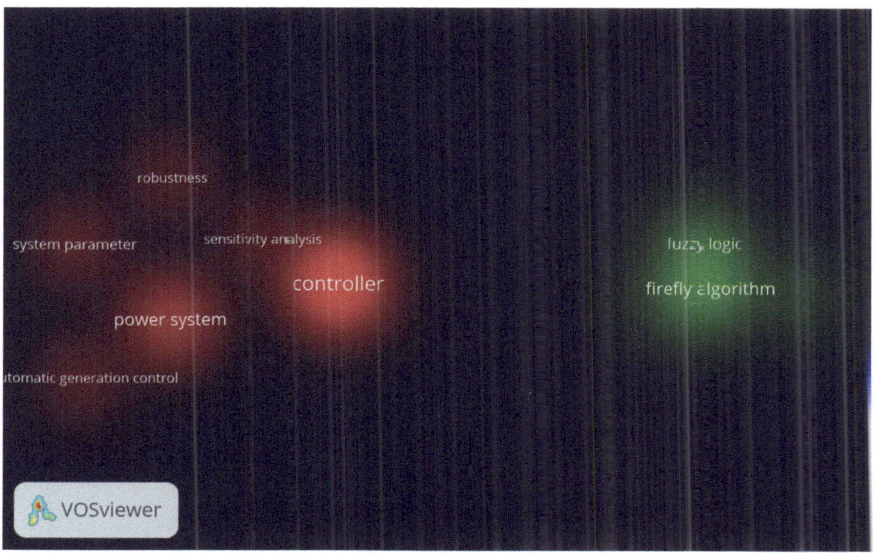


**Fig. 6** Cluster density of FA

complete updated references can be seen with the same search in WoS. In [14], was proposed a holistic optimization approach for inverted cart-pendulum control tuning, where a simplified Ant Colony Optimization method with a constrained Nelder-Mead algorithm (ACO-NM). Holistic optimization refers to a simultaneous tuning of the swing-up, stabilization and switching mode parameters. The holistic approach optimizes four controller structures, which include controllers that have never been tuned by a specific method besides by the trial-and-error method. Also, in [22] an improved ACO algorithm optimized fuzzy PID controller for load frequency control in multi area interconnected power systems was proposed. To demonstrate the robustness of the proposed control method, the authors implemented the sensitivity analysis under wide variation of operating conditions and system parameters. By other hand, in [105] was developed an improved ACO to design A2-C1 type fuzzy logic system. The design included parameters adjustment and rules selection, and the performance of the intelligent fuzzy system, which can be improved by choosing the most optimal parameters and reducing the redundant rules. In order to verify the feasibility of the algorithm, the intelligence fuzzy logic systems based on the algorithms are applied to predict the Mackey-Glass chaos time series. In [61] was developed a genetic algorithm (GA) and ACO, hybrid algorithm of GA and ACO to solve a multi-criteria supplier selection problem. It combines advantages of GA with great global converging rate and ACO with parallelism and effective feedback. Finally, in [95] was developed a method with fuzzy systems to dynamically update the parameters for the ACO and PSO algorithms. In the case of ACO, two fuzzy systems are designed for the Ant Colony System (ACS) algorithm variant. The first system adjusts the value for the pheromone evaporation parameter from the global pheromone trail update equation and the second system adjusts the values for the pheromone evaporation parameter from the local pheromone trail update equation. Figure 7, shows the number of publications by authors with at least two works.

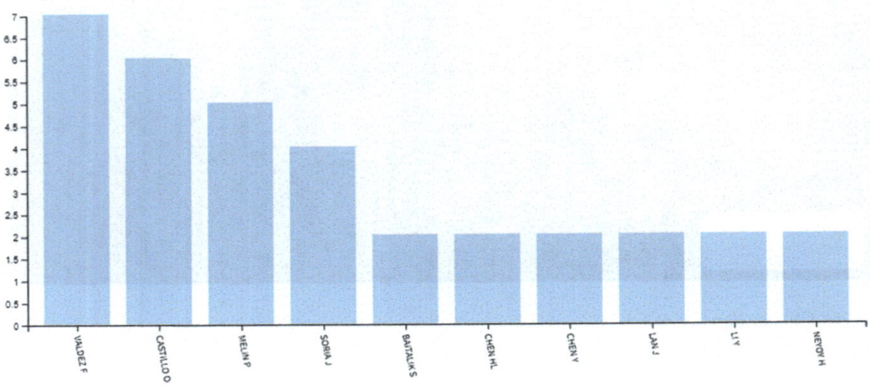


**Fig. 7** ACO with fuzzy by authors

Also, in Fig. 8 is shown the results obtained by countries, where it can be seen that India, China, Iran and Mexico have the major number of publications with the topic 'ACO with fuzzy logic'.

Also, of the same way and using the software VosViewer [72] an analysis with this search was calculated taking as reference the title and abstract fields and the main weight was by occurrences, we found a network with the works above described. Figure 10 shows the network visualization obtaining 7 clusters applying the method

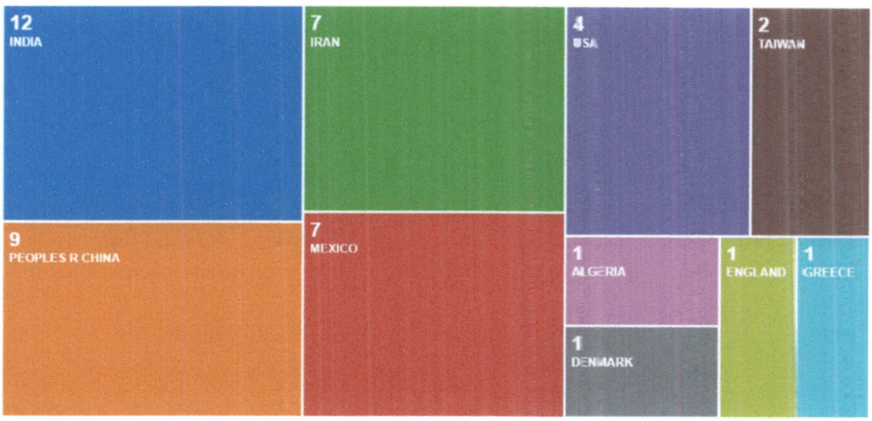


**Fig. 8** ACO with fuzzy logic by countries

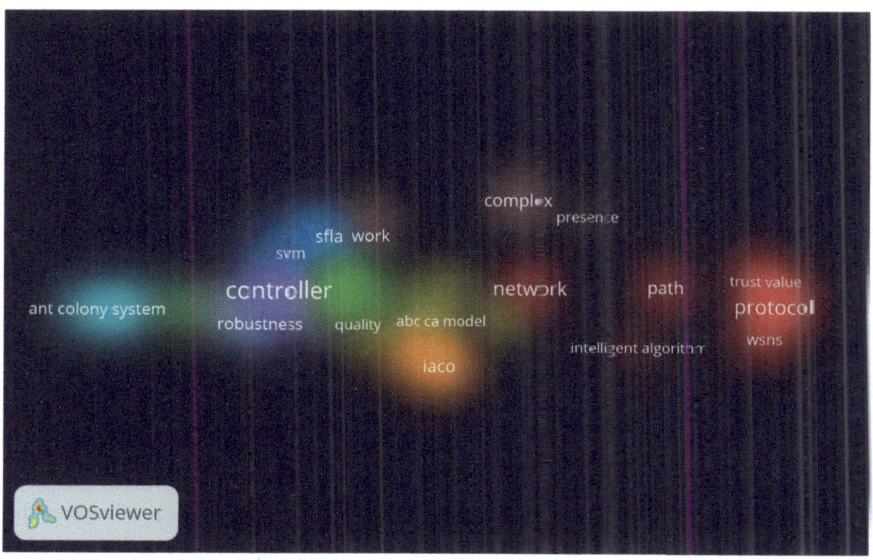


**Fig. 9** Cluster density of ACO

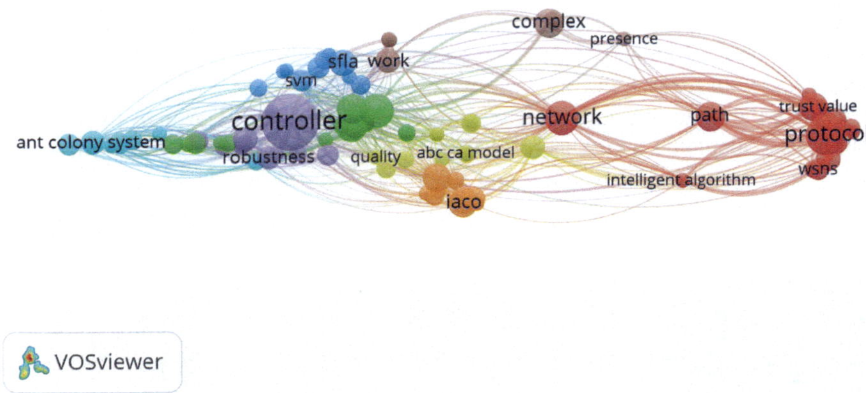


**Fig. 10** Network visualization with the search' ACO with fuzzy logic'

of normalization called association strength, with 675 links and a total link strength of 3762 with 92 items shown in Fig. 10. Figure 9 shows the cluster density for this search.

Other applications with ACO can be seen in [2, 3, 18, 21, 41, 58].

## *3.3 PSO Applications*

Here, are presented some applications with the PSO algorithm in fuzzy logic, neural networks and other works. PSO is an optimization method widely used in several areas, however, here we present the applications with fuzzy logic and other intelligent techniques. With the topic 'PSO with fuzzy logic' searching in WoS the results obtained are 398 publications related with this topic. As can be seen, the number of papers is greater than the two previous methods analyzed in this chapter. Also, only 5 papers are described, but the updated references can be consulted in WoS with the topic above described.

In [57], was proposed a fuzzy associative memory with auto encoding mechanisms using PSO and Differential evolution for the optimization of the association matrices studied. Also, in [73] was proposed an experimental estimation of hybrid adaptive neuro-fuzzy inference system-particle swarm optimization (ANFIS-PSO) based MPPT for PV grid integration under fluctuating sun irradiance. By other hand, in [36] was introduced a hybrid of genetic algorithm and particle swarm optimization to tune the parameters of different adaptive PID controllers. To evaluate the performance of the proposed hybrid optimization method on the different adaptive PID controllers, these controllers are applied to control the operation of one of the most difficult chemical processes, the divided wall distillation column. Also, in [83] proposed to build a new PSO-FS optimization platform identifying the optimal limits of gas ratios and corresponding rules for each fault type to enhance the diagnostic

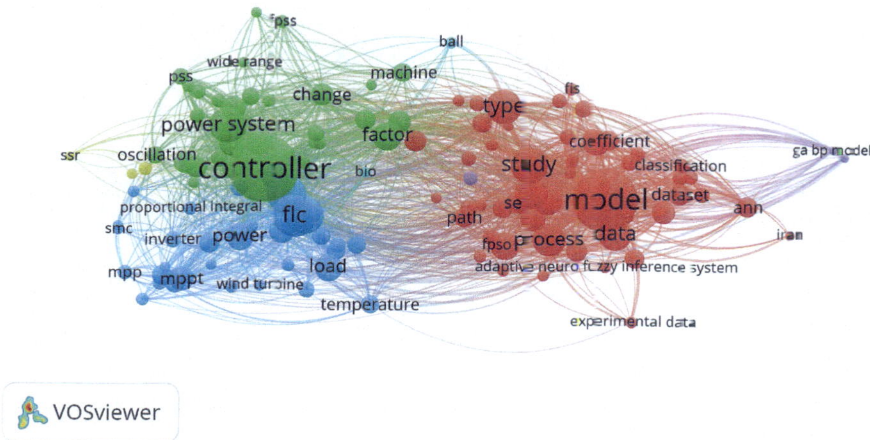


**Fig. 11** Network visualization with the search' PSO with fuzzy logic'

accuracies of Rogers' four-ratio and IEC 60599 code methods. In the proposed platform, PSO optimization technique specifies the gas ratio limits and corresponding rules where the fuzzy system identifies the fault types. Finally, with this topic in was proposed a discrete-time switched linear model of the PSO.

Also, of the same way and using the software VosViewer [72] an analysis with this search was calculated taking as reference the title and abstract fields and the main weight was by occurrences, we found a network with the works above mentioned. Figure 11 shows the network visualization obtaining 6 clusters applying the method of normalization called association strength, with 4601 links and a total link strength of 43,131 with 92 items shown in Fig. 11. Figure 12 shows the cluster density for this search.

Other works of PSO applied to neural networks, math functions and control can be reviewed in [8, 56, 59, 93].

## 3.4 BCO Applications

Also, we made a search in WoS, similar to other analyzed methods in this research with the topic 'BCO with fuzzy logic', however, we found only 8 papers with this method, maybe, because not many authors have tried to combine or apply these two techniques together. Here only 5 works are presented but the other papers can be consulted in WoS. Also, are cited other works where was used this algorithm.

In [20] was presented a comparison among the bee colony optimization (BCO), differential evolution (DE), and harmony search (HS) algorithms. In addition, for each algorithm, a type-1 fuzzy logic system (T1FLS) for the dynamic modification of the main parameters is presented. The dynamic adjustment in the main parameters

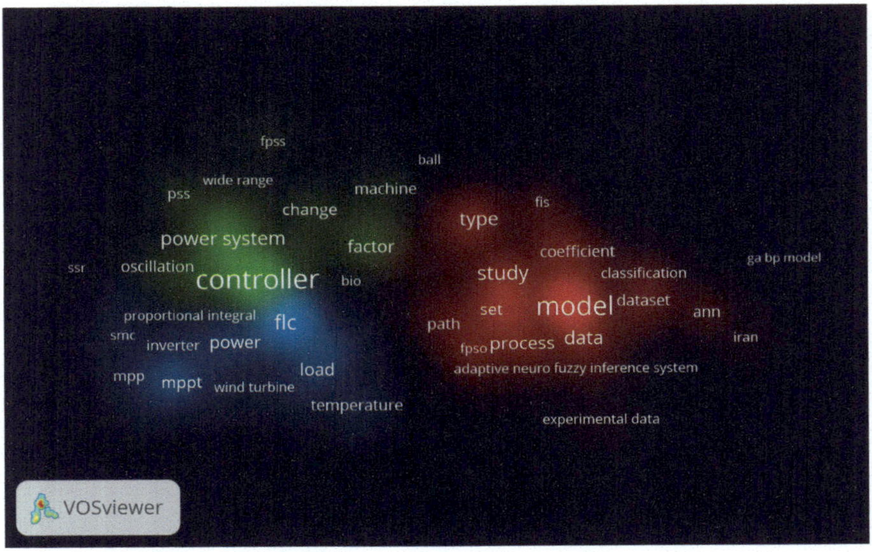


**Fig. 12** Cluster density of PSO

for each algorithm with the implementation of fuzzy systems aims at enhancing the performance of the corresponding algorithms. Each algorithm (modified and original versions) is analyzed and compared based on the optimal design of fuzzy systems for benchmark control problems, especially in fuzzy controller design. Also, in [24] a dynamic fuzzy-based dance mechanism for the bee colony optimization algorithm was proposed. In [19] a generalized type-2 fuzzy logic approach for dynamic parameter adaptation in bee colony optimization applied to fuzzy controller design. Also, in [7] a new fuzzy bee colony optimization with dynamic adaptation of parameters using interval type-2 fuzzy logic for tuning fuzzy controllers was introduced. Finally, Frumen et al. made a comparative study of type-2 fuzzy particle swarm, bee colony and bat algorithms in optimization of fuzzy controllers. In Fig. 13 is shown the number of paper by countries, highlighting Mexico with more number of works in this area.

Also, of the same way and using the software VosViewer [72] an analysis with this search was calculated taking as reference the title and abstract fields and the main weight was by occurrences, we found a network with the works above mentioned. Figure 14 shows the network visualization obtaining 4 clusters applying the method of normalization called association strength, with 201 links and a total link strength of 1092 with 34 items shown in Fig. 14. Figure 15 shows the cluster density for this search.

BCO in Traveling Salesman Person and other applications can be seen in [62–65, 89].

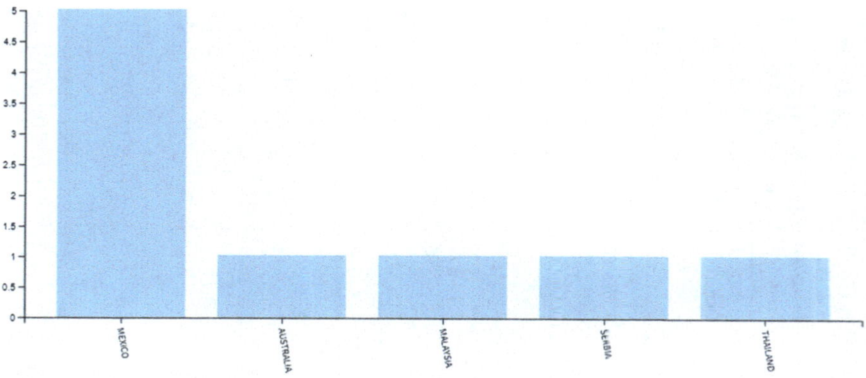


**Fig. 13** BCO with fuzzy logic by countries

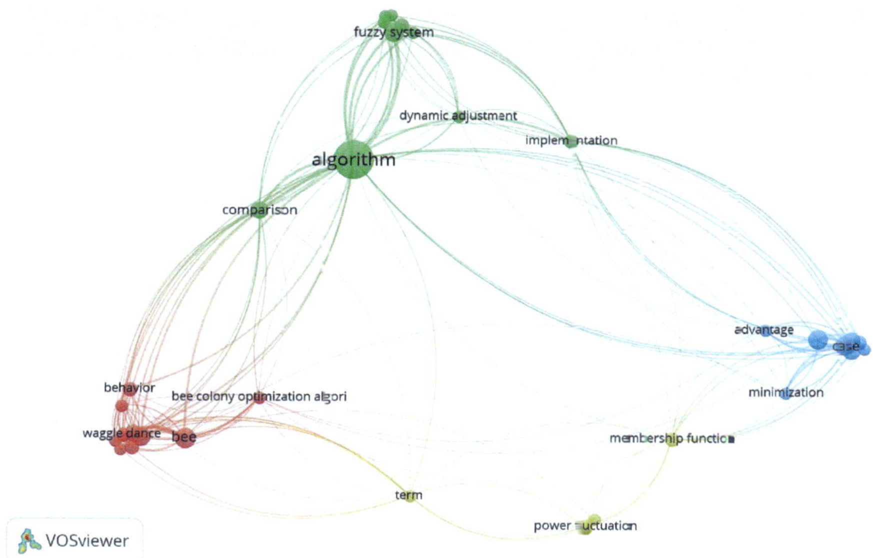


**Fig. 14** Network visualization with the search' BCO with fuzzy logic'

## 3.5 BA Applications

Musikapun and Pongcharoen [70] solved multistage, multi-machine, multi-product scheduling problems using bat algorithm, and they solved a class of non-deterministic polynomial time (NP) hard problems with a detailed parametric study. They also implied that the performance can be further improved by about 8.4% using an optimal set of parameters.

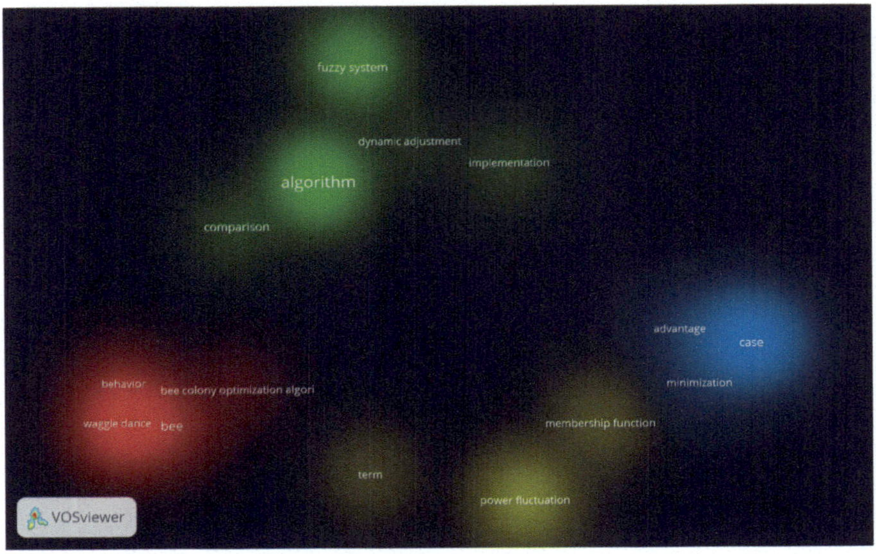


**Fig. 15** Cluster density of BCO

Yang [98] use the bat algorithm to study topological shape optimization in microelectronic applications so that materials of different thermal properties can be placed in such a way that the heat transfer is most efficient under stringent constraints. It can also be applied to carry out parameter estimation as an inverse problem. If an inverse problem can be properly formulated, then bat algorithm can provide better results than least-squares methods and regularization methods.

Lin et al. [60] presented a chaotic Levy flight bat algorithm to estimate parameters in nonlinear dynamic biological systems, which proved the effectiveness of the proposed algorithm.

Komarasamy and Wahi [53] studied K-means clustering using bat algorithm and they concluded that the combination of both K-means and BA can achieve higher efficiency and thus performs better than other algorithms.

Khan and Sahari [51] presented a study of a clustering problem for office workplaces using a fuzzy bat algorithm. Khan and Sahari [51] also presented a comparison study of bat algorithm with PSO, GA, and other algorithms in the context for e-learning, and thus suggested that bat algorithm has clearly some advantages over other algorithms.

Then, they [51] also presented a study of clustering problems using bat algorithm and its extension as a bi-sonar optimization variant with good results.

## 4 Conclusions

After making an exhaustive study of the different optimization algorithms based in swarm intelligence, specifically in animals and insects we can conclude the importance of these algorithms, and we can see as they are capable to improve the performance the many applications combining different techniques, which are usually required in complex optimization problems. For example, the Bee Colony Optimization is a meta-heuristic motivated by foraging behavior of honeybees. It represents general algorithmic framework that can be applied to various optimization problems in management, engineering, and control. ACO takes inspiration from the shortest path searching behavior of various ant species. PSO developed by Eberhart and Kennedy in 1995, inspired by the social behavior of bird flocking or fish schooling also is able for solving many complex optimization problems. Finally, BA, is the more recent method, inspired in the behavior of the micro bats, however is able to achieve very good results as the methods based on populations. In this paper, we only analyzed the most relevant proposed methods in the last years, however, there are other works with other algorithms where the researchers can find other similar metaheuristics. Therefore, in this study we focused on the methods based in swarm intelligence based on animal behavior. Recently, many researchers have been applying swarm intelligence in for solving complex applications and obtain better results than with conventional methods. Also, we can observe in this review that the optimization methods used with fuzzy systems, neural networks, etc., can be used to improve the results in many applications. The swarm intelligence is nowadays used by researchers to achieve the best results in the algorithms and obtain better results than the other techniques. The networks and cluster with the important topics used to make the calculus, give us a general idea of the researchers applying these methods in the world.

**Acknowledgements** We would like to express our gratitude to CONACYT, Tecnologico Nacional de Mexico/Tijuana Institute of Technology for the facilities and resources granted for the development of this research.

## References

1. M. Abdel-Basset, L.A. Shawky, Flower pollination algorithm: a comprehensive review. Artif. Intell. Rev. **52**(4), 2533–2557 (2019). https://dx.doi.org/10.1007/s10462-018-9624-4
2. F. Ahmadizar, H. Soltanpanah, Reliability optimization of a series system with multiple-choice and budget constraints using an efficient ant colony approach. Expert Systems with Applications **38**(4), 3640–3646 (2011). https://doi.org/10.1016/j.eswa.2010.09.018
3. J. Alcalá-Fdez, R. Alcalá, M.J. Gacto, F. Herrera, Learning the membership function contexts for mining fuzzy association rules by using genetic algorithms. Fuzzy Sets Syst. **160**(7), 905–921 (2009). https://doi.org/10.1016/j.fss.2008.05.012

4. A.M. Alhroob, W.J. Alzyadat, I.H. Almukahel, G.M. Jaradat, Adaptive fuzzy map approach for accruing velocity of big data relies on fireflies algorithm for decentralized decision making. IEEE Access **8**, 21401–21410 (2020)
5. P.J. Angeline, Evolutionary optimization versus particle swarm optimization: philosophy and performance differences, evolutionary programming VII. Lect. Notes Comput. Sci. **1447**, 601–610 (1998)
6. P.J. Angeline, Using Selection to Improve Particle Swarm Optimization, in *Proceedings 1998 IEEE World Congress on Computational Intelligence* (1998), pp. 84–89
7. G.A. Angulo, O. Castillo, A new fuzzy bee colony optimization with dynamic adaptation of parameters using interval type-2 fuzzy logic for tuning fuzzy controllers. Soft Comput. **22** (2016). https://doi.org/10.1007/s00500-016-2354-0
8. R. Argha, D. Diptam, C. Kaustav, Training artificial neural network using particle swarm optimization. Int. J. Adv. Res. Comput. Sci. Softw. Eng. Res. **3** (2013)
9. A. Askarzadeh, E. Rashedi, *Harmony Search Algorithm* (2017). https://doi.org/10.4018/978-1-5225-2322-2.ch001
10. N. Bacanin, T. Bezdan, E. Tuba, I. Strumberger, M. Tuba, optimizing convolutional neural network hyperparameters by enhanced swarm intelligence metaheuristics. Algorithms **13**, 67 (2020). https://doi.org/10.3390/a13030067
11. G. Beni, The concept of cellular robotic system, in *Proceedings of the 1988 IEEE International Symposium on Intelligent Control* (IEEE Computer Society Press, 1988), pp. 57–62
12. G. Beni, S. Hackwood, Stationary waves in cyclic swarms, in *Proceedings of the 1992 International Symposium on Intelligent Control* (IEEE Computer Society Press, 1992), pp. 234–242
13. G. Beni, J. Wang, Swarm intelligence, in *Proceedings of the Seventh Annual Meeting of the Robotics Society of Japan* (RSJ Press, 1989), pp. 425–428
14. M.J. Blondin, P.M. Pardalos, A holistic optimization approach for inverted cart-pendulum control tuning. Soft. Comput. **24**(6), 4343–4359 (2020). https://doi.org/10.1007/s00500-019-04198-7
15. E. Bonabeau, M. Dorigo, G. Theraulaz, *Swarm Intelligence* (Oxford University Press, 1997)
16. B. Bullnheimer, G. Kotsis, C. Strauss, Parallelization strategies for the ant system. *Kluwer Series on Applied Optimization* (1997), pp. 87–100
17. O. Castillo, P. Melin, Hybrid intelligent systems for time series prediction using neural networks, fuzzy logic, and fractal theory. IEEE Trans. Neural Netw. **13**(6), 1395–1408 (2002). https://dx.doi.org/10.1109/tnn.2002.804316
18. O. Castillo, H. Neyoy, M. Soriaj, García, F. Valdez, Dynamic fuzzy logic parameter tuning for ACO and its application in the fuzzy logic control of an autonomous mobile robot. Int. J. Adv. Robot. Syst. (2013)
19. O. Castillo, G.A. Angulo, A generalized type-2 fuzzy logic approach for dynamic parameter adaptation in bee colony optimization applied to fuzzy controller design. Inf. Sci. **460–461** (2017). https://doi.org/10.1016/j.ins.2017.10.032
20. O. Castillo, F. Valdez, J. Soria, G.A. Angulo, P. Ochoa, C. Peraza, Comparative study in fuzzy controller optimization using bee colony, differential evolution, and harmony search algorithms. Algorithms **12**, 9 (2018). https://doi.org/10.3390/a12010009
21. Y.H. Chang, C.W. Chang, C.W. Tao, H.W. Lin, Jin-Shiuh Taur, Fuzzy sliding-mode control for ball and beam system with fuzzy ant colony optimization. Expert Syst. Appl. **39**(3), 3624–3633 (2012)
22. G. Chen, Z. Li, Z. Zhang, S. Li, An improved ACO algorithm optimized fuzzy PID controller for load frequency control in multi area interconnected power systems. IEEE Access. **8**, 6429–6447 (2020)
23. H. Chen, X. Xu, L. Zhang, A new model for predicting sulfur solubility in sour gases based on hybrid intelligent algorithm. Fuel **262**, 116550 (2019). https://doi.org/10.1016/j.fuel.2019
24. S.S. Choong, L.P. Wong, C. Lim, A dynamic fuzzy-based dance mechanism for the bee colony optimization algorithm. Comput. Intell. **34**, 999–1024 (2018). https://doi.org/10.1111/coin.12159

25. S.C. Chu, P.W. Tsai, J.S. Pan, *Cat Swarm Optimization* (2006), pp. 854–858. https://doi.org/10.1007/11801603_94
26. A. Colorni, M., Dorigo, V. Maniezzo, Distributed optimization by ant colonies, in *Proceedings of the First European Conference on Artificial Life*, ed. by F. V.P. Bourgine (MIT Press, 1992), pp. 134–142
27. A. Colorni, M. Dorigo, V. Maniezzo, Genetic algorithms: a new approach to the timetable problem, eds.by In: M. Akgül, H.W. Hamacher, S.Tüfekçi Combinatorial Optimization. NATO ASI Series (Series F: Computer and Systems Sciences), vol **82**. (Springer, Berlin, Heidelberg, 1992). https://doi.org/10.1007/978-3-642-77489-8_14
28. M. Dorigo, Learning by probabilistic boolean networks, in *Proceedings of the IEEE International Conference on Neural Networks* (1994), pp. 887–891
29. M. Dorigo, G.D. Caro, Ant colony optimization: a new meta-heuristic. Proc. IEEE Congr. Evol. Comput. **2**, 1477 (1999)
30. M. Dorigo, G.D. Caro, The ant colony optimization meta-heuristic, in *New Ideas in Optimization* (1999), pp. 11–32
31. M. Dorigo, L. Gambardella, A study of some properties of ant-Q, in *Proceedings of the Fourth International Conference on Parallel Problem Solving From Nature* (1996), pp. 656–665
32. M. Dorigo, M. Birattari, T. Stützle, Ant colony optimization. IEEE Comput. Intell. Mag. **1**, 28–39 (2006). https://doi.org/10.1109/mci.2006.329691
33. M. Dorigo, E. Bonabeau, G. Theraulaz, Ant algorithms and stigmergy. Future Gener. Comput. Syst. **16**(8), 851–871 (2000). https://doi.org/10.1016/s0167-739x(00)00042-x
34. M. Dorigo, L.M. Gambardella, Ant colonies for the travelling salesman problem. Biosystems **43**(2), 73–81 (1997). https://dx.doi.org/10.1016/s0303-2647(97)01708-5
35. R.C. Eberhart, Kennedy, A new optimizer using particle swarm theory, in *Proceedings of Sixth International Symposium on Micro Machine and Human Science* (1995), pp. 33–43
36. E.M. El-Gendy, M.M. Saafan, M.S. Elksas, S.F. Saraya, F.F.G. Areed, Applying hybrid genetic–PSO technique for tuning an adaptive PID controller used in a chemical process. Soft. Comput. **24**(5), 3455–3474 (2020). https://doi.org/10.1007/s00500-019-04106-z
37. H. Fahim, W. Li, S. Javaid, M.M.S. Fareed, G. Ahmed, M.K. Khattak, Fuzzy logic and bio-inspired firefly algorithm based routing scheme in intrabody nanonetworks. Sensors **19**(24), 5526–5526 (2019). https://dx.doi.org/10.3390/s19245526
38. I. Fister, X.S. Yang, J. Brest, I. Fister, Modified firefly algorithm using quaternion representation. Expert Syst. Appl. **40**(18), 7220–7230 (2013). https://doi.org/10.1016/j.eswa.2013.06.070
39. C. Gallo, V. Capozzi, A simulated annealing algorithm for scheduling problems. J. Appl. Math. Phys. **7** (2019). https://doi.org/10.4236/jamp.2019.simann
40. A. Gandomi, A. Alavi, Krill Herd algorithm: a new bio-inspired optimization algorithm. Commun. Nonlinear Sci. Numer. Simul. IF = 2.806 **17**, 4831–4845 (2012). https://doi.org/10.1016/j.cnsns.2012.05.010
41. M.F. Ganji, M.S. Abadeh, A fuzzy classification system based on Ant Colony Optimization for diabetes disease diagnosis. Expert Syst. Appl. **38**(12), 14650–14659 (2011). https://dx.doi.org/10.1016/j.eswa.2011.05.018
42. Z.W. Geem, Novel derivative of harmony search algorithm for discrete design variables. Appl. Math. Comput. **199**(1), 223–230 (2008). https://dx.doi.org/10.1016/j.amc.2007.09.049
43. F. Glover, Tabu search—Part I. ORSA J. Comput. **1**(3), 190–206 (1989)
44. N. Goel, D. Gupta, S. Goel, *Performance of Firefly and Bat Algorithm for Unconstrained Optimization Problems* (Department of Computer Science, Maharaja Surajmal Institute of Technology GGSIP University C-4, 2013)
45. R. Greco, I. Vanzi, New few parameters differential evolution algorithm with application to structural identification. J. Traffic Transp. Eng. (Eng. Edn.) **6**(1), 1–14 (2019). https://doi.org/10.1016/j.jtte.2018.09.002
46. J.H. Holland, *Adaptation in Natural and Artificial Systems: An Introductory Analysis with Applications to Biology, Control and Artificial Intelligence* (University of Michigan Press, Ann Arbor, MI, 1975)

47. H.S. Hosseini, The intelligent water drops algorithm: a nature-inspired swarm-based optimization algorithm. Int. J. Bio-Inspired Comput. **1**(1/2), 71–71 (2009). https://dx.doi.org/10.1504/ijbic.2009.022775
48. A.S. Joshi, O. Kulkarni, G.M. Kakandikar, V.M. Nandedkar, Cuckoo search optimization a review. Mater. Today: Proc. **4**(8), 7262–7269 (2017). https://doi.org/10.1016/j.matpr.2017.07.055
49. D. Karaboga, B. Akay, A survey: algorithms simulating bee swarm intelligence. Artif. Intell. Review **31**(1–4), 61–85 (2009). https://doi.org/10.1007/s10462-009-9127-4
50. J. Kennedy, R.C. Eberhart, Particle swarm optimization, in *Proceedings of IEEE International Conference on Neural Networks* (1995), pp. 1942–1948
51. K. Khan, A. Sahai, A comparison of BA, GA, PSO, BP and LM for training feed forward neural networks in e-learning context. Int. J. Intell. Syst. Appl. **4**(7), 23–29 (2012). https://dx.doi.org/10.5815/ijisa.2012.07.03
52. S. Kirkpatrick, C.D. Gelatt, M.P. Vecchi, Optimization by simulated annealing, in *Science*. vol. 220 (American Association for the Advancement of Science (AAAS), 1983), pp. 671–680. https://doi.org/10.1126/science.220.4598.671
53. G. Komarasamy, A. Wahi, An optimized K-means clustering technique using bat algorithm. Eur. J. Sci. Res. **84**(2), 263–273 (2012)
54. K.N. Krishnanand, D. Ghose, Glowworm swarm based optimization algorithm for multimodal functions with collective robotics applications (2006). https://dx.doi.org/10.3233/mgs-2006-2301
55. E. Kuliev, V. Kureichik, Monkey search algorithm for ECE components partitioning. J. Phys.: Conf. Ser. **1015**, 042026 (2018). https://doi.org/10.1088/1742-6596/1015/4/042026
56. C. Li, T. Wu, Adaptive fuzzy approach to function approximation with PSO and RLSE. Expert Syst. Appl. **38**, 13266–13273 (2011)
57. L. Li, W. Pedrycz, T. Qu, Z. Li, Fuzzy associative memories with autoencoding mechanisms. Knowl.-Based Syst. **191**, 105090 (2020). https://doi.org/10.1016/j.knosys.2019.105090
58. Y.L. Li, Rong, The reliable design of one-piece flow production system using fuzzy ant colony optimization. Comput. Oper. Res. **36**(5), 1656–1663 (2009)
59. N.F.D. Lima, T.B. Ludermir, Frankenstein PSO applied to neural network weights and architectures, in *Evolutionary Computation (CEC)* (2011), pp. 2452–2456
60. J.H. Lin, C.W. Chou, C.H. Yang, H.L. Tsai, A chaotic Levy flightbat algorithm for parameter estimation in nonlinear dynamic biological systems. J. Comput. Inf. Technol. **2**(2), 56–63 (2012)
61. J. Luan, Z. Yao, F. Zhao, X. Song, A novel method to solve supplier selection problem: Hybrid algorithm of genetic algorithm and ant colony optimization. Mathem. Comput. Simul. **156**, 294–309 (2019). https://doi.org/10.1016/j.matcom.2018.08.011
62. P. Lučić, D. Teodorović, Bee system: modeling combinatorial optimization transportation engineering problems by swarm intelligence, in *Preprints of the RISTAN IV Triennial Symposium on Transportation Analysis* (Sao Miguel, 2001), pp. 441–445
63. P. Lučić, D. Teodorović, Transportation modeling: an artificial life approach, in *Proceedings of the 14th IEEE International Conference on Tools with Artificial Intelligence* (2002), pp. 216–223
64. P. Lučić, D. Teodorović, Vehicle routing problem with uncertain demand at nodes: the bee system and fuzzy logic approach, in *Fuzzy Sets in Optimization*, ed. by L.J. Verdegay (Springer, 2003), pp. 67–82
65. P. Lučić, D. Teodorović, Computing with bees: attacking complex transportation engineering problems. Int. J. Artif. Intell. Tools **12**(03), 375–394 (2003), https://dx.doi.org/10.1142/s0218213003001289
66. P. Manikannan, K. Udhayakumar, P. Pugazhendiran (2020). https://doi.org/10.17559/TV-20171029140308
67. G. Marković, D. Teodorović, V.A. Raspopović, Routing and wavelength assignment in all-optical networks based on the bee colony optimization. AI Commun. **20**, 273–285 (2007)

68. P. Melin, F. Olivas, O. Castillo, F. Valdez, J. Soria, José Mario García Valdez: Optimal design of fuzzy classification systems using PSO with dynamic parameter adaptation through fuzzy logic. Expert Syst. Appl. **40**(8), 3196–3206 (2013)
69. S. Mirjalili, The ant lion optimizer. Adv. Eng. Softw. **83**, 80–98 (2015). https://doi.org/10.1016/j.advengsoft.2015.01.010
70. P. Musikapun, P. Pongcharoen, Solving multi-stage multi-machine multi-product scheduling problem using bat algorithm, in *2nd International Conference on Management and Artificial Intelligence (IPEDR)*, vol. 35 (IACSIT Press, 2012), pp. 98–102
71. J. Ocenasek, J. Schwarz, Estimation of distribution algorithm for mixed continuous-discrete optimization problems, in *2nd Euro-International Symposium on Computational Intelligence* (2002), pp. 227–232
72. A. Perianes-Rodriguez, L. Waltman, N.J. van Eck, Constructing bibliometric networks: a comparison between full and fractional counting. J. Inf. **10**(4), 1178–1195 (2016). https://dx.doi.org/10.1016/j.joi.2016.10.006
73. N. Priyadarshi, S. Padmanaban, J.B. Holm-Nielsen, F. Blaabjerg, M.S. Bhaskar, An experimental estimation of hybrid ANFIS–PSO-based MPPT for PV grid integration under fluctuating sun irradiance. IEEE Syst. J. **14**(1), 1218–1229 (2020)
74. K.S. Rajesh, S.S. Dash, R. Rajagopal, Hybrid improved firefly-pattern search optimized fuzzy aided PID controller for automatic generation control of power systems with multi-type generations. Swarm Evol. Comput. **44**, 200–211 (2019). https://doi.org/10.1016/j.swevo.2018.03.005
75. E. Rashedi, H. Nezamabadi-pour, S. Saryazdi, GSA: a gravitational search algorithm. Inf. Sci. **179**(13), 2232–2248 (2009). https://doi.org/10.1016/j.ins.2009.03.004
76. A. Ratre, Taylor series-based compressive approach and firefly support vector neural network for tracking and anomaly detection in crowded videos, Avinash Ratre (Corresponding author). J. Eng. Res. **7**(4), 115–137 (2019)
77. R.G. Reynolds, An introduction to cultural algorithms, in *Proceedings of the 3rd Annual Conference on Evolutionary Programming* (1994), pp. 131–139
78. D. Rodrigues, L. Pereira, R. Nakamura, K. Costa, X. Yang, A. Souza, João Paulo Papa, A wrapper approach for feature selection based on Bat Algorithm and Optimum-Path Forest. Bauru, Brazil (2013)
79. Y. Shang, H. Nguyen, X. Bui, A novel artificial intelligence approach to predict blast-induced ground vibration in open-pit mines based on the firefly algorithm and artificial neural network. Nat. Resour. Res. **29**, 723–737 (2020)
80. S.W. Sharshir, M.E. Abd el aziz, M. Shafik, Enhancing thermal performance and modeling prediction of developed pyramid solar still utilizing a modified random vector functional link. Solar Energy **198**, 399–409 (2020). https://doi.org/10.1016/j.solener.2020.01.061
81. T. Stützle, *MAX-MIN Ant System for the Quadratic Assignment Problem* (FG Intellektik, FB Informatik, TU Darmstadt, Germany, 1997)
82. T. Stützle, F. Intellektik, T.U. Informatik, Darmstadt: an ant approach to the flow shop problem. Germany (1997)
83. I.B.M. Taha, A. Hoballah, S.S.M. Ghoneim, Optimal ratio limits of rogers' four-ratios and IEC 60599 code methods using particle swarm optimization fuzzy-logic approach. IEEE Trans. Dielectr. Electr. Insul. **27**(1), 222–230 (2020)
84. M.H.A. Talib, I.Z.M. Darus, P.M. Samin, Fuzzy logic with a novel advanced firefly algorithm and sensitivity analysis for semi-active suspension system using magneto-rheological damper. J. Ambient Intell. Hum. Comput. **10**(8), 3263–3278 (2019). https://dx.doi.org/10.1007/s12652-018-1044-4
85. W.J. Tang, Q.H. Wu, J.R. Saunders, Bacterial foraging algorithm for dynamic environments, in *IEEE Congress on Evolutionary Computation Sheraton Vancouver Wall Centre Hotel* (2006)
86. D. Teodorovic, M. Dell'orco, Bee colony optimization—A cooperative learning approach to complex transportation problems. *Proceedings of the 10th Mini-EURO Conference on Advanced OR and AI Methods in Transportation*, (Poznan, 2005), 13–16 September. pp. 51-60.

87. D. Teodorović, Transport modeling by multi-agent systems: a swarm intelligence approach. Transp. Plan. Tech **26**, 289–312 (2003)
88. D. Teodorović, M. Dell'Orco, Bee colony optimization—A cooperative learning approach to complex transportation problems, in *Advanced OR and AI Methods in Transportation. Proceedings of the 10th Meeting of the EURO Working Group on Transportation* (2005), pp. 51–60
89. D. Teodorović, P. Lučić, G. Marković, M.D. Orco, Bee colony optimization: principles and applications, in *Proceedings of the Eight Seminar on Neural Network Applications in Electrical Engineering – NEUREL*, ed, by B. Reljin, S. Stanković (2006), pp. 151–156
90. D. Teodorović, M. Šelmić, The BCO algorithm for the p median problem, in *Proceedings of the XXXIV Serbian Operations Research Conference* (2007)
91. D. Teodorović, Bee colony optimization (BCO), in *Innovations in Swarm Intelligence*, ed. by In: C.P. Lim, L.C. Jain, S. Dehuri (Springer, Berlin, Heidelberg, 2009), pp. 39–60. https://doi.org/10.1007/978-3-642-04225-6_3
92. F. Valdez, A review of optimization swarm intelligence-inspired algorithms with type-2 fuzzy logic parameter adaptation. Soft. Comput. **24**(1), 215–226 (2020). https://doi.org/10.1007/s00500-019-04290-y
93. F. Valdez, P. Melin, O. Castillo, Evolutionary method combining particle swarm optimisation and genetic algorithms using fuzzy logic for parameter adaptation and aggregation: the case neural network optimisation for face recognition (2010). https://dx.doi.org/10.1504/ijaisc.2010.032514
94. F. Valdez, P. Melin, O. Castillo, O. Montiel, A new evolutionary method with a hybrid approach combining particle swarm optimization and genetic algorithms using fuzzy logic for decision making (2008)
95. F. Valdez, J. Vázquez, F. Gaxiola, Fuzzy dynamic parameter adaptation in ACO and PSO for designing fuzzy controllers: the cases of water level and temperature control. Adv. Fuzzy Syst. **2018**, 1–19 (2018). https://doi.org/10.1155/2018/1274969
96. T. Qi Wu, M. Yao, J. Hua Yang, Dolphin swarm algorithm. Front. Inf. Technol. Electron. Eng. **17**, 717–729 (2016). https://doi.org/10.1631/fitee.1500287
97. X. Yang, *A New Metaheuristic Bat-Inspired Algorithm* (Trumpington Street, Cambridge CB2 1PZ, UK, 2010)
98. X. Yang, *Bat Algorithm: Literature Review and Applications* (School of Science and Technology. Middlesex University, The Burroughs, London NW4 4BT, United Kingdom, 2013)
99. X. Yang, S. Deb, Cuckoo search via Lévy flights, in *2009 World Congress on Nature Biologically Inspired Computing (NaBIC)* (2009), pp. 210–214
100. X.S. Yang, A new metaheuristic bat-inspired algorithm, in *Nature Inspired Cooperative Strategies for Optimization. Studies in Computational Intelligence*, vol. 284 (Springer, 2010), pp. 65–74
101. X.S. Yang, *Firefly Algorithms for Multimodal Optimization*, vol. 5792 (2010). https://doi.org/10.1007/978-3-642-04944-6_14
102. X.S. Yang, *Nature-Inspired Metaheuristic Algorithms* (2010)
103. X.S. Yang, *Nature-Inspired Optimization Algorithms* (Elsevier, Oxford, 2014). https://doi.org/10.1016/b978-0-12-416743-8.00016-6
104. X.S. Yang, S. Deb, Eagle strategy using Lévy Walk and Firefly algorithms for stochastic optimization, in *Studies in Computational Intelligence*, vol. 284 (2010), pp. 101–111. https://doi.org/10.1007/978-3-642-12538-6_9
105. Z. Zhang, T. Wang, Y. Chen, J. Lan, Design of type-2 fuzzy logic systems based on improved ant colony optimization. Int. J. Control Autom. Syst. **17** (2019). https://doi.org/10.1007/s12555-017-0451-1

# Optimization of Fuzzy Systems Through Metaheuristics in Control Systems

**Víctor Ortiz, Oscar Castillo, and Prometeo Cortés-Antonio**

**Abstract** Although there are modern computational methods that we can use to solve minimum or maximum problems in different areas, it is good to use classical or semi-there are various types of membership functions. This depends on the problem to be treated. The most used are triangular, trapezoidal, Gaussian among others. For this particular case within this writing we will use the triangular and Gaussian ones. As it is already known through optimization we can come up with the best solution for a particular problem. Since this does not work for all cases in general. For this reason, we will be using in this work Differential Evolution Algorithm (DEA) to optimize a fuzzy system, for the application of this in control as a inverted pendulum system. Similarly, the proposed methodology for optimizing the system is analyzed and explained.

**Keywords** Fuzzy system · Optimization · Fuzzy controllers · Bio-inspired algorithm · Differential evolution algorithm

## 1 Introduction

Fuzzy inference systems are popular today because they are based on the concept of fuzzy sets. The main idea is emulating human reasoning to have application in many fields of everyday life such as control, classification, data, expert systems, predictions, even. Precisely acting naturally. the multidisciplinary result is what determines its potential.

V. Ortiz · O. Castillo (✉) · P. Cortés-Antonio
Tijuana Institute of Technology, Tijuana, BC, Mexico
e-mail: ocastillo@tectijuana.mx

V. Ortiz
e-mail: victor.ortiz@tectijuana.edu.mx

P. Cortés-Antonio
e-mail: prometeo.cortes@tectijuana.edu.mx



P. Melin et al. (eds.), *Recent Advances of Hybrid Intelligent Systems Based on Soft Computing*, Studies in Computational Intelligence 915,
https://doi.org/10.1007/978-3-030-58728-4_17

Fuzzy systems have a fundamental structure, the first of which is the rules, the second is the dictionary of data that determines the membership function and practically the mechanism that ends fuzzy logical reasoning, which is the one that makes sense at the exit, which also can be called conclusion. [1, 2].

There are various types of membership functions. This depends on the problem to be treated. The most used are triangular, trapezoidal, gaussian, among others. For this particular case within this writing we will use the triangular and Gaussian ones [3].

On the other hand, there are bioinspired and optimization systems that emulate human behavior, but not in terms of rationing, but a way of life, its way of survival, the way of hunting and getting food, even the way to mutate and improve the breed. Doing those tasks in the best possible way with minimal effort is what we call optimization. Which can be to find a minimum a maximum it is enough with representing the alive beings informs of vectors or chromosomes to give it the sense of logical programming. [4].

There are currently many bio-inspired algorithms to mention a few examples: ant colony [5], shark nose [6], bees [7], swarm of particles [8], among others. In this writing we will particularly use the differential evolution [9].

In this investigation, a different way of optimizing is proposed, using the differential algorithm in its classic form, but instead of using gaussian functions, triangular functions will be used with their respective restrictions on their parameters. Checking if the Gaussian functions are the best to optimize this type of problem. Particularly in the inverted pendulum.

This investigation has the following sequence in Sect. 2 we will do a small but functional literature review, in Sect. 2 we will take a look at the proposed method, in Sect. 3 the details of the methodology used are presented and described, in Sect. 4 fuzzy logic controllers, in Sect. 5 the morning of the summary of the results is presented, in Sect. 6 the conclusions are reached which could be reached with the results obtained through the optimized inference system [10, 11].

## 2 Literature Review

You will find in this section the terminology and fundamental concepts so that we can understand and have a general and clearer idea about the method proposed in this research.

### *2.1 Differential Evolution Algorithm and Its Origins*

Rainer Storn and Kenneth Price proposed this method, which meaning comes from the Differential Evolution (DE) in nature [7] in 1995. But consolidated until 1997 in the Journal of Global optimization article. And the way of differentiating itself from other methods is its separable operations. It also deals with global optimization problems, but on continuous spaces. For better management it can be represented in the form of a vector, also known in other systems as chromosome or individual. In this case, it is determined to use an objective minimization function, that is, it is treated as an unrestricted optimization method. Defining it in another way to refer to its name, the main strategy of this algorithm is said to be the difference of two vectors of two parameters randomly selected to form the third vector of parameters, which is also called the solution vector.

But let us now see more clearly the formulation of the problem of this method. It has three fundamental characteristics. Optimize non-linear, non-differentiable and multi-objective functions [12].

- Parallelizable.
- Easy to use (that the parameters are easy to select or choose, as well as having a low computational cost)

Hay cuatro partes fundamentals que componen este algoritmo

- Initial population
- Mutation
- Crossing
- Selection.

*Subject to*:

$$F.O\ f(x) \tag{1}$$

$$G(x) \leq B \tag{2}$$

$$H(x) <= C \tag{3}$$

$$LB \leq x \leq UB \tag{4}$$

The next one equation represent the crossing state and we can say is the main equation of the algorithm.

$$v_{i,G+1}\left(X_{r1,G}\right) = X\_(r1, G) + F\left(X_{r2,G} - X_{r3,G}\right) \tag{5}$$

$x$ is a vector of n dimensions. $G$ $H$ are functions of x.
$B$ and $C$ are requirements and limitations. $LB$ y UB are the variables limitations.
$F$ is a scaling factor.

See below the application of the differential evolution process (see Fig. 1).

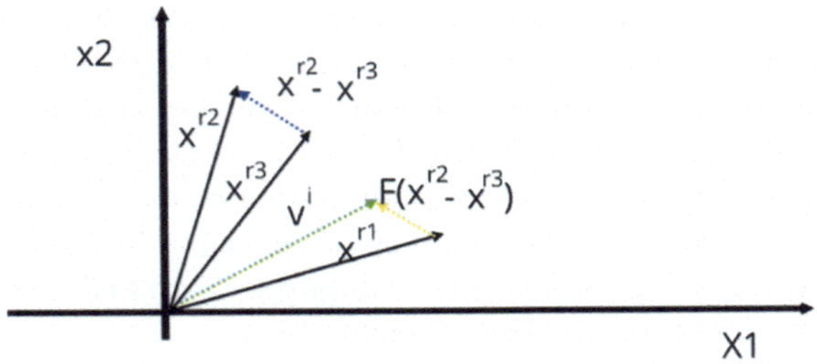


**Fig. 1** Example of a vector applying differential evolution algorithm

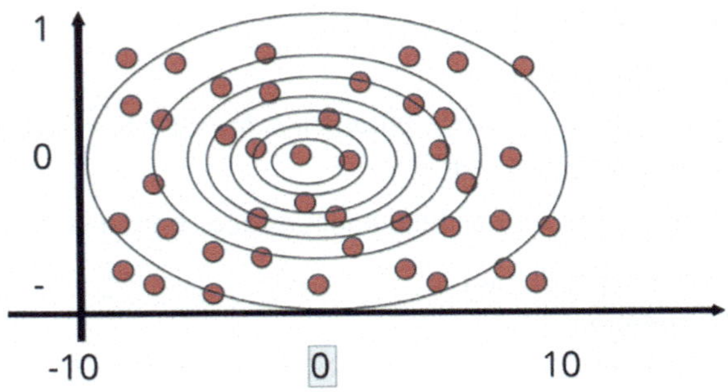


**Fig. 2** Initially uniformly distributed population

## 2.2 *Initial Population*

You must have a uniformly distributed random population, within both the upper and lower limits (see Fig. 2).

## 2.3 *Strategy*

As a population-based metaheuristic, it stands out for its easy use. But it was not only easy from the beginning, it took a lot of work to adapt, like most of this guy. One of the main goals he had to solve is diversity in populations or solutions. Another important feature is self-adaptation, which adds a touch of intelligence to this technique while you are searching [13].

**DEA Pseudo code**
A continuación se muestra el algoritmo de evolución diferencial en forma de Pseudocode (see Algorithm 1).

**Algorithm 1** Pseudo code Differential Evolutional Algorithm

Bigin
$G = 0$
*Create a random starting population* $\vec{x}_{i,G}$ $\forall_i$, $i = 1, \ldots NP$

*Evaluate* $f(\vec{x}_{i,G})\ \forall_i,\ i = 1, \ldots NP$
  *for* $G = 1$ *up to maximum generation do*
    for $i =$ up to NP do
      *choose randomly* $r_1 \neq r_2 \neq r_3$
      $J_{rand} = rand\ int\ (1, D)$
      *for* $j = 1$ *up to D do*
       if $rand_j[0, 1) < CRoJ = j_{rand} then$
      $u_{i,j,G+1} = x_{r3,j,G} + F(x_{r1,j,G} - x_{r2,j,G})$
      else
       $u_{i,j,G+1} = x_{i,j,G}$
      *end if*
   *end for*
   if $f(\vec{u}_i, G + 1) <= f(\vec{x}_i, G))$
      $\vec{x}_i, G + 1 = \vec{u}_i, G + 1$
    *end if*
  *end for*
  $G = G+1$
*end for*
End

## 2.4 Fuzzy Logic

One of the most important contributions made by Zadeh in 1965, from Berkeley California, work that was based on the theory of fuzzy sets. And that allows the representation of people's reasoning and How they handle uncertainty in their decisions Through fuzzy rules and linguistic terms.

This is practically the backbone of fuzzy sets such as: "very young", "young", "old", "very old", "very high" [14–16].

In this way, it challenges the conventional logic that suggests that there are no intermediate values between u yes or no, between a high or a low, between a young person or an old man, or even worse that there are no values between zero and one. However, it is not the true reasoning of people who have a classification between what is cold and what is hot, they think in many decimals between zero and one.

In other words, one person can be 100% young or therefore has equivalence 1 of the young group and another person can only be 70%, which gives a degree of membership of 0.7 in the group of young people. But the second person also already has 1° of membership in another set equal to 30% or 0.3.

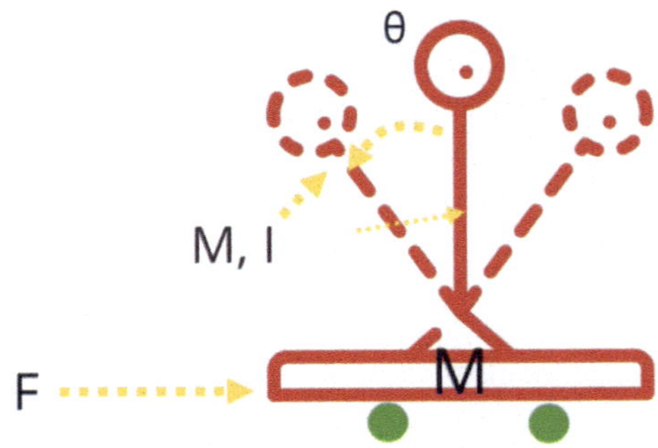


**Fig. 3** Inverted pendulum

## 2.5 Inverted Pendulum

This follows the same principle of the traditional pendulum but in-stead of the bar being suspended it is located at the top of the mobile car, whose main challenge is that by applying a force not necessarily similar, the car is capable of keeping the bar in balance.

Of course, different variables intervene [such as the limits of the car will move, the speed of the car in a linear way, the angular speed of the bar. Which we will describe in the form of equations later [17, 18].

A general diagram of the inverted pendulum is shown below (see Fig. 3).

## 2.6 Inverted Pendulum Dynamics

In particular, the work will focus on obtaining the best values for the fuzzy system. In Fig. 3 the used model is presented [19].

This method works for complex non-linear methods. Main characteristics of the pendulum lever remain balanced despite the fact that we apply force which will be dynamic on the car that supports it, causing the car to move on the X axis without letting the balance for this particular case we will have as a limit from −1.5 to 1.5. Below we will see a Table 1 where the used values are shown.

**Table 1** Used parameters

| Letter | Parameter | Value |
|---|---|---|
| M | Mass of the cart | 0.5 kg |
| m | Mass of the pendulum | 0.5 kg |
| b | Friction of the cart | 0.1 N/m/s |
| l | Length to pendulum center of mass | 0.3 m |
| I | Inertia of the pendulum | 0.006 kg * $m^2$ |

It should be mentioned that it is a difficult system to represent due to the physical characteristics it has. The following describes the differential equations for both the car and the lever and their degrees of freedom respectively. Applying the principles of Newton's second law, we have:

$$F = ma \tag{6}$$

$$M\ddot{x} + b\dot{x} + N = F \tag{7}$$

$$\ddot{x} = \frac{1}{M}\sum_{cart} fx = \frac{1}{M}(F - N - b\dot{x}) \tag{8}$$

$$\theta = \frac{1}{I}\sum_{pend} T = \frac{1}{I}(-Nl\cos\theta - Pl\sin\theta) \tag{9}$$

$$\theta = \frac{1}{I}\sum_{pend} T = \frac{1}{I}(-Nl\cos\theta - P\ell\sin\theta) \tag{10}$$

The forces between the car and the pendulum will also be included. That is required to model the X and Y components to the translation of the center of mass and its rotation.

$$m\ddot{x}_p = \frac{1}{I}\sum_{pend} F_x = N \tag{11}$$

$$\Rightarrow N = m\ddot{x}_p \tag{12}$$

$$m\ddot{y}_p = \sum_{pend} F_y = P - mg \tag{13}$$

$$\Rightarrow P = m\left(\ddot{y}_p + g\right) \tag{14}$$

To represent the derivatives of $\ddot{x}_p$ y $\ddot{y}_p$ as follows

$$x_p = x + l\sin\theta \tag{15}$$

$$\dot{x}_p = \dot{x} + l\theta\sin\theta \tag{16}$$

$$\ddot{x}_p = \dot{x} + l\dot{\theta}^2\sin\theta + l\ddot{\theta}\cos\theta \tag{17}$$

On the other hand, the equations of component Y:

**Table 2** Plant states

| Parameter |
|---|
| Pendulum angle<br>Angular velocity |
| Cart position coordinate |
| Pendulum angle from vertical |

$$y_p = -l\cos\theta \tag{18}$$

$$\dot{y}_p = l\dot{\theta}\sin\theta \tag{19}$$

$$\ddot{y}_p = l\dot{\theta}^2\cos\theta + l\ddot{\theta}\,\sin\theta \tag{20}$$

They can be substituted in N and P (Table 2).

$$N = \dot{x} + l\dot{\theta}^2\sin\theta + l\ddot{\theta}\,\cos\theta \tag{21}$$

$$P = l\dot{\theta}^2\cos\theta + l\ddot{\theta}\,\sin\theta \tag{22}$$

Figure 4 as shown as below, represents the method used (see Fig. 4).

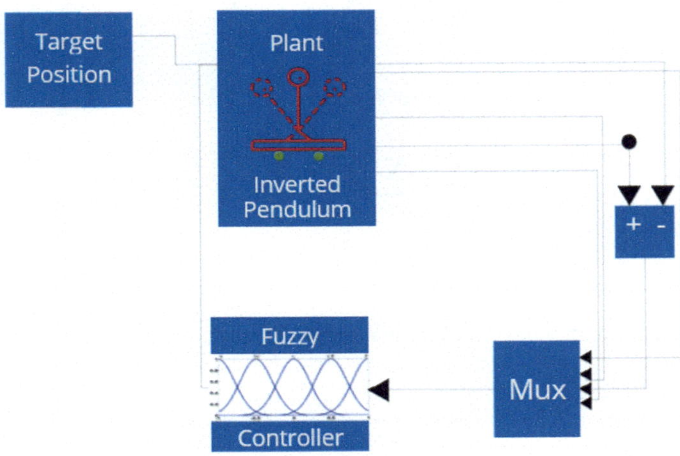


**Fig. 4** Inverted pendulum simulink diagram

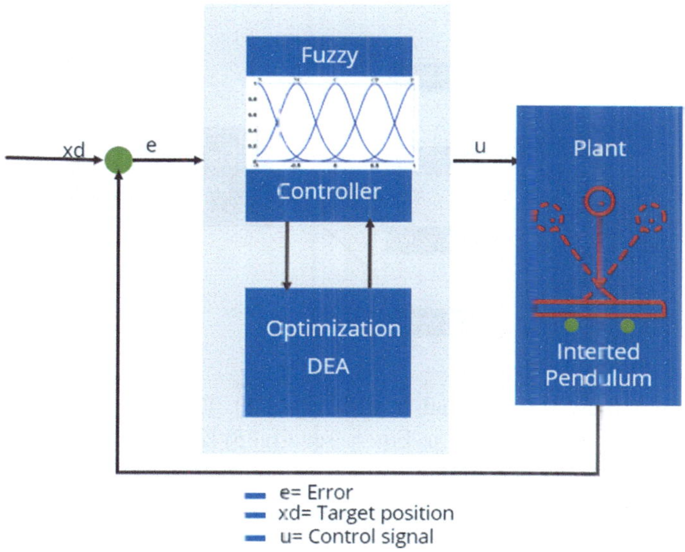


**Fig. 5** System control blocks diagram

## 3 Methodology

Beginning from the fuzzy system created with convenient membership functions and with random parameters we will optimize these parameters using the differential evolution algorithm to later send them to be tested in the inverted pendulum plant, returning the root mean square error and we will see how close we are to the objective function. This will become an iterative cycle until the lowest possible error is obtained or that makes it the stop criterion [20].

On the other hand, there are the parameters of the application method for which we will use the values recommended in the existing literature, since for now we will not focus on finishing the best values for those parameters, as shown in Fig. 3 (see Fig. 5).

## 4 Fuzzy Logic Controller

These rules of the fuzzy system describe a relation between the linguistic variables based on the membership functions, taking into account the inputs and outputs [21–23].

A fuzzy Mamdani type system is designed containing two linear error (ev) and angular error (ew) inputs, which comprise a range of [−1 1], which contain three membership functions, two trapezoidal at the ends and one triangular in the middle.

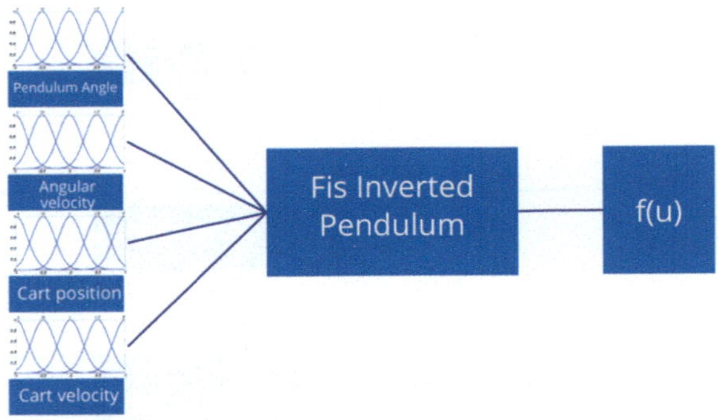


**Fig. 6** Inverted pendulum fuzzy system

So it has two outputs Torque 1 and Torque 2, [−1 1]. They contain three triangular membership functions, as shown in Fig. 6 (see Fig. 5) [24].

This fuzzy controller has 16 rules, which are shown a below:

1. If (pendulum_Angle is low) and (Angular_Velocity is Low) and (Cart_position is Low) and (Cart_velocity is Low) then (out is outmf1) (1)
2. If (pendulum_Angle is low) and (Angular_Velocity is Low) and (Cart_position is Low) and (Cart_velocity is Hight) then (out is outmf2) (1)
3. If (pendulum_Angle is low) and (Angular_Velocity is Low) and (Cart_position is Hight) and (Cart_velocity is Low) then (out is outmf3) (1)
4. If (pendulum_Angle is low) and (Angular_Velocity is Low) and (Cart_position is Hight) and (Cart_velocity is Hight) then (out is outmf4) (1)
5. If (pendulum_Angle is low) and (Angular_Velocity is Hight) and (Cart_position is Low) and (Cart_velocity is Low) then (out is outmf5) (1)
6. If (pendulum_Angle is low) and (Angular_Velocity is Hight) and (Cart_position is Low) and (Cart_velocity is Hight) then (out is outmf6) (1)
7. If (pendulum_Angle is low) and (Angular_Velocity is Hight) and (Cart_position is Hight) and (Cart_velocity is Low) then (out is outmf7) (1)
8. If (pendulum_Angle is low) and (Angular_Velocity is Hight) and (Cart_position is Hight) and (Cart_velocity is Hight) then (out is outmf8) (1)
9. If (pendulum_Angle is hight) and (Angular_Velocity is Low) and (Cart_position is Low) and (Cart_velocity is Low) then (out is outmf9) (1)
10. If (pendulum_Angle is hight) and (Angular_Velocity is Low) and (Cart_position is Low) and (Cart_velocity is Hight) then (out is outmf10) (1)
11. If (pendulum_Angle is hight) and (Angular_Velocity is Low) and (Cart_position is Hight) and (Cart_velocity is Low) then (out is outmf11) (1)
12. If (pendulum_Angle is hight) and (Angular_Velocity is Low) and (Cart_position is Hight) and (Cart_velocity is Hight) then (out is outmf12) (1)
13. If (pendulum_Angle is hight) and (Angular_Velocity is Hight) and (Cart_position is Low) and (Cart_velocity is Low) then (out is outmf13) (1)
14. If (pendulum_Angle is hight) and (Angular_Velocity is Hight) and (Cart_position is Low) and (Cart_velocity is Hight) then (out is outmf14) (1)

15. If (pendulum_Angle is hight) and (Angular_Velocity is Hight) and (Cart_position is Hight) and (Cart_velocity is Low) then (out is outmf15) (1)
16. If (pendulum_Angle is hight) and (Angular_Velocity is Hight) and (Cart_position is Hight) and (Cart_velocity is Hight) then (out is outmf16) (1).

## 5 Results and Discussion

A large number of experiments were carried out using the differential evolution method with the parameters detailed below, including and including the restrictions that are handled in the membership functions (see Fig. 7) (Table 3).

The mean square error was used as the aptitude function, as see Eq. (8), with which it is intended to minimize the error of the desired trajectory [25] (Tables 4, 5 and 6).

$$MSE = \frac{1}{2}\sum_{i=1}^{n}\left(\hat{Y}_i - Y_i\right)^2 \tag{22}$$

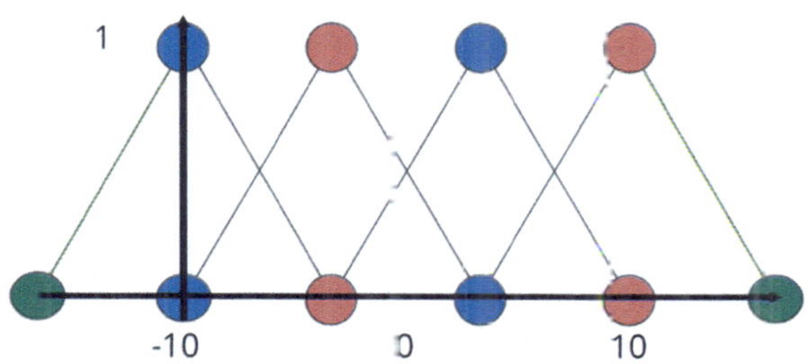


**Fig. 7** Restrictions used in membership functions

**Table 3** Table shows the algorithm parameters

| Parameter | Value |
|---|---|
| Maximun number of iterations | 1000 |
| Population size | 100 |
| Lower bound of scaling factor | 0.2 |
| Upper bound of scaling factor | 0.8 |
| Crossover probability | 0.2 |

**Table 4** Original membership function details

| FIS variable | MF | Parameters |
|---|---|---|
| In1 | in1mf1 | [0.3 2 −0.3 0] |
| | in1mf2 | [0.3 2 0.3 0] |
| In2 | in2mf1 | [1 2 −1 0] |
| | in2mf2 | [1 2 1 0] |
| In3 | in3mf1 | [3 2 −3 0] |
| | in3mf2 | [3 2 3 0] |
| In4 | in4mf1 | [3 2 −3 0] |
| | in4mf2 | [3 2 3 0] |

**Table 5** Original method outputs plots

| Output | Parameters |
|---|---|
| outmf1 | [41.37 10.03 3.162 4.288 0.3386] |
| outmf2 | [40.41 10.05 3.162 4.288 0.2068] |
| outmf3 | [41.37 10.03 3.162 4.288 0.3386] |
| outmf4 | [40.41 10.05 3.162 4.288 0.2068] |
| Outmf5 | [38.56 10.18 3.162 4.288 −0.04893] |
| Outmf6 | [37.6 10.15 3.162 4.288 −0.1807] |
| Outmf7 | [38.56 10.18 3.162 4.288 −0.04893] |
| Outmf8 | [37.6 10.15 3.162 4.288 −0.1807] |
| Outmf9 | [37.6 10.15 3.162 4.288 0.1807] |
| Outmf10 | [38.56 10.18 3.162 4.288 0.04891] |
| Outmf11 | [37.6 10.15 3.162 4.288 0.1807] |
| Outmf12 | [38.56 10.18 3.162 4.288 0.04892] |
| Outmf13 | [40.41 10.05 3.162 4.288 −0.2068] |
| Outmf14 | [41.37 10.03 3.162 4.288 −0.3386] |
| Outmf15 | [40.41 10.05 3.162 4.288 −0.2068] |
| Outmf16 | [41.37 10.03 3.162 4.288 −0.3386] |

**Table 6** Differential Evolution optimized membership function details

| FIS variable | MF | Parameters |
|---|---|---|
| Pendulum angle | Low | [−0.2389 −0.1596 −0.08023] |
| | High | [−0.4336 0.1051 0.6438] |
| Angular velocity | Low | [−0.1444 0.03231 0.2089] |
| | High | [0.09108 0.4496 0.8081] |
| Cart position | Low | [−2.833 −0.2695 2.294] |
| | High | [−1.527 0.3161 2.159] |
| Cart velocity | Low | [−1.728 −0.7008 0.3259] |
| | High | [−0.6306 0.734 2.98] |

## 6 Conclusions and Future Work

By way of conclusion, it can be highlighted that without affecting the outputs, only optimizing the type of functions and their parameters we can simulate the same system used in the original fuzzy system, not necessarily being better in results but being able to get closer to the desired values in a shorter time of time. On the other hand, it can be tested in a future optimizer with a newer optimization method or even adding fuzzy logic of type 2 [26–30, 31] or applying the presented ideas in other kinds of problems [32–39].

**Acknowledgements** We are very grateful to CONACYT (National Council of Science Technology). It is also very important to mention the TecNM (National Technological Institute of Mexico) and Tijuana Institute Technology.

## References

1. X.-S. Yang, M. Karamanoglu, Swarm intelligence and bio-inspired computation: an overview, in *Swarm Intelligence and Bio-Inspired Computation* (2013), pp. 3–23
2. P. Gupta, R. Cambini, S.S. Appadoo, Recent advances in optimization theory and applications (2018). https://doi.org/10.1007/s10479-018-2984-yDO
3. D. Pamučar, D, L. Vasin, P. Atanasković, M. Miličić, Planning the city logistics terminal location by applying the green-median model and Type-2 neurofuzzy network. Comput. Intell. Neurosci. (2016)
4. D.S. Pamučar, D. Božanic´, N. Komazec, Risk assessment of natural disasters using fuzzy logic system of type 2. Manag. J. Sustain. Bus. Manag. Solut. Emerg. Econ. **21**, 23–34 2017
5. S. Sremac, I. Tanackov, M. Kopić, D. Radović, ANFIS model for determining the economic order quantity. Decis. Mak. Appl. Manag. Eng. **1**, 1–12 (2018)
6. E.H. Mamdani, Application of fuzzy algorithms for control of simple dynamic plant. Proc. Inst. Electr. Eng. **121**, 1585 (1974)
7. L.A. Zadeh, The concept of a linguistic variable and its application to approximate reasoning—I. Inf. Sci. **8**, 199–249 (1975)
8. R. Kothamasu, S.H. Huang, Adaptive Mamdani fuzzy model for condition-based maintenance. Fuzzy Sets Syst. **158**, 2715–2733 (2007)
9. S. Akkurt, G. Tayfur, S. Can, Fuzzy logic model for the prediction of cement compressive strength. Cem. Concr. Res. **34**, 1429–1433 (2004)
10. O. Castillo, L. Amador-Angulo, J.R. Castro, M. Garcia-Valdez, A comparative study of type-1 fuzzy logic systems, interval type-2 fuzzy logic systems and generalized type-2 fuzzy logic systems in control problems. Inf. Sci. **354**, 257–274 (2016)
11. M.I. El-Hawwary, A.L. Elshafei, H.M. Emara, H.A.A. Fattah, Adaptive fuzzy control of the inverted pendulum problem. IEEE Trans. Control Syst. Technol **14**, 1135–1144 (2006)
12. E. Ontiveros, P. Melin, O. Castillo, High order α-planes integration: a new approach to computational cost reduction of General Type-2 Fuzzy Systems. Eng. Appl. Artif. Intell. **74**, 186–197 (2018)
13. L. Zheng, D. Peng, Y. Sun, S. Gao, Research of model identification for control system based on improved differential evolution algorithm, in *Advanced Computational Methods in Life System Modeling and Simulation*, volume 761, M. Fei, S. Ma, X. Li, X. Sun, L. Jia, Z. Su (Springer, Singapore, 2017), pp. 282–293

14. P. Ochoa, O. Castillo, J. Soria, Differential evolution using fuzzy logic and a comparative study with other metaheuristics, in *Nature-Inspired Design of Hybrid Intelligent Systems* (Springer, Cham, Switzerland, 2017), pp. 257–268
15. P. Ochoa, O. Castillo, O, J. Soria, Type-2 fuzzy logic dynamic parameter adaptation in a new Fuzzy Differential Evolution method, in *Proceedings of the 2016 Annual Conference of the North American Fuzzy Information Processing Society (NAFIPS)*, El Paso, TX, USA, 31 October–4 November 2016, pp. 1–6
16. X. Zhang, X. Zhang, Shift based adaptive differential evolution for PID controller designs using swarm intelligence algorithm. Clust. Comput. **20**, 291–299 (2017)
17. H. Beirami, A.Z. Shabestari, M.M. Zerafat, Optimal PID plus fuzzy controller design for a PEM fuel cell air feed system using the self-adaptive differential evolution algorithm. Int. J. Hydrogen Energy **40**, 9422–9434 (2015)
18. M. Salehpour, A. Jamali, A. Bagheri, N. Nariman-zadeh, A new adaptive differential evolution optimization algorithm based on fuzzy inference system. Eng. Sci. Technol. Int. J. **20**, 587–597 (2017)
19. D. Pamučar, S. Ljubojević, D. Kostadinović, B. Đorović, Cost and risk aggregation in multi-objective route planning for hazardous materials transportation—A neuro-fuzzy and artificial bee colony approach. Expert Syst. Appl. **65**, 1–15 (2016)
20. M. Konar, A. Bagis, Performance comparison of particle swarm optimization, differential evolution and artificial bee colony algorithms for fuzzy modelling of nonlinear systems. Elektron. Elektrotech. **22**, 8–13 (2016)
21. M. Dell'Orco, M. Marinelli, M.G. Altieri, Solving the gate assignment problem through the fuzzy bee colony optimization. Transp. Res. Part C Emerg. Technol. **80**, 424–438 (2017)
22. O. Castillo, L. Amador-Angulo, A generalized type-2 fuzzy logic approach for dynamic parameter adaptation in bee colony optimization applied to fuzzy controller design. Inf. Sci. (2017)
23. J.S.R. Jang, C.T. Sun, E. Mizutani, Neuro-fuzzy and soft computing: a computational approach to learning and machine intelligence, Prentice Hall, Upper Sadler, NJ, USA (1997)
24. L. Amador-Angulo, O. Mendoza, J. Castro, A. Rodríguez-Díaz, P. Melin, O. Castillo, Fuzzy sets in dynamic adaptation of parameters of a bee colony optimization for controlling the trajectory of an autonomous mobile robot. Sensors **16**, 1458 (2016)
25. R. Martínez-Soto, O. Castillo, L.T. Aguilar, Optimization of interval type-2 fuzzy logic controllers for a perturbed autonomous wheeled mobile robot using genetic algorithms. Inf. Sci **179**(13), 2158–2174 (2009)
26. G.M. Mendez, O. Castillo, Interval type-2 TSK fuzzy logic systems using hybrid learning algorithm, in *The 14th IEEE International Conference on Fuzzy Systems*, FUZZ'05 (2005), pp. 230–235.
27. P. Melin, C.I. González, J.R. Castro, O. Mendoza, O. Castillo, Edge-Detection method for image processing based on generalized Type-2 fuzzy logic. IEEE Trans. Fuzzy Syst. **22**(6), 1515–1525 (2014)
28. C.I. González, P. Melin, J.R. Castro, O. Castillo, O. Mendoza, Optimization of interval type-2 fuzzy systems for image edge detection. Appl. Soft Comput. **47**, 631–643 (2016)
29. C.I. González, P. Melin, J.R. Castro, O. Mendoza, O. Castillo, An improved Sobel edge detection method based on generalized type-2 fuzzy logic. Soft. Comput. **20**(2), 773–784 (2016)
30. E. Ontiveros, P. Melin, O. Castillo, High order α-planes integration: a new approach to computational cost reduction of General Type-2 Fuzzy Systems. Eng. Appl. of AI **74**, 186–197 (2018)
31. O. Castillo, *Type-2 Fuzzy Logic in Intelligent Control Applications* (Springer, 2012)
32. P. Melin, O. Castillo, Intelligent control of complex electrochemical systems with a neuro-fuzzy-genetic approach. IEEE Trans. Ind. Electron. **48**(5), 951–955
33. L. Aguilar, P. Melin, O. Castillo, Intelligent control of a stepping motor drive using a hybrid neuro-fuzzy ANFIS approach. Appl. Soft Comput. **3**(3), 209–219 (2003)

34. P. Melin, O. Castillo, Adaptive intelligent control of aircraft systems with a hybrid approach combining neural networks, fuzzy logic and fractal theory. Appl. Soft Comput. **3**(4), 353–362 (2003)
35. P. Melin, J. Amezcua, F. Valdez, O. Castillo, A new neural network model based on the LVQ algorithm for multi-class classification of arrhythmias. Inf. Sci. **279**, 483–497 (2014)
36. P. Melin, O. Castillo, *Modelling, Simulation and Control of Non-linear Dynamical Systems. An Intelligent Approach Using Soft Computing and Fractal Theory* (CRC Press, USA and Canada, 2002)
37. P. Melin, D. Sánchez, O. Castillo, Genetic optimization of modular neural networks with fuzzy response integration for human recognition. Inf. Sci. **197**, 1–19 (2012)
38. M.A. Sanchez, O. Castillo, J.R. Castro, P. Melin, Fuzzy granular gravitational clustering algorithm for multivariate data. Inf. Sci. **279**, 498–511 (2014)
39. D. Sanchez, P. Melin, Optimization of modular granular neural networks using hierarchical genetic algorithms for human recognition using the ear biometric measure. Eng. Appl. Artif. Intell. **27**, 41–56 (2014)

# Optimal Design of Interval Type-2 Fuzzy Tracking Controllers of Mobile Robots Using a Metaheuristic Algorithm

**Felizardo Cuevas, Oscar Castillo, and Prometeo Cortes-Antonio**

**Abstract** This paper describes a modification of a biological inspired algorithm based on shark behavior, the Shark Smell Optimization (SSO), for the optimization of the membership functions parameters for the fuzzy controllers of autonomous mobile robots. SSO is a metaheuristic technique based on the behavior presented by sharks in nature, which can be used for solving optimization problems. First, SSO is used to optimize benchmark control problems. Second, the traditional SSO is tested with the optimization of the membership function's parameters of type-1 fuzzy controllers. Third, tests are also performed with the Interval Type-2 Fuzzy Logic Controller. The comparison of results between the controller optimized with SSO and the controller optimized with WDO demonstrates that the proposed method shows better performance in the optimal design of fuzzy controllers.

**Keywords** Mobile robots · Type-2 fuzzy systems · Type-1 fuzzy logic controller · Type-2 fuzzy logic controller · Shark smell optimization · Wind-driven optimization

## 1 Introduction

The main goal of optimization is to find the best solution for a set of possible solutions to a particular problem. There are cases in which the problem solving space is too large and this may result in a time to find the solution becoming unaffordable. In other cases, there are different computational intelligence areas that provide a set of techniques

F. Cuevas · O. Castillo (✉) · P. Cortes-Antonio
Tijuana Institute of Technology, Tijuana, BC, Mexico
e-mail: ocastillo@tectijuana.mx

F. Cuevas
e-mail: felizardo.cuevas@tectijuana.edu.mx

P. Cortes-Antonio
e-mail: prometeo.cortes@tectijuana.edu.mx


P. Melin et al. (eds.), *Recent Advances of Hybrid Intelligent Systems Based on Soft Computing*, Studies in Computational Intelligence 915,
https://doi.org/10.1007/978-3-030-58728-4_18

to solve search and optimization problems [1, 2]. Such techniques can provide very competitive results, although not necessarily the best alternatives. Algorithms based on populations have become a new paradigm of collective intelligence and can find the best or near to best solutions to optimization problems with reasonable times and costs [3, 4].

Many scientists in the field of artificial intelligence have taken these algorithms as a research topic. Swarm intelligence, can be viewed as a set of metaheuristic techniques of artificial intelligence that are based on the analysis of systems with collective behavior [5–7]. These systems are present in nature, generally in a self-organized manner [8]. Many systems based on search and optimization algorithms have been applied to solve a wide range of problems.

In [9] an algorithm inspired by the movement of stars, galaxies, and superclusters of galaxies under the force of gravity is proposed. In this work, different fuzzy systems were designed for the dynamic adaptation of the $c_3$ and $c_4$ parameters to measure the performance of the algorithm with seven mathematical functions with a different number of dimensions. This method increases the efficiency of the algorithm by providing multiple cycles of exploration and exploitation, thus increasing the chances of accurately finding a global minimum.

In [10] a new method for dynamic parameter adaptation in particle swarm optimization (PSO) is proposed. PSO is a metaheuristic inspired on social behaviors, which is very useful in optimization problems. This paper proposes an improvement to the convergence and diversity of the swarm in PSO using fuzzy logic. Fuzzy rules were used to control the key parameters in PSO to achieve the best possible dynamic adaptation of these parameter values.

In [11] a modification to the bee colony optimization algorithm (BCO), with a fuzzy approach to dynamically change its parameters, was proposed. This is a metaheuristic technique inspired by the behavior presented by bees in nature, which can be used for solving optimization problems. First, the traditional BCO was tested with the optimization of fuzzy controllers. Second, a modification of the original method is presented by including fuzzy logic to dynamically change the main parameter values of the algorithm during execution. Third, the proposed modification of the BCO algorithm with the fuzzy approach is used to optimize benchmark control problems.

In [12, 13] the BCO technique was used to find the optimal distribution of the membership functions in the design of fuzzy controllers. BCO is used specifically for tuning membership functions of the fuzzy controller for trajectory stability in an autonomous mobile robot. Two types of perturbations were added to the model for the Generalized Type-2 Fuzzy Logic System to better analyze its behavior under uncertainty and this shows better results when compared to the original BCO.

In [14] a modification of the bee colony optimization algorithm (BCO), with a fuzzy approach to dynamically change its parameters, was proposed. This is a metaheuristic technique inspired on the behavior presented by bees in nature, which can be used for solving optimization problems. First, the traditional BCO was tested with the optimization of fuzzy controllers. Second, a modification of the original method was presented by including fuzzy logic to dynamically change the main parameter values of the algorithm during execution. Third, the proposed modification to the BCO algorithm with the fuzzy approach was used to optimize benchmark control problems.

In [14–16] the Differential Evolution (DE) algorithm uses fuzzy logic for dynamic parameter adaptation of the mutation parameter (F). This modification of the algorithm is called Fuzzy Differential Evolution algorithm (FDE).

Recent work like in [5] have shown that an IT2FS can obtain better results when compared with type-1 fuzzy systems. Proposed approach is used to achieve the best possible dynamic adaptation of these membership function parameter values. Previously, the same problem was solved by using type-1 fuzzy logic [8, 11], but now in this paper we are now proposing the use of IT2FL. The use of IT2FL allows us to know if the use of a fuzzy interval type 2 system (IT2FS) improves the quality of the results of our approach. The present paper aims at analyzing the performance of T1 FLC and IT2 FLC to find the advantages and disadvantages of this controllers against the others and based on this, we compare the performance of controllers in noise environments. In this case, the main contribution of this work is the implementation of a method to dynamically determine the optimal parameter values of a fuzzy controller using the shark smell optimization algorithm (SSO). In this work, the SSO is applied to different benchmark control problems. There are previous works by researchers who use the same SSO metaheuristic algorithm [17, 18], with the difference that in this document, SSO is used for the dynamic adjustment of the fuzzy controller parameters, where for the first time it is used in the adjustment of controller MFs fuzzy. In this case, the algorithm is only used to adjust the parameters of membership functions of a fuzzy controller, but not the rules.

This paper is organized as follows. The basic concepts and operation of the SSO algorithm are described in Sect. 2, the proposed methodology is explained in Sect. 3, the case studies used in this work are described in Sect. 4, the experimental results with type-1 fuzzy logic, type-2 fuzzy logic and the statistical comparison between the two algorithms SSO (Shark Smell Optimization) and WDO (Wind Driven Optimization) respectively are described in Sect. 5. Finally, in Sect. 6 the conclusions are presented.

## 2 Shark Smell Optimization (SSO)

This section describes the model proposed by Oveis Abedinia, Nima Amjady and Ali Ghasemi in 2004, based on the observation of intelligent behavior of sharks searching for food or prey [19–21]. Shark Smell optimization (SSO) is a meta-heuristic algorithm [19, 20] that belongs to the class of nature-inspired algorithms. This method (SSO) uses a technique similar to the way the sharks look for food in nature and how they use their location and optimization methods to find optimal routes towards the source of food [20]. Natural systems have become currently an important source of ideas and models for the development of many artificial systems [21]. Various biological and natural processes have inspired these types of algorithms.

## 2.1 Behavior and Structure of SSO

Natural systems have become currently an important source of ideas and models for the development of many artificial systems. The mathematical model of the shark search process consists of three main components shown below [19, 22]:

1. Initialization of the food source (Population of solutions).
2. Shark search (Determination of food location).
3. Scout sharks (exploration phase).

## 2.2 Background on SSO

This method was proposed in [19]. The shark smell optimization algorithm (SSO) is based on the shark smelling abilities for localizing the source of food. In sharks' movement, the concentration of the smell is an important factor to guide the shark to the prey. In other words, the shark moves in the way with higher smell concentration. Figure 1 presents the movement of shark to the odor source based on its concentration. This characteristic is used in the SSO algorithm to find the best solution to an optimization problem. In this algorithm, several assumptions are considered, which have been presented in [19]. The following steps briefly explain the algorithm (for a minimization problem).

**Initialization**
A population of initial solutions is randomly generated for an optimization problem in a feasible search area (sea). A source (prey) represents the optimal solution whilst the quality of the solution is represented by the smell strength at a particular position. According to [22] the initial solution is given as follows:

$$X_i^1 = \left[X_1^1, X_2^1, X_3^1, \ldots X_{NP}^1\right] \quad \text{NP} = \text{Population size} \tag{1}$$

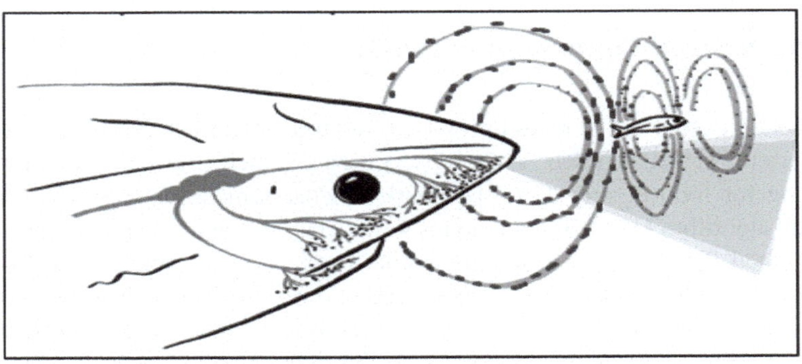

**Fig. 1** Shark's movement to the blood odor source and towards the prey

where there $i$th initial position vector $X_i^k$, i.e. $i$th initial candidate solution for the optimization problem, is as follows [Eq. (2)]:

$$X_{i,j}^1 = \left[X_{i,1}^1, X_{i,2}^1, X_{i,3}^1, \ldots X_{i\,ND}^1\right] \quad \text{ND} = \text{Number of decision variables} \qquad (2)$$

where $X_{i,j}^k$ is the $J$th dimension of the $i$th shark position; $ND$ is the number of decision variables in an optimization problem. The smell Intensity in each position indicates its closeness to the prey.

**Forward Movement**

As the blood disperses into the water, the shark will move towards the target with a velocity '$V$', oriented by the smell of the stronger odor particles, hence leading to a high-quality solution. In correspondence with the position vector, each velocity vector has dimensional component elements:

$$V_i^1 = \left[v_{i,1}^1, v_{i,2}^1, v_{i,3}^1, \ldots v_{i,ND}^1\right] \qquad (3)$$

By increasing the odor concentration, the velocity of the shark will increase. In each stage the magnitude of $V_{ij}^1$, is given as follows:

$$\left|\mathrm{v}_{ij}^k\right| = \min\left[\left|\eta_k.R_1.\nabla(OF)\right|_{x_{ij}^k} + \left|\propto_k .R_2.\mathrm{v}_{ij}^{k-1}\right., \left|\beta_k.\mathrm{v}_{ij}^{k-1}\right|\right] \qquad (4)$$

$$\mathrm{i} = 1, 2, \ldots, \mathrm{NP} \quad \mathrm{j} = 1, 2, \ldots, \mathrm{ND} \quad \mathrm{k} = 1, 2, \ldots, \mathrm{k}_{\max}$$

where:

| | |
|---|---|
| $\beta_k$ | is a velocity limiter ratio for stage k. |
| $\eta_k$ | is an element in [0 1]. |
| $\propto_k$ | is the inertia coefficient. |
| $\nabla(OF)$ | the gradient of the objective function. |

$R_1$ and $R_2$ are random values, which gives more randomness to the search when determining the velocity reached by the gradient function and the to broaden the search of the algorithm. The momentum rate $\propto_k$ becomes constant for stage k (number of stages for shark's forward movement) and the velocity is dependent from its former state. The considered sign for the value of $\mathrm{v}_{ij}^k$ depends on the direction of the selected term of the minimum operator.

The velocity vector will determine the new position during the forward movement of the shark given by:

$$\mathbf{Y}_{\mathbf{i}}^{\mathbf{k+1}} = x_i^k + v_i^{k-1}.\Delta t_k \qquad (5)$$

where:

$\mathbf{Y}t_k$ time interval which is assumed to be 1.

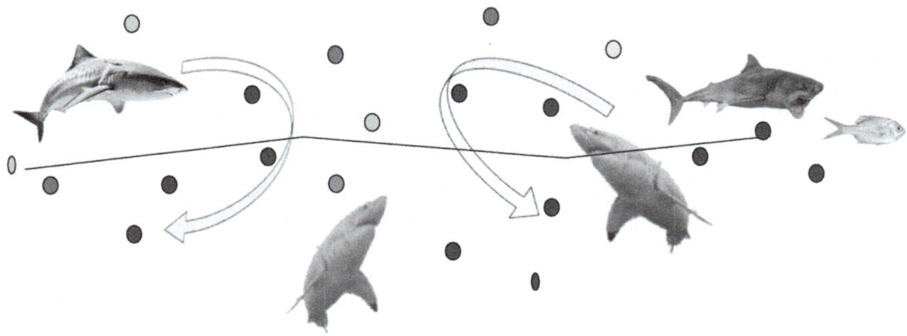

**Fig. 2** Shark's rotational movement

$Y_i^{k+1}$ new position.
$x_i^k$ actual position.
$v_i^{k-1}$ previous velocity.

**Rotational Movement**
The rotational movement allows the shark to identify the stronger odor particles when it moves forward and this enables a local search within the SSO algorithm. As can be noted from Fig. 2, the rotation of the shark is on a closed contour and not necessarily a circle. From the optimization viewpoint, the shark implements a local search in each stage to find better candidate solutions. This is modeled by the equation below:

$$Z_i^{k+1,m} = Y_i^{k+1} + R_3.Y_i^{k+1} \quad \mathrm{m} = 1, 2, \ldots, \mathrm{M} \quad \mathrm{k} = 1, 2, \ldots, \mathrm{k_{max}} \tag{6}$$

The rotational movement allows the shark to identify the stronger odor particles when it moves forward and this enables a local search in the SSO algorithm.

$R_3$ random number with uniform distribution in the range [−1 1].
$M$ indicates number of points in the local search of each stage.

**Updating of the Shark Position**
If the shark finds a point with stronger smell during the rotational movement; it will follow this point and continue the search path from it, as shown in Fig. 2. This characteristic is implemented in the SSO algorithm as follows:

$$X_i^{k+1} = argmin\{OF(Y_i^{k+1}), OF\left(Z_i^{k+1,1}\right), \ldots, OF\left(Z_i^{k+1,M}\right) \tag{7}$$

$$\mathrm{i} = 1, 2, \ldots, \mathrm{NP} \quad \mathrm{k} = 1, 2, \ldots, \mathrm{k_{max}} \quad m = 1, 2, \ldots, M$$

$X_i^{k+1}$ the next position of the Shark.
$Y_i^{k+1}$ forward movement.

$Z_i^{k+1,M}$ rotational movement with range of random number generation.

The objective function *OF* should be minimized. The cycle will continue until k reaches the minimum value (best individual) in the given population in a particular search space, which is defined for the particular optimization problem. SSO has a number of parameters, like the other metaheuristic optimization methods, including population size NP and number of stages $k_{max}$ as well as $\propto$, $\beta$ and $\eta$ for each stage. We have empirically observed that SSO works well with these values: of $r_i = 0.75$, $\propto = 0.1$ and $\beta = 3$. However, these parameters can be fine-tuned for each optimization problem separately.

## 3 Dynamic Setting of Parameters Membership Functions of the Fuzzy Controller with the SSO

This section is focused on describing the dynamic adjustment of the parameters of fuzzy controllers. Methodologies similar to this proposal are described in [23–25]. The main goal in choosing the SSO algorithm is because there are few research works published of this algorithm, there also exist other variants [17, 18] and we consider a good idea to use it and analyze if good results can be achieved when compared with respect to other algorithms.

### *3.1 Dynamic Adaptation of Fuzzy Controller Parameters with the Proposed Methodology*

For the development of the proposed method, we establish the following sequence of steps in the SSO algorithm:

1. 40 dimensions are needed to establish the position of the points to optimize in the MFs.
2. We set a lower and upper limit of −1 to 1 normalizing range in the membership functions in fuzzy input and output sets.
3. The objective function is based on the controller's fuzzy inference system.

The methodology to optimize the parameters of the membership functions with the shark smell algorithm, evaluating the fuzzy controller in the plant and the results obtained in the evaluation with respect to the desired path, is illustrated in Fig. 3.

In Fig. 4, we graphically show the sequence of steps of the proposed algorithm.

The pseudocode of the proposed algorithm is presented below.

Step 1: Randomly generate the initial population of n blood particles for the MF parameters. To start, a set of random source odor particles is generated, where each row (vector) represents a blood particle. The particle has in its memory a position.

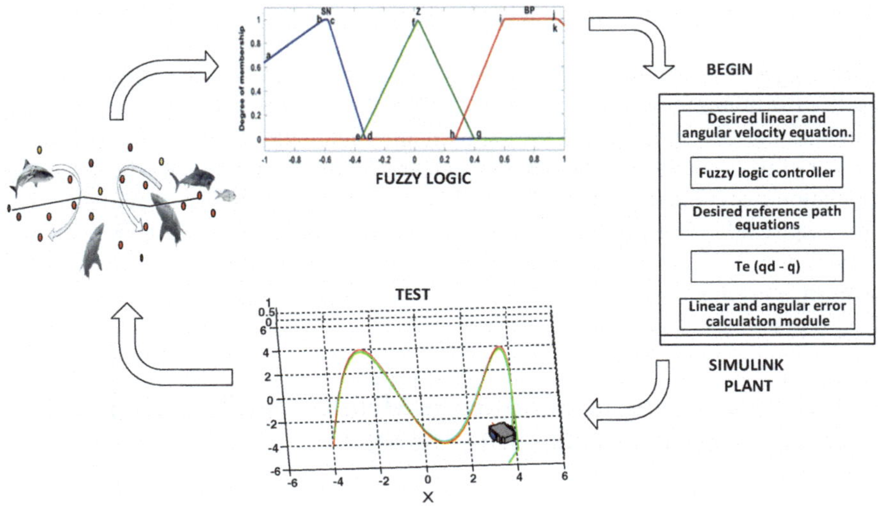


**Fig. 3** Procedure for optimization of the parameters of the membership functions

and that position is a possible solution to the problem, which in this case represents the values of the membership function parameters of the fuzzy controller. The initial population must contain possible position candidate solutions that satisfy the constraints. Set k = 0, and evaluate the fitness value of the initial populations using Eq. (2). In this case 40 dimensions are needed to establish the position of the parameters to optimize in the MFs. We set a lower and upper limit of −1 to 1 by normalizing the range in the membership functions in the fuzzy input and output sets.

Step 2: Select the best position for the neighborhood search. The selected odor particle memory contains the best position found so far and to define the fitness of each shark position the equation of the mean square error is used, which is Eq. (3). The objective function is based on the error between the real values and the output of the fuzzy controller.

Step 3: Evaluate the fitness of each particle and identify each of the parameters of the MFs using SSO.

Step 4: Represent the new value of the MF parameters from each shark position.

Step 5: Select the fittest positions of each shark.

Step 6: Move dynamically each of the MF parameters of the Inputs-Outputs.

Step 7: Check if the new fitness Function value is better than the previous Fitness Function value. If this is satisfied, update final position of each of the FIS membership function parameters.

Step 8: Check the stopping criteria. If satisfied, terminate the search, else K = K + 1 up $K_{\max}$, show and save the best values found and return to evaluate new solutions.

We propose the following fuzzy control architecture to use dynamic variation of parameters at runtime, as is shown in Fig. 3. The fuzzy logic controller design is

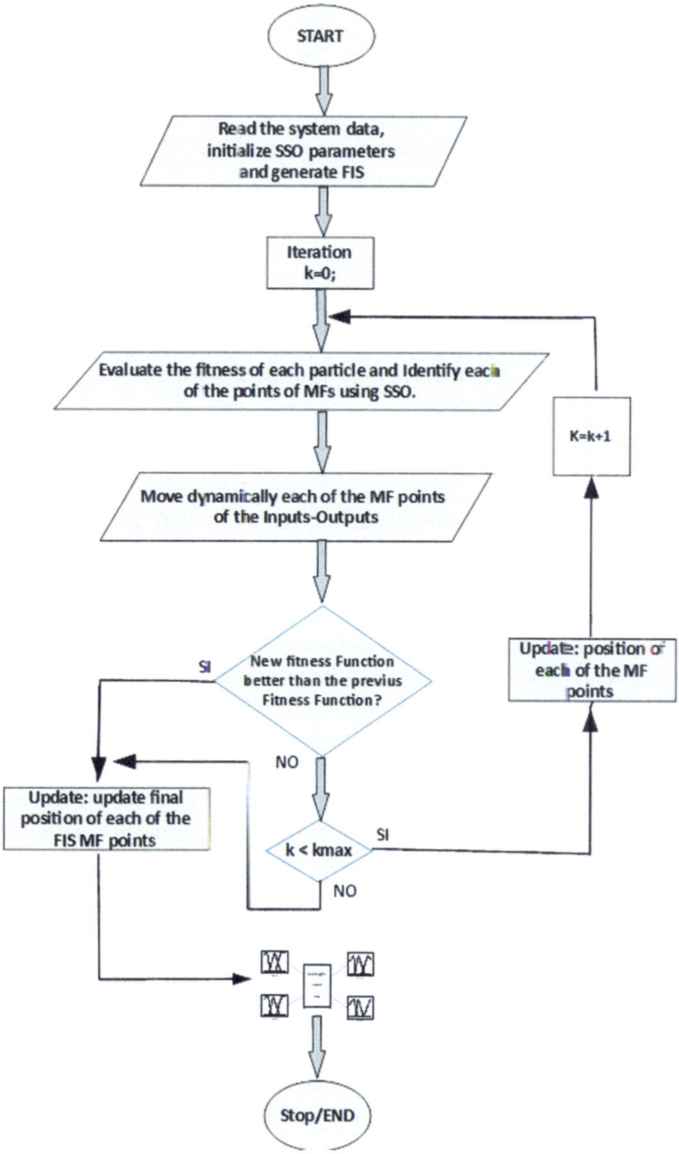


**Fig. 4** Procedure to adjust the fuzzy controller parameters using SSO

**Table 1** Comparison of the methods

| Characteristics | WDO | SSO |
|---|---|---|
| Population | Air package | Shark Smell |
| New speed | $u_{t+1}^i$ | $v_{ij}^k$ |
| Current speed | $u_t^i$ | $v_i^k$ |
| Actual position | $x_t^i$ | $x_i^k$ |
| Next position | $x_{t+1}^i$ | $x_{i+1}^k$ |
| Better experience | – | $Y_i^{k+1}$ |
| Best group experience | – | $z_i^{k+1,M}$ |
| Increase | $T$ | **k** |
| Uniform random numbers between 0 and 1 | – | $r_1, r_2, r_3$ |

*WDO* Wind-driven optimization; SSO: shark smell optimization

of the Mamdani type and has 2 inputs and 2 outputs where the parameters of the Membership Functions are dynamically adjusted with 9 rules that are chosen with prior knowledge of the problem within the exploitation and exploration in search spaces.

To better explain the difference between the used method, with respect to one of the most used metaheuristics: particle swarm optimization (PSO) which was developed by Kennedy and Eberhart in [26, 27], PSO has biological inspiration in nature, to be more specific in the behavior of birds, where each bird represents a particle. A comparison is made among PSO, WDO and SSO illustrated in Table 1. Considering that the SSO algorithm is also based on PSO, the following comparison table is made, and a Clarification is made that PSO is not currently used in this methodology, but rather only used in this table for comparison purposes.

## 3.2 *Fuzzy Logic Controller*

The term fuzzy logic was introduced with the 1965 proposal of fuzzy set theory by Zadeh [28]. Fuzzy logic had however been studied since the 1920 s, as multiple-valued logic—notably by Łukasiewicz and Tarski [29]. The use of linguistic variables can model human ways of thinking, and it is much easier to program systems based on the logic of human reasoning. By combining fuzzy logic with other artificial intelligence techniques, this mixture of techniques is able to achieve something that previously seemed impossible: the autonomy of mobile robots. So today, robots currently help with household chores, and even medical areas, in all areas of the industry, robots play a very important role in manufacturing processes. The good performance and efficiency in the work environment, creates the need for software and hardware to be updated and optimized regularly, as discussed in [30, 31].

We propose a fuzzy logic controller optimization; this fuzzy controller is of Mamdani type and is applied to create a smooth response in a mobile robotic platform, instead of a response with oscillations that could be produced with traditional hard logic. Other metaheuristics and combinations among them have also been used hybridizing with fuzzy logic in current robotics as in [32, 33]. Also Castillo and Melin et al. [5, 34, 35] describe the applications of the method. Therefore, for designing the fuzzy system of Mamdani type, with dynamical adjustment of parameters of fuzzy controller, the two measures described as linear velocity error $\epsilon_{lv}$ and angular velocity error $e_{wv}$ were considered as inputs. The therefore, for designing the fuzzy system of Mamdani type, with dynamical adjustment of parameters of fuzzy controller, the two measures described as linear velocity error $e_{lv}$ and angular velocity error $e_{wv}$ were considered as inputs. The system has the fuzzy outputs; Parmotor 1 and Parmotor 2. In regards to the inputs of the fuzzy system, the error variables have by themselves a defined range of possible values which range from $-50$ to 50, but with the linear velocity error $e_{lv}$ and angular velocity error $e_{wv}$, we perform a normalization of the values of these to have values between $-1$ and 1. Equation 8 shows how the normalization of the linear velocity error is performed and Eq. 9 shows how the normalization of the angular velocity error is obtained.

$$\boldsymbol{e}_{lv} = \frac{\mathbf{V}_{Ref} - \mathbf{V}_{\mathrm{Actual}}}{50} \tag{8}$$

$$\boldsymbol{e}_{wv} = \frac{\mathbf{w}_{Ref} - \mathbf{w}_{Actual}}{50} \tag{9}$$

The design of the input variables can be appreciated in Figs. 5 and 6, which show the $e_{lv}$ and $e_{wv}$ inputs respectively, where each input is granulated into three membership functions, trapezoidal at the ends and triangular at the middle. For the output variables, as mentioned above, we performed a normalization of the values of these to have values between $-1$ and 1, so that the output variables were designed using this range of values. Each output is granulated into three triangular membership

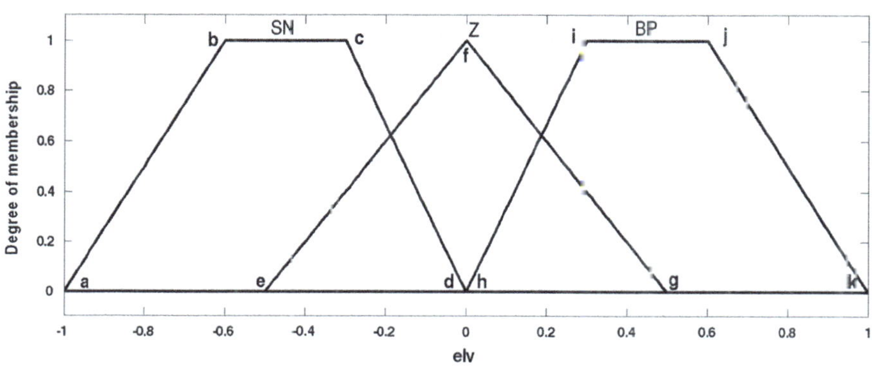


**Fig. 5** Input 1: Linear velocity error $e_{lv}$

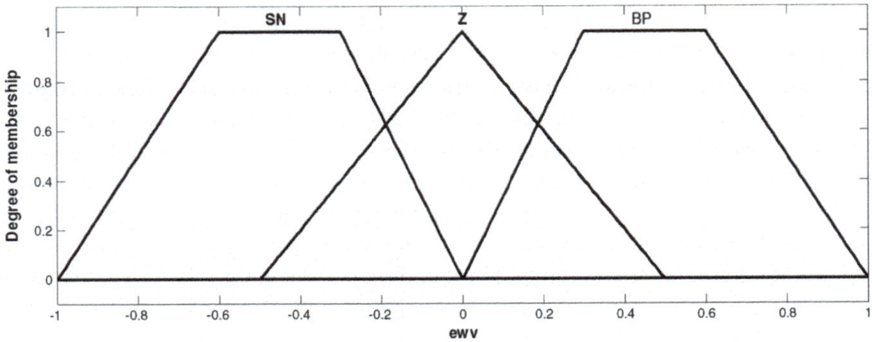


**Fig. 6** Input 2: Angular velocity error $e_{wv}$

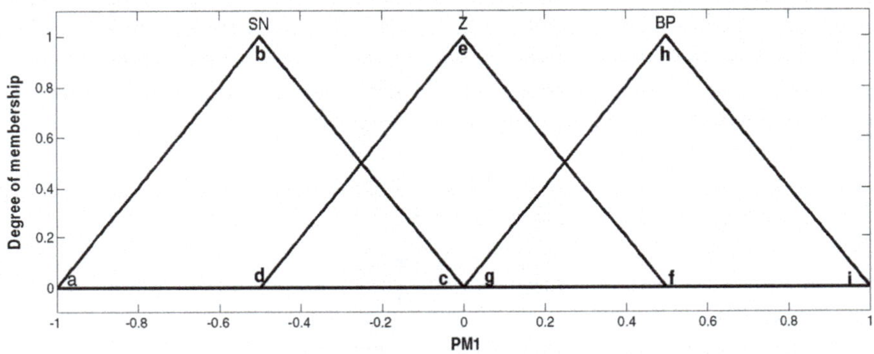


**Fig. 7** Output 1: Parmotor 1 $PM_1$

functions, the design of the output variables can be find in Figs. 7 and 8, for $PM_1$ and $PM_2$ respectively. The fuzzy system has the linear and angular velocity errors as inputs, as shown.

In Fig. 9. To design the rules of each fuzzy system, it was decided that in early iterations the SSO algorithm must explore and eventually exploit.

Taking into account other variables such as linear velocity error and angular velocity error for example, when $e_{lv}$ is small negative (SN) and $e_{wv}$ is zero (Z), that is, that the smell particles are close together, we must use exploitation, and when $e_{lv}$ is big positive (BP) and $e_{wv}$ is big positive (BP) we must use exploration, this in search of the candidate solutions. The fuzzy controller has 9 rules, which are shown below in Table 2 [36].

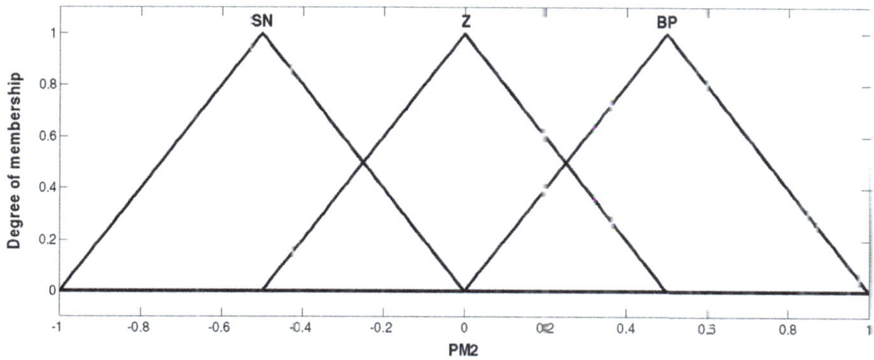


**Fig. 8** Output 2: Parmotor 2 $PM_2$

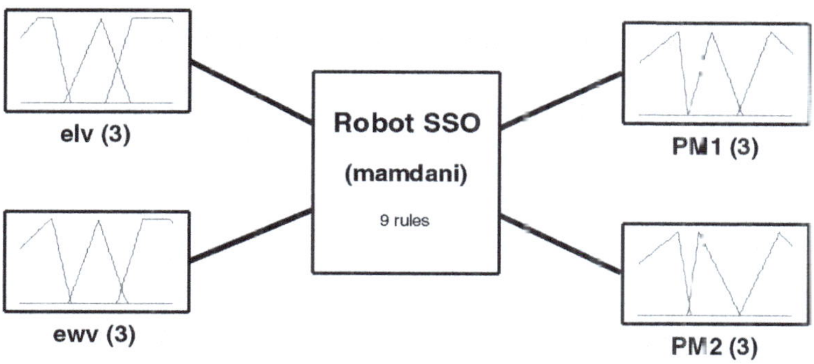


**Fig. 9** Fuzzy system for the robot controller. Type-1 FLC

**Table 2** Fuzzy system robot controller

| Rules | Antecedent | | Consequent | |
|---|---|---|---|---|
| No. | elv | ewv | Parmotor 1 $PM_1$ | Parmotor 2 $PM_2$ |
| 1 | SN | SN | NS | NS |
| 2 | SN | Z | NS | Z |
| 3 | SN | BP | NS | BP |
| 4 | Z | SN | Z | NS |
| 5 | Z | Z | Z | Z |
| 6 | Z | BP | Z | BP |
| 7 | BP | SN | BP | NS |
| 8 | BP | Z | BP | Z |
| 9 | BP | BP | BP | BP |

If-then fuzzy rules

### 3.3 Interval Type-2 Fuzzy Logic

The concepts of type-1 and type-2 fuzzy logic systems was developed by Zadeh between 1965 and 1975 [37, 38]. This section presents a concise overview of IT2FLS with the intention of providing basic knowledge of how the IT2FLS works to achieve its objective.

First, we break down the overall behavior of the robot that we call "behavior", which may have the task of following a line. In Fig. 10, a controller is designed to determine the control action of the mobile robot. These basic behaviors have as a control structures a controller with type-2 fuzzy logic formed from a set of fuzzy IF-THEN rules [39, 40]. The actions are monitored by the controller's rule set to select the appropriate actions that are transmitted to the system.

We know that the T2FS is located in a region built by expanding a main type-1 mem-bership function (T1MF). T2FS is obtained by using fuzzy sets to granulate the input domains of the base line T1FS with a footprint of uncertainty (FOU) as shown in Figs. 11, 12, 13 and 14. Consequently, the T1MF is extended to T2MF by adding the FOU to represent the uncertainty of the problem. The variability in each of the actions is modeled with interval type-2 fuzzy sets (IT2FS). The T2FS linguistic input variables and their ranges are used for path tracking, as shown in Figs. 11 and 12, with two outputs, as shown in Figs. 13 and 14.

Therefore, membership functions have values distributed according to the uncertainty. As well as those that belongs to the antecedents also in the consequent parts. From the design, three MFs, the $e_{lv}$ input (SN: Small Negative, Z: Zero, BP: Big Positive), three MFs for the $e_{wv}$ input (SN: Small Negative, Z: Zero, BP: Big Positive), three MFs for the $PM_1$ and $PM_2$ outputs (SN: Small Negative, Z: Zero, BP: Big Positive), are built. Nine rules were developed for the T2FLS for path tracking.

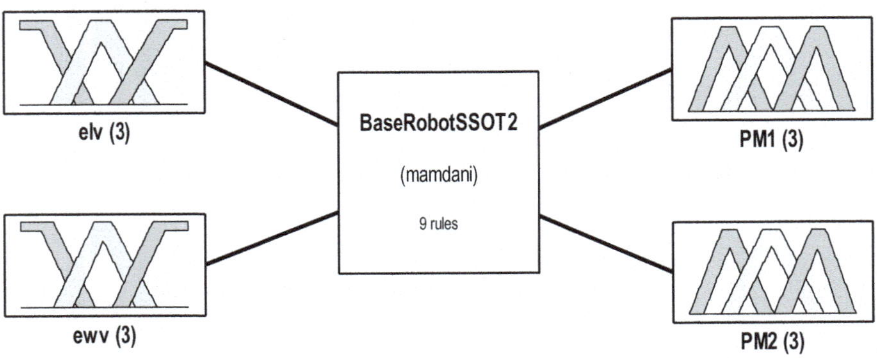


**Fig. 10** It2flc architecture

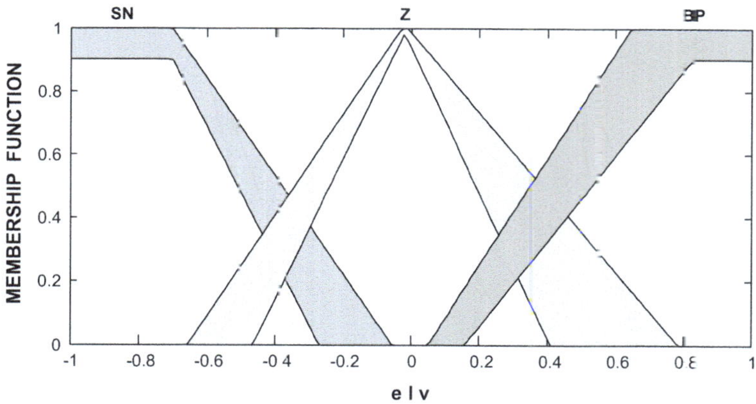


**Fig. 11** Footprints of IT2 MFs for input $e_{lv}$

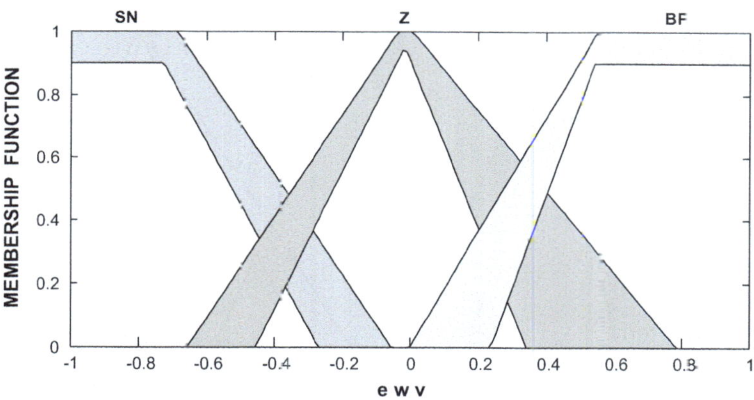


**Fig. 12** Footprints of IT2 MFs for input $\boldsymbol{e_{wv}}$

The number of rules (9) is determined by expert knowledge. Fuzzy rules are obtained from the combination of two inputs with three membership functions. The rules are used to control the parmotors of wheels right and left, some of these rules are as presented in Table 2.

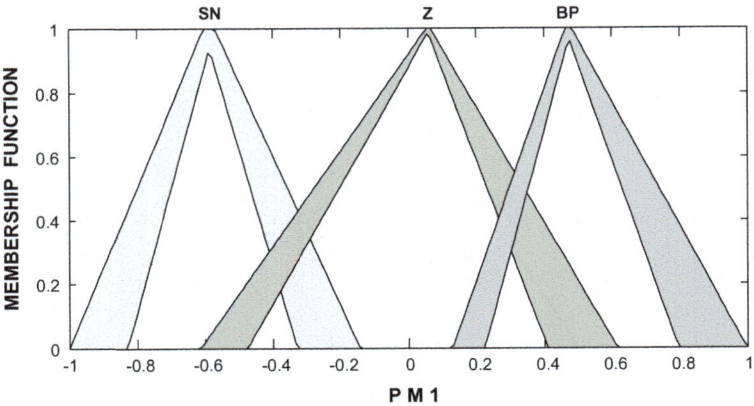


**Fig. 13** Footprints of IT2 MFs for output $PM_1$

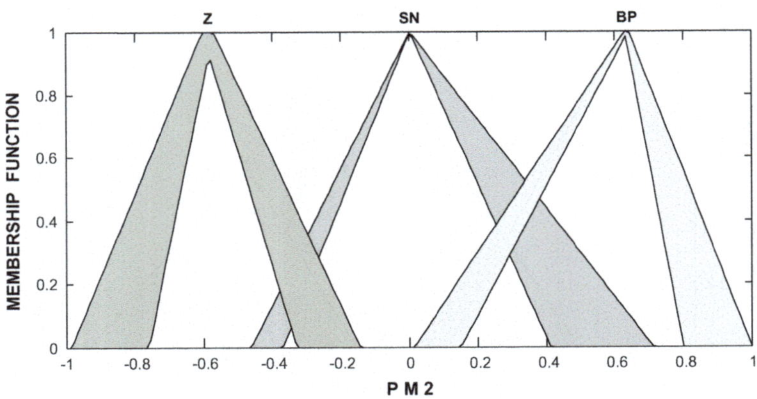


**Fig. 14** Footprints of IT2 MFs for the output $PM_2$

# 4 Study Cases

## *4.1 Case 1: Benchmark Functions*

Also for the comparison of the proposed method with respect to other methods, we considered benchmark mathematical functions, defined in [41], which are a total of 27 functions and in each one it is tried to find values that give us the global minimum of each function. In Fig. 15 there is a sample of the functions that are used. The problem that is considered (Benchmark mathematical functions) [23] is illustrated in Fig. 15, and the proposed methodology is as shown in Fig. 16, where we can notice

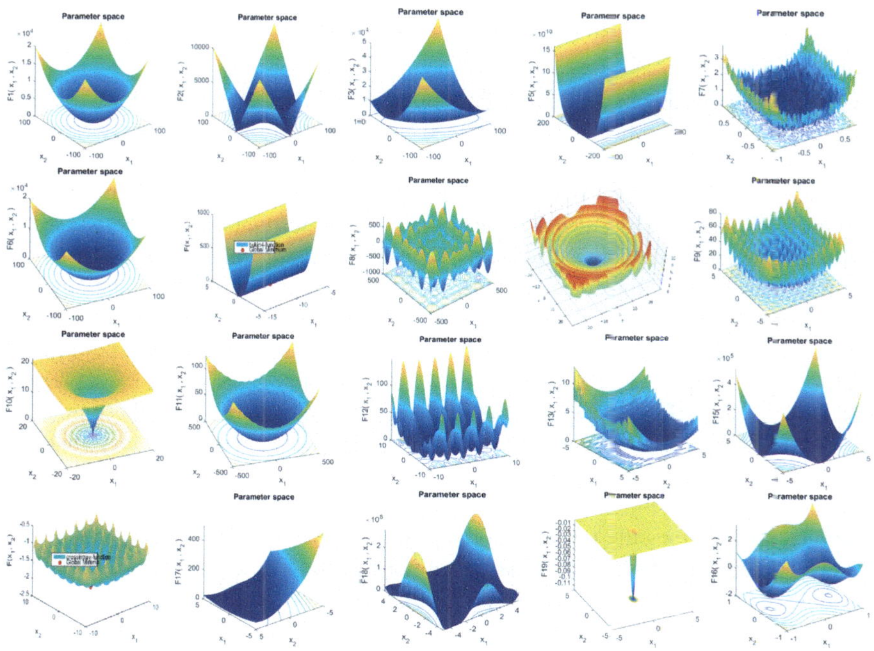

Fig. 15 Benchmark mathematical functions

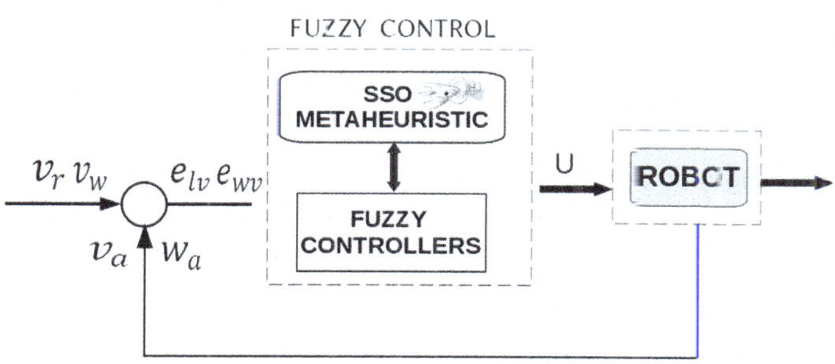


Fig. 16 Proposed methodology

that fuzzy parameters are adjusted by SSO, and in turn this "SSO" searches for the optimal parameters for the membership functions of the fuzzy input-output sets.

### 4.2 Case: Fuzzy Controller of an Autonomous Mobile Robot

We propose the optimization for a fuzzy controller of an Autonomous Mobile Robot of differential type, using the proposed SSO for this specific problem; below, the fuzzy controller is explained in more detail. The controller for the autonomous mobile robot is as used in [39–42]. The main objective of the controller is the better search of a desired trajectory, and it has 9 rules that define the relationships between the linguistic variables of the fuzzy system. The linguistic variables are as follows: in the first input, it is the linear velocity error ($e_{lv}$); the second input, it is the angular velocity error ($e_{wv}$). The two inputs have the same linguistic values: big positive (BP), zero (Z), and small negative (SN), with the same membership functions in both fuzzy input sets. The outputs are Parmotor 1 ($PM_1$) and Parmotor 2 ($PM_2$) they represent the change of direction when the robot's wheels interact, and each of these outputs has three triangular functions. Figure 17 represents the desired trajectory that the robot should follow. In this figure, the Y-axis shows the desired Y displacement, and the X-axis represents the X displacement.

Figure 18 describes the red line as the desired trajectory, and the real trajectory of the robot in blue, and as we can note, the robot is not lost and follows closely the desired trajectory, since it generates an error of $4.64 \times 10^{-03}$ and a standard deviation of $1.03 \times 10^{-02}$.

## 5 Simulation Results

In the case of the results obtained with the optimization of benchmark functions using the shark smell method, this is presented in the first part of this section. The results with the optimization of the membership functions of the fuzzy controller used with the autonomous mobile robot platform are optimization generates a vector

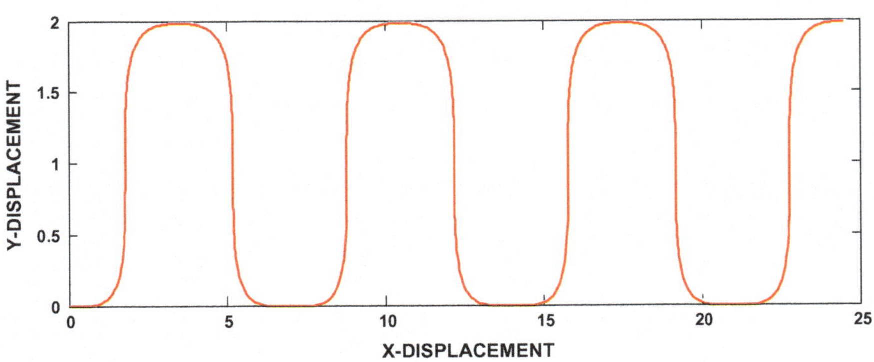


**Fig. 17** Desired trajectory normalized axis

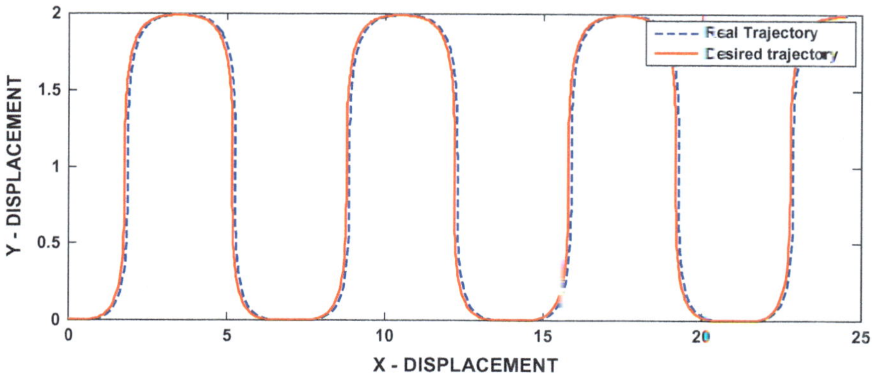


**Fig. 18** Obtained mean square error (MSE)

of values that are presented in the second part of the section. This optimization generates a vector of values that are defined as the dynamic parameters of the optimized membership functions.

## 5.1 *Case 1 Results: Benchmark Functions*

Table 3 shows the parameters used in the optimization of benchmark functions. Table 4 column 1 shows the used functions; column 2 represents the minimum of the functions; column 3 shows the obtained average; column 4 represent the standard deviations respectively obtained with the SSO; To conclude columns 5 and 6 show the results of the SSO method. This was evaluated 30 times for each of these functions with the same parameters to obtain the averages and standard deviations. Table 4 above shows the optimization results of benchmark functions where the minimum of functions $f_{15}$, $f_{16}$, $f_{17}$, and $f_{18}$ is different from zero. With the evaluation with different minima, the performance of the optimization in the function with the SSO can be better observed.

**Table 3** Parameters used in the experiments

| Population | Iterations | Dimensions |
|---|---|---|
| 100 | 500 | 30 |

**Table 4** Results for 40 dimensions

| $F_{cn}$ | $f_{min}$ | WDO | | SSO | |
|---|---|---|---|---|---|
| | | Average | Std. dvtn | Average | Std. dvtn |
| $f_1$ | 0 | $3.6 \times 10^{-27}$ | $3.9 \times 10^{-27}$ | $1.6 \times 10^{-13}$ | $6.9 \times 10^{-14}$ |
| $f_2$ | 0 | $3.2 \times 10^{-14}$ | $2.7 \times 10^{-14}$ | $3.7 \times 10^{-06}$ | $3.0 \times 10^{-07}$ |
| $f_3$ | 0 | $2.4 \times 10^{-20}$ | $2.0 \times 10^{-20}$ | $2.7 \times 10^{-13}$ | $3.1 \times 10^{-14}$ |
| $f_4$ | 0 | $5.9 \times 10^{-13}$ | $6.2 \times 10^{-13}$ | $2.2 \times 10^{-13}$ | $4.1 \times 10^{-14}$ |
| $f_5$ | 0 | 28.569 | $7.6 \times 10^{-02}$ | $1.8 \times 10^{-13}$ | $4.7 \times 10^{-16}$ |
| $f_6$ | 0 | $1.7 \times 10^{-02}$ | $4.5 \times 10^{-03}$ | $5.7 \times 10^{-05}$ | $3.1 \times 10^{-06}$ |
| $f_7$ | 0 | $1.1 \times 10^{-02}$ | $3.0 \times 10^{-02}$ | $1.3 \times 10^{-12}$ | $9.4 \times 10^{-15}$ |
| $f_9$ | 0 | −118.27 | 0 | $4.1 \times 10^{-13}$ | $1.0 \times 10^{-14}$ |
| $f_{10}$ | 0 | $9.5 \times 10^{-15}$ | $1.2 \times 10^{-14}$ | $6.9 \times 10^{-08}$ | $3.5 \times 10^{-09}$ |
| $f_{11}$ | 0 | 0 | 0 | $3.8 \times 10^{-13}$ | $3.3 \times 10^{-29}$ |
| $f_{15}$ | 0.0 | $3.0 \times 10^{-04}$ | $2.2 \times 10^{-07}$ | $3.1 \times 10^{-06}$ | $4.7 \times 10^{-07}$ |
| $f_{16}$ | 1.0 | 1.0316 | $6.8 \times 10^{-16}$ | 1.07 | $3.4 \times 10^{-16}$ |
| $f_{17}$ | 0.3 | 0.3979 | $6.8 \times 10^{-05}$ | $6.0 \times 10^{-12}$ | $3.7 \times 10^{-15}$ |
| $f_{18}$ | 3 | 7.783 | $3.6 \times 10^{-15}$ | $2.5 \times 10^{-01}$ | $3.8 \times 10^{-02}$ |

## *5.2 Case 2 Results: Dynamic Adjustment of Fuzzy Controller Parameters*

Table 5 shows the parameters used in the WDO and SSO methods. Tables 6 and 7 show the 30 experiments performed to obtain the best optimized fuzzy system, using the MSE as metric, where it can be observed that the best error found is of 0.00169, and in general, the values only varied from $10^{-02}$ to $10^{-03}$.

In Fig. 19, the dynamic adjustment that was made in the parameters of the membership functions of the inputs of the fuzzy controller is observed, where the uncertainty that is generated between each of them can be observed, and with this, the error that shows the fuzzy controller in the simulation is considerably improved. Figures 20, 21, 22 and 23 shows the optimized movement of the parameters of the membership functions of the inputs $e_{vl}$, $e_{wl}$ and the outputs $PM_1$ and $PM_2$, respectively. In addition, as can be noted from Fig. 19, the overlap between existing functions helps the robot to have a better tracking of the desired trajectory, in comparison with the other method used in this work. It can be noted in Tables 6 and 7, that there is a variation in the results obtained with the SSO method, since it has the best MSE of 0.00000126, but the worst is 0.05150. Unlike WDO where there is a big variation in the results; the best MSE with 0.000019, the worst of 0.057094. For this reason, on average, this

**Table 5** Parameters used WDO and SSO method for fuzzy controller

| Populatión | Iteratións | Dimensions |
|---|---|---|
| 30 | 1500 | 40 |

**Table 6** Results of the experiments with the fuzzy controller, obtained with SSO

| Experiment | MSE | Experiment | MSE |
|---|---|---|---|
| 1 | **$1.2679 \times 10^{-06}$** | 16 | $6.3590 \times 10^{-04}$ |
| 2 | $1.3972 \times 10^{-05}$ | 17 | $1.3007 \times 10^{-03}$ |
| 3 | $6.7674 \times 10^{-04}$ | 18 | $4.7000 \times 10^{-03}$ |
| 4 | $1.1398 \times 10^{-05}$ | 19 | $8.4020 \times 10^{-03}$ |
| 5 | $3.7205 \times 10^{-05}$ | 20 | $4.7000 \times 10^{-03}$ |
| 6 | $2.4143 \times 10^{-05}$ | 21 | $1.4700 \times 10^{-02}$ |
| 7 | $2.9927 \times 10^{-06}$ | 22 | $5.1500 \times 10^{-02}$ |
| 8 | $5.0183 \times 10^{-05}$ | 23 | $7.0000 \times 10^{-03}$ |
| 9 | $2.9581 \times 10^{-05}$ | 24 | $3.7010 \times 10^{-03}$ |
| 10 | $5.6985 \times 10^{-06}$ | 25 | $5.9000 \times 10^{-03}$ |
| 11 | $6.5592 \times 10^{-06}$ | 26 | $2.7000 \times 10^{-02}$ |
| 12 | $4.0882 \times 10^{-05}$ | 27 | $8.2560 \times 10^{-04}$ |
| 13 | $1.8416 \times 10^{-06}$ | 28 | $6.5050 \times 10^{-04}$ |
| 14 | $1.1593 \times 10^{-05}$ | 29 | $2.8000 \times 10^{-03}$ |
| 15 | $3.6720 \times 10^{-06}$ | 30 | $5.4490 \times 10^{-04}$ |

**Table 7** Results of the experiments with the fuzzy controller, obtained with WDO

| Experiment | MSE | Experiment | MSE |
|---|---|---|---|
| 1 | **$1.90 \times 10^{-05}$** | 16 | $1.64 \times 10^{-04}$ |
| 2 | $1.90 \times 10^{-05}$ | 17 | $1.64 \times 10^{-04}$ |
| 3 | $5.40 \times 10^{-05}$ | 18 | $2.60 \times 10^{-04}$ |
| 4 | $5.40 \times 10^{-05}$ | 19 | $2.93 \times 10^{-04}$ |
| 5 | $6.20 \times 10^{-05}$ | 20 | $4.33 \times 10^{-04}$ |
| 6 | $6.20 \times 10^{-05}$ | 21 | $1.62 \times 10^{-03}$ |
| 7 | $6.90 \times 10^{-05}$ | 22 | $1.65 \times 10^{-03}$ |
| 8 | $8.70 \times 10^{-05}$ | 23 | $6.21 \times 10^{-03}$ |
| 9 | $8.70 \times 10^{-05}$ | 24 | $6.35 \times 10^{-03}$ |
| 10 | $9.10 \times 10^{-05}$ | 25 | $7.73 \times 10^{-03}$ |
| 11 | $9.80 \times 10^{-05}$ | 26 | $1.67 \times 10^{-02}$ |
| 12 | $1.05 \times 10^{-04}$ | 27 | $1.67 \times 10^{-02}$ |
| 13 | $1.15 \times 10^{-04}$ | 28 | $2.65 \times 10^{-02}$ |
| 14 | $1.38 \times 10^{-04}$ | 29 | $2.92 \times 10^{-02}$ |
| 15 | $1.60 \times 10^{-04}$ | 30 | $5.7 \times 10^{-02}$ |

method can be the best at optimizing this specific problem. Table 8 summarized the results of the optimization methods.

It can be noted in Tables 6 and 7, that there is less variation in the results obtained with the SSO method, since it has the best MSE of 0.00000126, but the worst is of

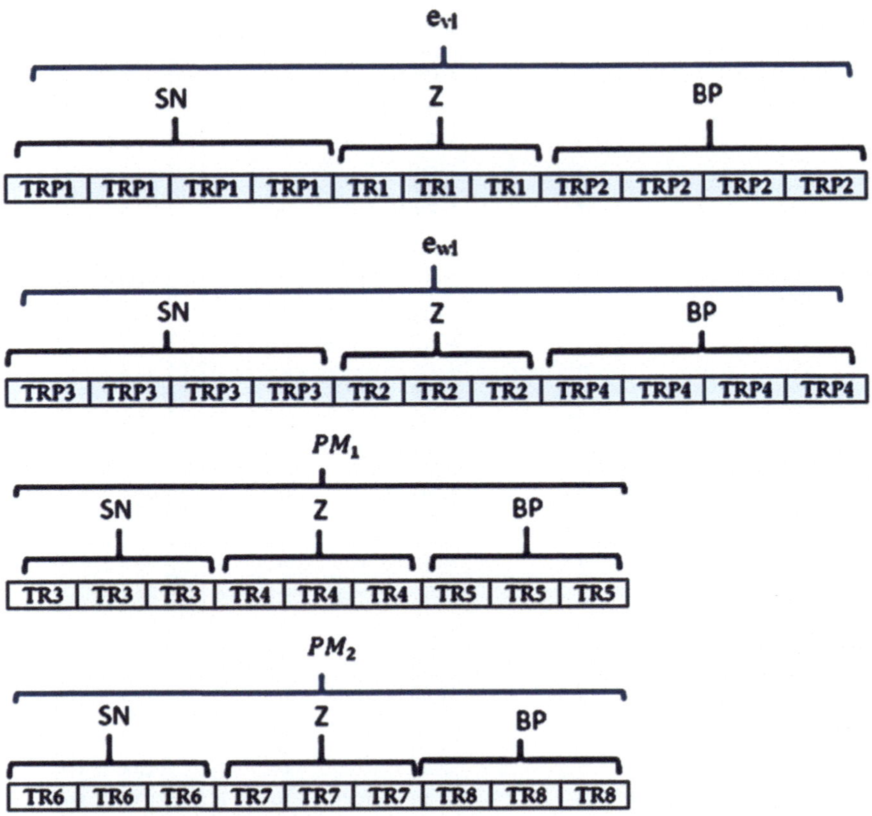


**Fig. 19** Particle vector for optimizing the membership functions of the fuzzy controller

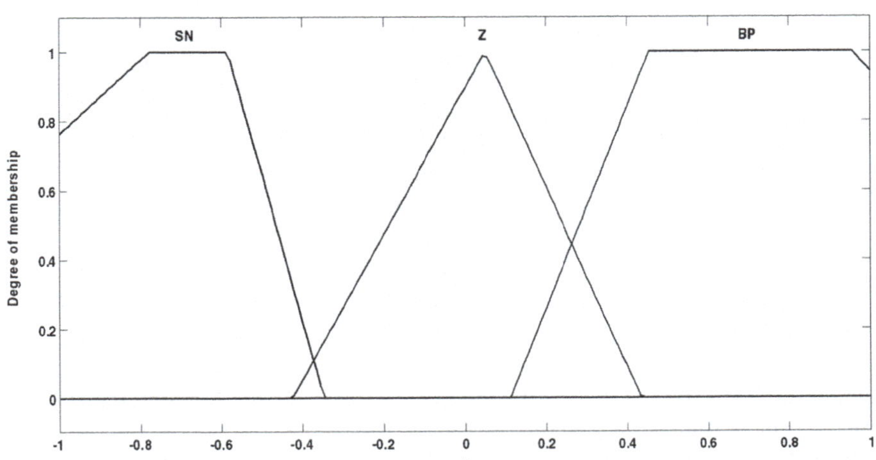


**Fig. 20** Optimized Input 1: Linear velocity error (evl)

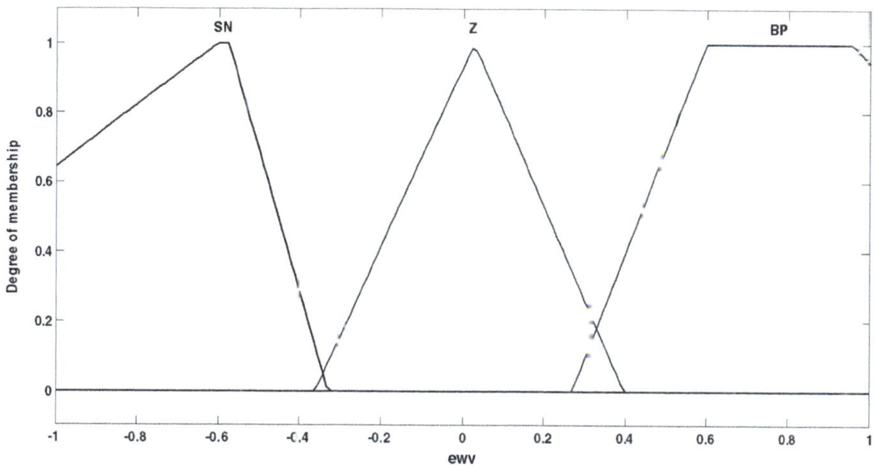


**Fig. 21** Optimized Input 2: Angular velocity error (ewl)

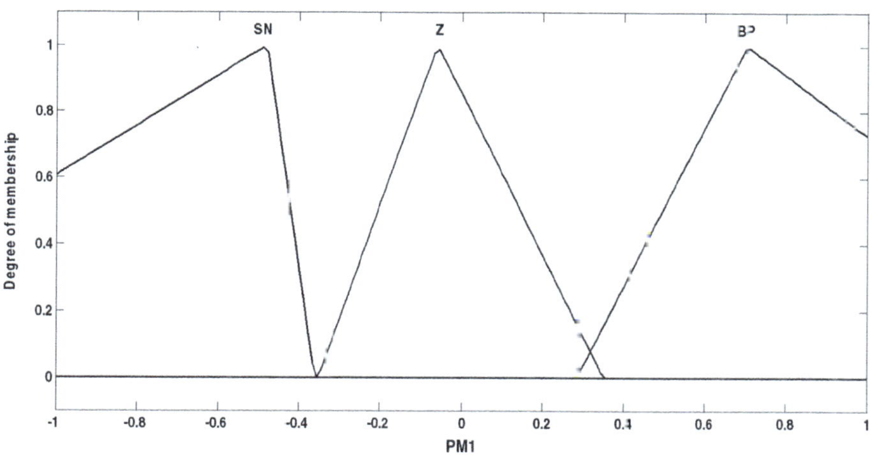


**Fig. 22** Optimized Output 1: $\boldsymbol{PM_1}$

0.05150. Unlike WDO where there is a variation in the results; the best MSE with 0.000019, the worst of 0.057094. For this reason, on average, the SSO method can be considered the best at optimizing this specific problem. Table 8 summarizes the results of the optimization methods.

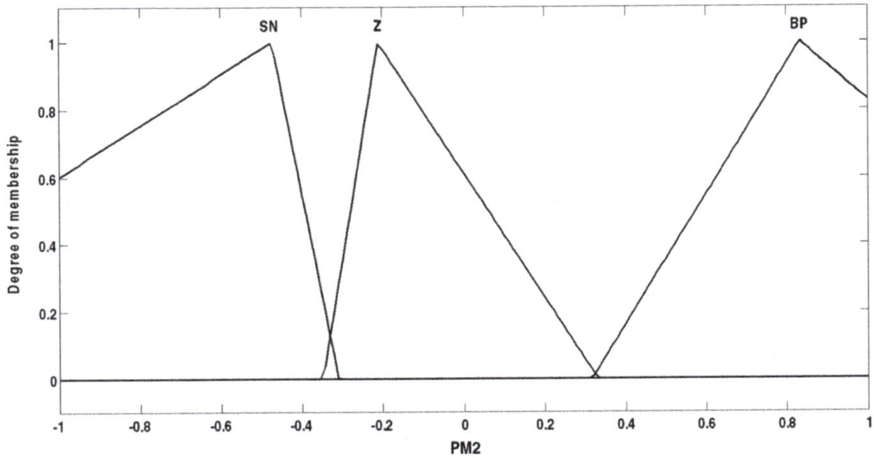


**Fig. 23** Optimized Output 2: $PM_2$

**Table 8** Results of the optimization methods

| | WDO | SSO |
|---|---|---|
| Average | $6.00 \times 10^{-03}$ | $4.6 \times 10^{-03}$ |
| Standard deviation | $1.26 \times 10^{-02}$ | $1.10 \times 10^{-02}$ |

# 6 Conclusions

The proposed approach implements a type-1 and type-2 fuzzy controller that produces the parmotor forces in the wheels for path tracking, with the ability for path tracking that arise in the environment so that the robot can travel the path without losing it and with the least possible error. In addition, it is worth mentioning that in the path following by the robot with T1FLC and T2FLC, it is possible to observe how the values of the parmotor 1 and parmotor 2 are changing when analyzing the results with the integral parameters applied and the times of path. Generally, the IT2 FLC gains are smaller than those of the baseline T1FLC, and they decrease as the FOU increases. Because smaller IT2 gains generate smaller control signals, this explains why IT2FLCs may be more robust to disturbances and why their response slower [42, 43]. The findings provide theoretical explanation for the experimental observations suggesting that an IT2 fuzzy controller is better able to cope with uncertainty and eliminate disturbances. For future work, our goal is to test our focus on other types of problems that have more uncertainty, varying the MFs of the input-output fuzzy sets, and then testing the true power of fuzzy type 2 interval systems by doing more experimentation with other optimization methods. Also, we are thinking in considering which SSO parameters to be adjusted dynamically. In addition, include other metrics, such as stability metrics in more complex problems.

## References


1. A.R. Yildiz, Comparison of evolutionary-based optimization algorithms for structural design optimization. Eng. Appl. Artif. Intell. **26**(1), 327–333 (2013)
2. J. Vesterstrm, R. Thomsen, A comparative study of differential evolution, particle swarm optimization, and evolutionary algorithms on numerical benchmark problems, in *Proceedings of the 2004 Congress on Evolutionary Computation*, CEC2004 (2004)
3. J. Perez, F. Valdez, O. Castillo, O. Roeva, Bat algorithm with parameter adaptation using Interval Type-2 fuzzy logic for benchmark mathematical functions, in *2016 IEEE 8th International Conference on Intelligent Systems, IS 2016—Proceedings* (2016), pp. 120–127
4. J. Pérez, F. Valdez, Castillo, Modification of the bat algorithm using type-2 fuzzy logic for dynamical parameter adaptation, in *Studies in Computational Intelligence*, vol. 667 (2017), pp. 343–355
5. O. Castillo, P. Melin, Experimental study of intelligent controllers under uncertainty using type-1 and type-2 fuzzy logic. Stud. Fuzziness Soft. Comput. **223**, 121–132 (2008)
6. E.H. Mamdani, S. Assilian, An experimental in linguistic synthesis with a fuzzy logic controller. Int. J. Man Mach. Stud. **7**(1), 1–13 (1975)
7. N. Karnik, N.N.J.M. Mendel, Q. Liang., Type-2 fuzzy logic systems. IEEE Trans. Fuzzy Syst. **7**, 643–658 (1999)
8. F. Olivas, F. Valdez, O. Castillo, C.I. Gonzalez, G. Martinez, P. Melin, Ant colony optimization with dynamic parameter adaptation based on interval type-2 fuzzy logic systems. Appl. Soft. Comput. J. **53**(C), 74–87 (2017)
9. E. Bernal, O. Castillo, J. Soria, F. Valdez, Galactic swarm optimization with adaptation of parameters using fuzzy logic for the optimization of mathematical functions, in *Studies in Computational Intelligence*, vol. 749 (2018), pp. 131–140
10. F. Olivas, F. Valdez, O. Castillo, Fuzzy classification system design using PSO with dynamic parameter adaptation through fuzzy logic. Stud. Comput. Intell. **574**, 29–47 (2015)
11. C. Caraveo, F. Valdez, O. Castillo, Optimization of fuzzy controller design using a new bee colony algorithm with fuzzy dynamic parameter adaptation. Appl. Soft Comput. J. **43**, 131–142 (2016)
12. L. Amador-Angulo, O. Mendoza, J.R. Castro, A. Rodríguez-Díaz, P. Melin, O. Castillo, Fuzzy sets in dynamic adaptation of parameters of a bee colony optimization for controlling the trajectory of an autonomous mobile robot. Sensors (Switzerland) **16**(9), 1458 (2016)
13. A.A. Zamani, E. Bijami, F. Sheikholeslam, B. Jafrasteh, Optimal fuzzy load frequency controller with simultaneous auto-tuned membership functions and fuzzy control rules. Turkish J. Electr. Eng. Comput. Sci. (2014)
14. P. Ochoa, O. Castillo, J. Soria, Fuzzy differential evolution method with dynamic parameter adaptation using type-2 fuzzy logic, in *2016 IEEE 8th International Conference on Intelligent Systems, IS 2016—Proceedings* (2016), pp. 113–118
15. P. Ochoa, O. Castillo, J. Soria, Differential evolution with dynamic adaptation of parameters for the optimization of fuzzy controllers (2014), pp. 275–288
16. P. Ochoa, O. Castillo, J. Soria, Differential evolution algorithm with interval type-2 fuzzy logic for the optimization of the mutation parameter, in *Studies in Computational Intelligence*, vol. 749 (2018), pp. 55–65
17. M. Ahmadigorji, N. Amjady, A multiyear DG-incorporated framework for expansion planning of distribution networks using binary chaotic shark smell optimization algorithm. Energy (2016)
18. M. Bagheri, A. Sultanbek, O. Abedinia, M.S. Naderi, M.S. Naderi, N. Ghadimi, Multi-objective shark smell optimization for solving the reactive power dispatch problem, in *Proceedings—2018 IEEE International Conference on Environment and Electrical Engineering and 2018 IEEE Industrial and Commercial Power Systems Europe, EEEIC/I and CPS Europe 2018* (2018)
19. G. Oveis Abedinia Nima Amjady Ali, A new metaheuristic algorithm based on shark smell optimization. Complexity **21**(5), 97–116 (2014)

20. O. Abedinia, N. Amjady, A. Ghasemi, A new metaheuristic algorithm based on shark smell optimization. Complexity (2016)
21. N. Gnanasekaran, S. Chandramohan, P.S. Kumar, A. Mohamed Imran, Optimal placement of capacitors in radial distribution system using shark smell optimization algorithm. Ain Shams Eng. J. (2016)
22. O. Abedinia, N. Amjady, Short-term wind power prediction based on hybrid neural network and chaotic shark smell optimization. Int. J. Precis. Eng. Manuf. Green Technol. (2015)
23. M. Bidar, H.R. Kanan, Modified firefly algorithm using fuzzy tuned parameters, in *13th Iranian Conference on Fuzzy Systems, IFSC 2013* (2013)
24. J. Pérez, F. Valdez, O. Castillo, A new bat algorithm with fuzzy logic for dynamical parameter adaptation and its applicability to fuzzy control design. Stud. Comput. Intell. **574** (2015)
25. L. Amador-Angulo, O. Castillo, Optimization of the type-1 and type-2 fuzzy controller design for the water tank using the bee colony optimization, in *2014 IEEE Conference on Norbert Wiener in the 21st Century: Driving Technology's Future, 21CW 2014—Incorporating the Proceedings of the 2014 North American Fuzzy Information Processing Society Conference, NAFIPS 2014, Conference Proceedings* (2014), pp. 1–8
26. R. Eberhart, J. Kennedy, New optimizer using particle swarm theory, in *Proceedings of the International Symposium on Micro Machine and Human Science* (1995)
27. R.C. Eberhart, Y. Shi, Particle swarm optimization: Developments, applications and resources, in *Proceedings of the IEEE Conference on Evolutionary Computation, ICEC* (2001)
28. L.A. Zadeh, Knowledge representation in fuzzy logic. IEEE Trans. Knowl. Data Eng. **1**(1), 89–100 (1989)
29. R. Giles, Łukasiewicz logic and fuzzy set theory. Int. J. Man. Mach. Stud. (1976)
30. S. Galichet, L. Foulloy, Fuzzy controllers: synthesis and equivalences. IEEE Trans. Fuzzy Syst. **3**(2), 140–148 (1995)
31. Z. Zhao, M. Tomizuka, S. Isaka, Fuzzy gain scheduling of PID controllers. IEEE Trans. Syst. Man Cybern. ieeexplore.ieee.org
32. S.E. Oltean, M. Dulau, R. Puskas, Position control of Robotino mobile robot using fuzzy logic, in *2010 IEEE International Conference on Automation, Quality and Testing, Robotics, AQTR 2010—Proceedings* vol. 1 (2010), pp. 366–371
33. F. Cuevas, O. Castillo, Design and implementation of a fuzzy path optimization system for omnidirectional autonomous mobile robot control in real-time, in *Fuzzy Logic Augmentation of Neural and Optimization Algorithms: Theoretical Aspects and Real Applications* (Springer, Cham), pp. 241–252
34. P. Melin, L. Astudillo, O. Castillo, F. Valdez, M. Garcia, Optimal design of type-2 and type-1 fuzzy tracking controllers for autonomous mobile robots under perturbed torques using a new chemical optimization paradigm. Expert Syst. Appl. **40**(8), 3185–3195 (2013)
35. O. Castillo, P. Melin, W. Pedrycz, Design of interval type-2 fuzzy models through optimal granularity allocation. Appl. Soft Comput. J. **11**(8), 5590–5601 (2011)
36. L. Astudillo, P. Melin, O. Castillo, Chemical optimization algorithm for fuzzy controller design, in *Springer Briefs in Applied Sciences and Technology* (2014)
37. L.A. Zadeh, Fuzzy sets. Inf. Control (1965)
38. L.A. Zadeh, Fuzzy sets as a basis for a theory of possibility. Fuzzy Sets Syst. (1978)
39. F. Cuevas, O. Castillo, P. Cortes-Antonio, Towards an Adaptive Control Strategy Based on Type-2 Fuzzy Logic for Autonomous Mobile Robots, in *IEEE International Conference on Fuzzy Systems*, vol. 2019-June (2019)
40. F. Cuevas, O. Castillo, P. Cortes, Towards a control strategy based on type-2 fuzzy logic for an autonomous mobile robot, in *Hybrid Intelligent Systems in Control, Pattern Recognition and Medicine* (Springer, Cham, 2020), pp. 301–314
41. M. Molga, C. Smutnicki, Test functions for optimization needs. Test Funct. Optim. Needs (2005)
42. L. Astudillo, P. Melin, O. Castillo, Optimization of type-2 and type-1 fuzzy tracking controllers for an autonomous mobile robot under perturbed torques by means of a chemical optimization paradigm. Stud. Fuzziness Soft. Comput. **294**, 3–26 (2013)

43. C. Caraveo, F. Valdez, O. Castillo, A new meta-heuristics of optimization with dynamic adaptation of parameters using type-2 fuzzy logic for trajectory control of a mobile robot. Algorithms, **10**(3) (2017)

The manufacturer's authorised representative in the EU is Springer Nature Customer Service Centre GmbH, Europaplatz 3, 69115 Heidelberg, Germany. If you have any concerns regarding our products, please contact ProductSafety@springernature.com

Printed and bound by CPI Group (UK) Ltd, Croydon, CR0 4YY
30/09/2026
02225506-0007